GREG COWIE

KU-759-068

AN INTRODUCTION TO PHYSICAL OCEANOGRAPHY

ATLANTIS *under sail in the North Atlantic.* [*Photograph courtesy U.S. Navy.*]

AN INTRODUCTION TO PHYSICAL OCEANOGRAPHY

BY WILLIAM S. von ARX, Sc.D.

Physical Oceanographer, Woods Hole Oceanographic Institution
Professor of Oceanography, Massachusetts Institute of Technology

ADDISON–WESLEY PUBLISHING COMPANY
Advanced Book Program
Reading, Massachusetts
London · Amsterdam · Don Mills, Ontario · Sydney · Tokyo

William S. von Arx
An Introduction to Physical Oceanography

This book is in the
ADDISON–WESLEY SERIES IN THE EARTH SCIENCES

and is contribution Number 1245,
from the Woods Hole Oceanographic Institution.

First printing, 1962
Second printing, 1964
Third printing, with corrections, 1967
Fourth printing, 1969
Fifth printing, 1971
Sixth printing, with additional corrections, 1974
Seventh printing, 1977
Eighth printing, 1979

Library of Congress Catalog Card No. 61–5026
International Standard Book No. 0-201-08174-1

Copyright © 1962, 1974 by Addison–Wesley Publishing Company, Inc.
Published simultaneously in Canada.

All rights reserved. No part of this publication may be reproduced, stored in a retrieval system, or transmitted, in any form or by any means, electronic, mechanical, photocopying, recording, or otherwise, without the prior written permission of the publisher, Addison–Wesley Publishing Company, Inc., Advanced Book Program, Reading, Massachusetts 01867, U.S.A.

Printed in the United States of America

ABCDEFGHIJ-HA-79

Preface

An understanding of the physical nature of the oceans is an essential and often inspiring part of any complete study of the earth. With the attention now being given to the earth sciences, a growing number of students are developing a sincere interest in the oceans as well as the atmosphere and solid earth. This interest can be difficult to satisfy because, even more than is the case in geology and meteorology, oceanography is not a discipline. Rather it is a field in which many disciplines find relevance. This makes knowledge of the oceans the valid concern of many students, mostly well advanced in their undergraduate training or beginning graduate work, who are specializing in a variety of related but actually quite different fields.

The language of physical oceanography is primarily that of physics with a rather special twist that is familiar to dynamical meteorologists. The concepts of physical oceanography are strange to others and especially difficult for those without some training in applied mathematics. A teacher can be hard pressed to select a level of physical terminology and symbolism that is equally rewarding to all who sit before him, and it is too time-consuming to give the necessary series of orientation lectures. Clearly this is the function of the written word. There is no text, including this one, that has sufficient flexibility to smooth out all dissimilarities of background through preparatory reading; but some difficulties can be eased. That is the object of this book.

In its broad plan the text is divided into three parts treating certain aspects of the geophysical situation of the oceans, the physical concepts and interpretative theories related to the ocean-atmosphere system, and the practical side of the science. Relatively little descriptive oceanography is included since this is so clearly treated in other books and articles to which references are given. Mathematics is used only sparingly and in parallel with verbal or pictorial description.

In preparing this text I have drawn information from many published sources and from many informal discussions with my colleagues. Care has been taken to express grateful acknowledgment for each use of published material. Efforts to do the same with unpublished material are certain to

v

be deficient in number but not sincerity. Some of those who have given me counsel, kind criticism and outright instruction to the enormous benefit of this book are Robert S. Arthur, Alan J. Faller, Frederick C. Fuglister, Columbus O'D. Iselin, Hans A. Panofsky, Alfred C. Redfield, Hunter Rouse, Victor P. Starr, Melvin E. Stern, Henry M. Stommel, T. Ferris Webster, and George Veronis. I am particularly indebted to Hans A. Panofsky for his contributions to Chapter 4 on elementary fluid mechanics and to Henry M. Stommel, George L. Pickard, and Gordon A. Riley for reviewing the whole text and offering many valuable suggestions.

There is also a considerable amount of time and effort expended in the preparation of a book by other persons. I would like to thank Charles E. Spooner, Robert W. Allen, and Georg Boltz for making photographic reproductions, John W. Stimpson and Joseph E. Banks for preparing many of the illustrations, Miss Helen F. Phillips for typing each of the several drafts of the manuscript, and Mrs. Ann J. Martin for editorial assistance, especially in connection with the references given in Appendix A. I am also indebted to the Office of Naval Research for supporting this endeavor and permitting the royalties that may accrue from publication to be used in furthering the educational programs of the Woods Hole Oceanographic Institution.

W. S. v.A.

Woods Hole, Massachusetts

Contents

Introduction

Oceanography, an assemblage of many sciences oriented toward a study of the earth's oceans, is principally concerned with various aspects of sea water: its motions and chemical constituents, its physical properties and behavior, its relationships to the solid earth, the atmosphere, and to living organisms of all kinds, its economic and technical potentialities, its role as a part of the earth's outer covering. Where sea water wets the solid crust, oceanography enters the domain of geology. Where it reflects sunlight, is distilled into the atmosphere, or exerts a drag on the winds, oceanography is joined with meteorology. Where marine forms of life exist or land forms migrate by way of the sea, oceanography merges with biology. And where man must combat or find uses for the sea or sea water itself, oceanography is allied with engineering and technology. Oceanography is concerned with the sea as a major part of the human environment. It is a relatively unspecialized field, and a large number of disciplines find application within its boundaries.

In the traditional view oceanography is subdivided into three main branches: physical oceanography, chemical oceanography, and biological oceanography. But this classical subdivision is restrictive. Today it is equally appropriate to include marine meteorology, submarine geology and geophysics, and certain branches of engineering as fundamental parts of the science.

Physical oceanography involves two major activities: (1) a direct observational study of the oceans and the preparation of synoptic charts of oceanic properties, and (2) a theoretical study of the physical processes which might be expected to lead to the observed behavior of the oceans. The first is a branch of physical geography and the second is a branch of theoretical physics. Neither can stand without chemical and biological information as a part of the description of the oceans and as a cross-check on the validity of physical reasoning.

The chemist's role in the study of the oceans is nearly central, being as indispensable to physical as it is to biological studies. Through chemical analysis we have learned that the composition of the salt dissolved in sea water is, in a broad way, nearly the same regardless of its total concentration,

1

which suggests that the whole world ocean is well mixed, and we infer that this mixture must be sustained by various classes of currents. Chemistry also provides a method for determining the density of sea water. From a knowledge of sea-water density as a function of depth, it has become possible to compute the field of relative motion in the sea that contributes to the mixing process.

Near shore the "constancy of composition" of sea water cannot be assumed to hold even approximately. Dissolved inorganic and organic materials from rivers, industrial wastes, and ground water constitute an important environmental influence on the productivity of organisms in the sea. These problems make the work of the chemist nearly indistinguishable from that of the biologist.

Biologists often work in association with both chemists and physicists to determine the relations between organisms and their environment. Part of the biologist's problem is to determine the organic and inorganic intake of organisms, the cycles by which these materials are returned to supply future generations, and the paths these materials take through the food cycle of the sea. Lately radiochemical evidence has been used to trace substances through the food chain as well as to study the characteristics of predation and the diseases of organisms. When the source of radioactive material is localized in space and time, some inferences can be drawn concerning the migrations of the larger pelagic forms of life.

These interrelated investigations have bearing on the fishing industry, which is at present the most important economic contribution of the oceans. Economic oceanography, however, leads beyond fishery resources to consideration of waste-disposal problems, the recovery of metals and salts from the sea, the extraction of power from waves and tides, a study of the motion of ships in a seaway and of the management of ships so as to take advantage of favorable currents and weather, the conversion of sea water to potable water, and the construction of well-drilling platforms which will withstand the onslaught of waves and the erosion of currents. The engineering difficulties attending the construction of offshore platforms have received recent public notice, but in the background there is the quiet triumph of the submarine telegraph companies which have maintained lines on the sea bed for more than a century. This achievement was based on a scientific study of the sea undertaken to satisfy an economic need. Such motivation has been mainly responsible for progress in the science from the earliest times.

CHAPTER 1

Early Explorations and Ideas

The history of physical oceanography is concerned with efforts to penetrate both the geographic and intellectual unknown. Because exploration of the unknown cannot be planned in advance, it is characteristic of the history of science that a chronological review of discovery reflects confusion. Each advance in knowledge is attained either by accident or through some small, inspired insight into the nature of things. The memorable contributions arise mainly from controversy or from the efforts of those individuals whose work, however it was inspired, has provided insights into the nature of phenomena still to be observed.

Study of the sea seems always to have been prompted by a practical rather than an abstract curiosity about the natural world. Old documents on seamanship show that the first need of the mariner, then as now, has been to know his position. Early navigators probably relied on soundings a good deal but soon learned to avoid the treacherous shoals and currents near shore. The maritime commerce of the Phoenicians and Greeks made the Mediterranean and adjacent seas reasonably well known even before the time of Homer (ca. 850 B.C.). However, it took time and extraordinary courage to venture westward into the broad reaches of the Atlantic beyond the Straits of Gibraltar and eastward into the Indian Ocean. The Homeric conception of the world, about 1000 B.C., included only the land immediately adjacent to the Mediterranean, surrounded by Oceanus, the indefinite land-encircling ocean which lay everywhere beyond the frontiers of knowledge.

Africa remained a puzzle for many centuries. About 600 B.C., according to Herodotus, King Necho of Egypt sent an expedition manned with Phoenician sailors down the Red Sea and along the east coast of Africa. The ship is said to have returned to the Mediterranean, three years later, by way of Gibraltar. This expedition should have established that Africa is a separate continent, but the chronicle was rejected by scholars of that time. In the light of present knowledge, a westward passage around the

3

coast of Africa was entirely feasible because it would have been aided by favorable winds and ocean currents during the dry monsoon as far south as the Cape of Good Hope. The trip north on the west coast of Africa would have been made largely under beam winds with a weak following current.

By 500 B.C. trade by sea in the Mediterranean had reached the point where it was profitable for pirates to be abroad, and competition among city-states assumed sufficient intensity for the bolder traders to seek new markets beyond the Straits of Gibraltar. Hanno, a Carthaginian, sailed southward along the west coast of Africa as far perhaps as Cape Palmas, Liberia. Beyond this, the map of the world drawn in 450 B.C. by Herodotus shows the Mediterranean and southern Asiatic world confronted by unknown oceans on the south and the European world bordered on the north by unknown lands.

Hecataeus' map of the world (ca. 500 B.C.) was very little expanded beyond the Homeric map and included only a small portion of southern Asia (Fig. 1–1). Probably most of this was filled in by the expedition (329–325 B.C.) eastward to the Indus River led by Alexander the Great, which revealed the relationships of several bodies of water to one another, namely, the Caspian Sea, Persian Gulf, and Arabian Sea, and to the then known world around the Red Sea and Mediterranean.

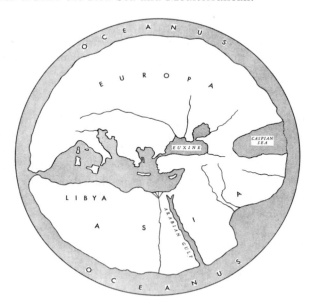

Fig. 1–1.   Hecataeus' map (ca. 500 B.C.). [From M. R. Cohen and I. E. Drabkin, 1948, *A Source Book in Greek Science*, New York: McGraw-Hill.]

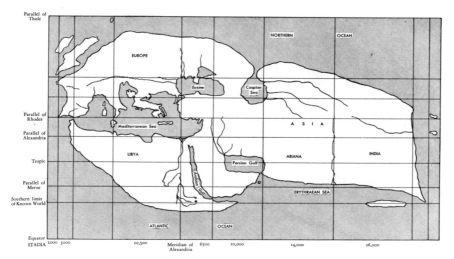

Fig. 1–2. Eratosthenes' map. [From M. R. Cohen and I. E. Drabkin, 1948, *A Source Book in Greek Science*, New York: McGraw-Hill.]

In the fourth century B.C. the astronomer Pytheas sailed from Massilia through the Gibraltar Straits on a voyage of discovery, fortified with reasonably accurate astronomical determinations of latitude. On his journey he is said to have explored Great Britain and investigated the tin ores of Cornwall. The rise and fall of the great tides of the English Channel were ascribed by Pytheas to the influence of the moon, although he knew nothing of the principles of gravitation. He also gave the name of Thule to a land to the north which may have been either the Scandinavian peninsula or Iceland. On the map drawn about 300 B.C. by Dinaearchus of Messina, a pupil of Aristotle, parallels of latitude are used for the first time.

The Greek mathematician and astronomer Eratosthenes (ca. 250 B.C.), who lived about 200 years after the time of Pytheas, determined the circumference of the earth with what turned out to be remarkable accuracy. From this he computed the ratio of the inhabited portion to that which remained to be explored. He also produced a map of the world (Fig. 1–2) which was not greatly improved until the time of Ptolemy in 150 A.D. The map by Ptolemy (Fig. 1–3) represents the Indian Ocean as an enclosed sea bounded on the south by land stretching from Africa to China.

The Indian Ocean was the first to be used for trade but, strangely, was one of the last to be thoroughly explored. The ancients carried on a brisk trade between the Mediterranean and the East by way of the Red Sea and Indian Ocean. This sea traffic was much influenced by the monsoon. The wet monsoon of the Northern Hemisphere summer permitted the ships of the Greek and Arab traders to penetrate the Arabian Sea and Bay of

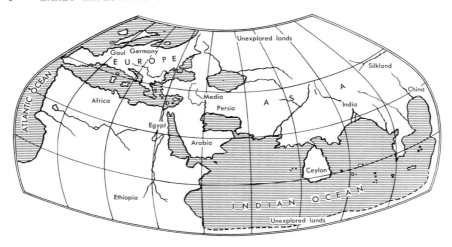

Fig. 1–3. Ptolemy's map. [From D. E. Smith, 1923, *History of Mathematics*, Vol. 1, Boston: Ginn.]

Bengal, and then during the period of the dry monsoon of the Northern Hemisphere winter to find favorable winds for the homeward journey. It was the reversing currents associated with the monsoon winds of the northern Indian Ocean that favored this traffic, made fruitful by the ready markets for oriental products in the Roman world. With the fall of Rome the trade dwindled, but unwittingly the knowledge that the Greeks and Romans possessed had been deposited with Arabian scholars for safekeeping during the Dark Ages. By this accident the Ptolemaic view of the world was kept intact until the eleventh century A.D. when, as a by-product of the Crusades, Western civilization was re-educated concerning its own past.

With the lapse of nearly fifteen centuries in the Christian era, men again became generally inquisitive about the natural world. At this time Portuguese mariners re-explored the west coast of Africa, venturing beyond the terminus of Hanno's expedition. The Cape of Good Hope was not only reached but rounded by Bartholomeu Diaz in 1488. Africa was circumnavigated from west to east in 1497 by Vasco da Gama, who believed in and found a sea route to India. Christopher Columbus headed west with the same object in view and stumbled upon the Antilles of the New World in October, 1492. By the sixteenth century the uncomfortable fact of the earth's rotundity was proved beyond question by Magellan's expedition in 1519–1522. In the course of this voyage Magellan is supposed to have attempted a sounding in the Low Archipelago of the Pacific but failed to reach bottom with a few hundred fathoms of sounding line. This may be the first deep-sea sounding ever attempted.

Fig. 1–4. Captain James Cook. [From J. Murray and J. Hjort, 1912, *The Depths of the Ocean*, London: Macmillan.]

Magellan sailed from Spain with five ships to make his way through the Straits of Magellan (the nature of which is awesomely described by Captain Joshua Slocum in his classic book *Sailing Alone around the World*) and crossed the Pacific to meet death in April, 1521, at the hands of the aborigines of Cebu in the Philippines. Survivors of his expedition returned to Spain in September, 1522.

Nearly six decades after Magellan, Sir Francis Drake found the gap between Tierra del Fuego and the mainland of Antarctica, the Drake Passage. This provided the closing link in the discovery of the Southern (Antarctic) Ocean. It remained to be known, however, whether or not land lay to the south. It was Captain James Cook (Fig. 1–4) who suggested that an Antarctic continent existed.

Captain Cook was commissioned to lead a scientific party to the South Pacific to observe the transit of Venus in 1769 and subsequently, in the year 1772, to circumnavigate the Antarctic Ocean at about latitude 60°S. Development of knowledge concerning the continent of Antarctica has led to a steadily decreasing estimate of its area. Even recent seismic evidence shows that certain parts of Antarctica are covered with ice so thick that the rocks supporting it must lie below present sea level.

With the completion of Cook's circumnavigation, the bold outline of the world ocean was known. The depth of the sea, however, remained to be explored. The first expedition to measure the vertical extent of the ocean and one of the last to map its southern boundaries was led by Captain James C. Ross in the ships *Erebus* and *Terror* during the years 1839 to 1843.

Fig. 1–5. Sir Charles Wyville Thomson. [From W. A. Herdman, 1923, *Founders of Oceanography and Their Work*, London: Edward Arnold.]

Fig. 1–6. H.M.S. *Challenger* shortening sail to sound. [From J. Murray and J. Hjort, 1912, *The Depths of the Ocean*, London: Macmillan.]

Fig. 1–7. Matthew Fontaine Maury. [From C. L. Lewis, 1927, *Matthew Fontaine Maury, Pathfinder of the Seas*, Annapolis: U.S. Naval Institute.]

A few years before the Ross expedition, the *Beagle* (1831 to 1836), with Charles Darwin aboard, made its famous voyage in which so much new knowledge of the "natural history" of the ocean islands was obtained. Darwin also looked into the geologic structure and possible origins of ocean islands. This voyage ushered in the succession of cruises devoted to scientific study of the natural history and philosophy of the seas which culminated in the efforts of Sir Charles Wyville Thomson (Fig. 1–5) in *Lightning* (1868), *Porcupine* (1869 to 1870), and finally in the *Challenger* expedition of 1873 to 1876.

The *Challenger* expedition (Fig. 1–6) represented the first systematic attempt to examine the depth and breadth of the world's ocean from the chemical, physical, and biological points of view.* Among many other things, the *Challenger* expedition placed a scientific foundation under the empirical studies of the ocean circulation begun 20 years earlier by Matthew Fontaine Maury (1806–1873) (Fig. 1–7).

As a result of physical injury while a lieutenant in the U.S. Navy, Maury was forced to spend most of his naval career ashore in charge of the Depot of Charts and Instruments. This personal misfortune brought to Maury's attention piles of logbooks containing numerous daily observations of the

---

* The detailed account of the *Challenger* expedition is given in a narrative prepared by the expedition's historian Sir John Murray, *Report on the Scientific Results of the Voyage of H.M.S. Challenger During the Years 1873–76*. This chronicle fills but two of some fifty volumes published as a consequence of the *Challenger* expedition.

Fig. 1–8.   Vilhelm F. K. Bjerknes. [From *Weatherwise*, 11(3), 1957.]

wind and currents of the oceans, which he assembled on charts.   From these he discovered that the average ocean currents are related to the average winds and can be characterized as great eddylike motions that change somewhat with the seasons.   This discovery not only made it possible to manage slow-moving sailing ships so as to take advantage of favorable winds and currents, but laid the foundations of "the physical geography of the sea."   In 1855 Maury published a summary of his studies under that title.   These findings, together with those of the *Challenger* expedition, provided the nucleus of present understanding of the architecture of the oceans.   Study of the physical mechanisms that bring these features into existence has been the main concern of physical oceanography ever since.

Some of the major contributions to this effort came from the school of thought stimulated by the Norwegian theoretical physicist Vilhelm F. K. Bjerknes (Fig. 1–8).   In 1898# Bjerknes published a paper which provided a basis for determining the field of motion in the sea from measurements of the vertical and horizontal distributions of pressure.   Currents and volumes of water moving in the oceans are difficult to measure directly because there are no convenient reference marks at sea that are assuredly at rest.   The practical methods for applying dynamical principles at sea were developed during the first quarter of the present century by Bjørn Helland-Hansen, J. W. Sandstrøm, and several other Norwegian, Swedish,

---

See Appendix A.

and German oceanographers whose names will become familiar in following chapters.

Near the turn of the century two other major steps were taken in the formulation of the modern point of view. Maury's observation of the close relationship between surface winds and ocean surface currents was given a physical explanation by V. W. Ekman in 1905. On the basis of Ekman's idealized model and the growing collection of reliable observations of the oceans at all depths, it was possible in 1907 for O. Krümmel to publish the first of a two-volume work, *Handbuch der Ozeanographie* (1907, 1911), giving physical justification for many of his views on the ocean structure and circulations.

At the same time that progress was being made in physical oceanography, some chemical techniques were developed as needed to determine the density and pressure of sea water by methods superior to the old hydrometer techniques used on the *Challenger*. Otto Pettersson suggested that if the chloride concentration in sea water could be assumed to represent the total salt content, and if the water temperature were also known, it would be possible to compute the density of sea water with the necessary accuracy for dynamical computations. This suggestion was made in Stockholm in 1899 at the first international conference for the exploration of the sea. In the three-year interval before the second conference in 1901, the relationships between sea-water density and total salt, and between total salt and chloride concentrations, had been measured and tabulated by an extraordinarily active group working in Copenhagen. The tables prepared by Martin Knudsen for the computation of salinity from measurement of chloride concentration are still in use. The standards of salinity prepared under the direction of Knudsen and later by Fridtjof Nansen are still carefully preserved and sparingly used. A supplemental supply of standard sea water has been provided by F. M. Soule and C. A. Barnes, mainly for work in connection with the International Ice Patrol.

With the *Titanic* disaster of 1912, steps were taken to provide an international iceberg warning system. By 1914 this program was put into effect as the International Ice Patrol. In 1926 E. H. Smith* showed how Bjerknes' principles could be applied to the circulation of the Labrador Sea and Baffin Bay areas adjacent to the North Atlantic shipping lanes. An oceanographic survey group under Smith's and later Soule's (1933) scientific direction, supported by contributions from many nations, has not only made an extraordinary study of high-latitude circulations but has supplied the major burden of proof that the indirect methods proposed by the Norwegian group are valid for practical purposes.

---

* E. H. Smith, 1926, "A Practical Method for Determining Ocean Currents." *Bull. U.S. Coast Guard*, 14, 50 pp.

Fig. 1–9. *Discovery I.* [From *Discovery Papers*, Vol. 1, 1929, Cambridge University Press; now issued by the National Institute of Oceanography.]

In the icy waters at the other end of the world, another major oceanographic effort was undertaken to study the whale fishery of the Antarctic. This enterprise has resulted in a surprisingly profound understanding of the Antarctic in spite of its remoteness. In 1925 the *Discovery* Committee in England, acting on behalf of the Falkland Island Dependencies, commissioned *Discovery I* (Fig. 1–9) and later *Discovery II* (Fig. 1–10) and *William Scoresby* to engage in marine and whale fishery surveys with the support of the taxes collected from whaling ships using British ports.

Further work of excellence has been carried out in all oceans by the American nonmagnetic ship *Carnegie* (Fig. 1–12) and in the South Atlantic by the cruises of the German *Meteor* (Fig. 1–11). These and many lesser expeditions have provided the framework for modern descriptive knowledge of the deep oceans.

Fig. 1–10. *Discovery II.* [From *Discovery Reports*, Vol. 1, 1929, Cambridge University Press; now issued by the National Institute of Oceanography.]

Fig. 1–11. *Meteor.* [From *Wissenschaftliche Ergebnisse der Deutschen Atlantischen Expedition auf dem Forschungs- und Vermessungsschiff "Meteor" 1925–27*, Bd. I, A. Defant, editor, Berlin: Walter de Gruyter.]

Fig. 1–12. *Carnegie*, by the wind in the South Pacific. [From J. Harland Paul, 1932, *The Last Cruise of the Carnegie*, Baltimore: Williams and Wilkins.]

Concurrent with the progress of ocean-wide surveys was the more intensive work of the fisheries oceanographers in western Europe and eastern North America. These energetic men believed that the direction of greatest promise lay in extending to the open sea those special lines of inquiry which were provided with a sound theoretical basis and could be tested in relatively small areas. Henry B. Bigelow (Fig. 1–13) summarized the problems of the time (1931) in his book *Oceanography; Its Scope, Problems, and Economic Importance* and laid the groundwork for an era of modern research centered in the United States. Although many of the problems he cited still have not been solved, his philosophy has prospered. Under

Fig. 1–13.   Henry Bryant Bigelow.   [WHOI Archives.]

Fig. 1–14.   *Atlantis*.   [Photograph by D. M. Owen.]

his guidance the Woods Hole Oceanographic Institution was founded in 1930, the research ship *Atlantis* (Fig. 1–14) built and commissioned. The work of *Atlantis* has been directed mainly by Columbus O'D. Iselin toward a study of the morphology of the circulation of the western North Atlantic and its periodic as well as secular changes.

In a comparatively short time, enough data were assembled to invite theoretical inquiry. Comparisons of these and other observations with theoretical models of the ocean circulation were soon made possible by the vigorous adaptation of the principles of fluid mechanics to oceanographic problems in the gifted mind of Carl-Gustav Rossby. Rossby's bold approach to a study of the basic phenomena of fluid motion on a rotating earth has raised many more problems than it has settled, and the restless upheaval continues to the great benefit of all concerned.

On the other side of the American continent, a marine biological station established in 1891 in connection with the Department of Zoology of the University of California became in 1925 the Scripps Institution of Oceanography under the direction of T. Wayland Vaughan, who was succeeded in 1936 by Harald U. Sverdrup (Fig. 1–15). It was Sverdrup, together with M. W. Johnson and R. H. Fleming, who wrote in 1942 a compendious book, *The Oceans, Their Physics, Chemistry, and General Biology*, which brought together much of the world's knowledge of the whole science. This book joined the old European and growing American points of view and probably helped to shift the center of gravity of oceanographic research effort from the war-torn and economically hard-pressed European laboratories to those of North America. Only recently have the European laboratories begun to prosper, along with those of Japan, Germany and, if reports are correct, those of Soviet Russia.

The influence of British institutions on the origins and present progress of oceanography has been profound. The Admiralty, acting on the recommendation of the Royal Society, in 1873 sent out the *Challenger* expedition. The institution that has come to be known as the Liverpool Observatory and Tidal Institute was founded in 1845 at the time Maury was engrossed in his famous work. The Plymouth Marine Laboratory, opened in 1888, is still active. To these, since the war, has been added the National Institute of Oceanography, which under the direction of G. E. R. Deacon has assumed a prominent position among laboratories concerned with the modern approach to physical study of the oceans.

The basic attitude in modern oceanography is to supplement physical deductions from Newtonian principles with direct observation as closely and as often as possible. This approach need not be remote from practical problems. Close knitting of the faculties of physical imagination with those of direct observation is a highly effective scientific procedure. The older inductive method of global surveys followed by years of pains-

Fig. 1–15.   Harald Ulrick Sverdrup. [Photograph by Paul Williams.]

taking data analysis is not yet outmoded but is rapidly giving way to deft inquiry based on a keener appreciation of the nature of the physical world.

Long before an observational basis was established for the scientific study of the oceans, the principles of theoretical physics and mathematical analysis were being formulated by the mathematicians, astronomers, engineers, and occasional physicians and clerics of the seventeenth, eighteenth, and nineteenth centuries. The roots of theoretical physical oceanography press deeply into all aspects of a physical description of the earth and are strongly nourished by Newtonian principles, mainly as given in the *Principia Mathematica Philosophiae Naturalis* of 1687. This great study plus some basic principles of fact and logic, such as Euclidian geometry (300 B.C.), the hydrostatic principle of Archimedes (287?–212 B.C.), the records of Babylonian and Egyptian astronomy which have reached us through the scholarship of the Alexandrian Greeks and the spoils of the Crusades, almost complete the record of progress toward modern understanding of the earth up to the beginning of the seventeenth century. But from the Newtonian era onward, the record of achievement seems almost explosive; perhaps because it is connected with our time, or because this era really has been a period of extraordinarily successful preoccupation with the human environment. Some of the history of these developments is contained in the chapters that follow and more will be found in Appendix A. References to Appendix A are identified in the text by the symbol "#."

SUPPLEMENTARY READING

BIGELOW, H. B., 1931, *Oceanography*, Boston: Houghton Mifflin.

BOWDITCH, N., 1958, *American Practical Navigator*, Parts 1 and 6, U.S. Navy Hydrographic Office Publication No. 9.

BROWN, H., *et al.*, 1959, *Oceanography 1960 to 1970*, Washington, D. C.: National Academy of Sciences—National Research Council.

BROWN, H. S., 1954, *The Challenge of Man's Future*, New York: Viking.

COLLINDER, Per, 1955, *A History of Marine Navigation*, New York: St. Martin's Press.

DEACON, G. E. R., H. U. SVERDRUP, H. STOMMEL, and C. W. THORNTHWAITE, 1955, "Discussions on the Relationships Between Meteorology and Oceanography," *J. Mar. Res.*, 14:499–515.

GARRATT, G. R. M., 1950, *One-Hundred Years of Submarine Cables*, London: His Majesty's Stationery Office.

GOULD, LT. CDR. R. T., 1923, *The Marine Chronometer, Its History and Development*, London: J. D. Potter.

TAYLOR, E. G. R., 1957, *The Haven-Finding Art; A History of Navigation from Odysseus to Captain Cook*, New York: Abelard-Schuman, Ltd.

VAUGHAN, T. W., *et al.*, 1937, *International Aspects of Oceanography, Oceanographic Data and Provisions for Oceanographic Research*, Washington, D. C.: National Academy of Sciences.

On Geological and Astronomical Backgrounds

According to a casual definition, the earth can be regarded as being composed of three materials, rock, water, and air, arranged in three layers —the lithosphere, hydrosphere, and atmosphere. The earth is also an astronomical body, and lacking this condition very few of the present concerns of geologists, oceanographers, and meteorologists would remain. The effects of the nonuniform distribution of sunlight over the earth and the equally energetic but more uniform radiation of earth heat into space, acting with rotational and gravitational forces, produce a complex of interdependent fluid phenomena which characterize the world as we know it. In this chapter we will review some of the facts and hypotheses concerning the geological and astronomical setting of the world ocean that bear upon its properties and phenomena.

The geometry of the oceans

The oceans represent only 1/790 of the volume of the earth but are conspicuous because they are so extraordinarily thin in comparison with their horizontal dimensions. Indeed, on a globe they are well represented by the thickness of varnish or at most by the thickness of the paper on which a map is printed, since they average only 1/1680 of the earth's radius in depth. Oceanographers are accustomed to think of the ocean in terms of diagrams which are strongly exaggerated in the vertical dimension. This habit can lead to misconceptions. But there is no simple alternative, for sections of oceans drawn to the natural scale and printed in a book the width of this one might extend from margin to margin but from top to bottom be hardly more than a thin line. Under most circumstances a drawing to natural scale could be fitted on the edge of a page.

In spite of the almost vanishing thickness of the ocean layer compared with its width, the average depth of the oceans is approximately 4 km.

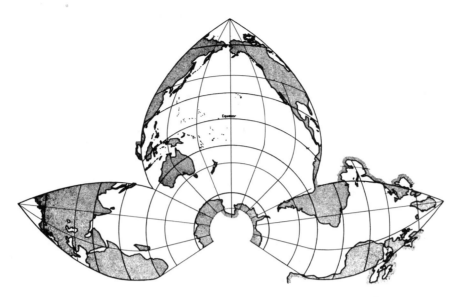

Fig. 2–1. The world ocean in Bartholomew's petal projection. [Adapted from Plate 2 of *The Times Atlas of the World*, Vol. 1.]

According to recent estimates, the total area of the globe is 510,100,934 km^2, of which the land area is 148,847,000 km^2 (29%), while that of the sea is 361,254,000 km^2 or about 71% of the whole. The volume of sea water on the earth is close to 1,369,000,000 km^3 and comprises more than 98% of the hydrosphere.

The *world ocean* is composed of three branches extending northward from a circumpolar ring around the Antarctic continent (Fig. 2–1). The Atlantic branch is the longest, stretching from the Antarctic Ocean to the Arctic Sea. The Pacific branch reaches northward to the Bering Strait and, by the shallowness of this passage, is effectively cut off from the Arctic basin. In spite of this abbreviation, the Pacific Ocean is so wide it covers nearly one-third of the globe. The Indian Ocean branch is the shortest, covering roughly one-seventh of the earth's surface. This seventh, however, lies mainly in the tropics and is adjacent to the largest land mass on the earth, so that the continentality of the land exerts a marked influence on its circulation.

The Atlantic Ocean system, covering one-fifth of the earth, is in effect a broad channel whose properties are subject to the gradual variation of climate with latitude from south to north. The floor of the Atlantic arm is divided by an extensive mid-ocean ridge (Fig. 2–2). The North Atlantic, almost at the end of this arm, is a kind of cul-de-sac, but it is the area that

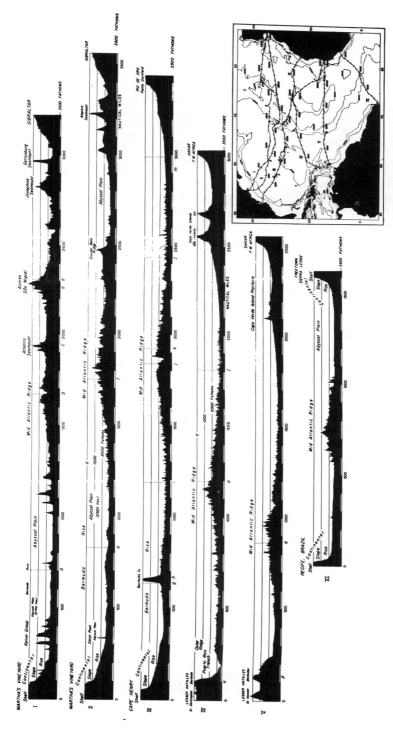

Fig. 2-2. Six topographic profiles across the North Atlantic basin. Vertical exaggeration 40:1. [From B. C. Heezen, M. Tharp, and M. Ewing, 1959, *The Floors of the Oceans, I. The North Atlantic*, Geological Society of America Special Paper 65.]

Fig. 2–3. Splintered berg off Melville Bay. [Courtesy International Ice Patrol.]

has been studied most intensively. This may be unfortunate, because the generalizations drawn from studies of the North Atlantic waters may not be wholly valid for other ocean situations.

As the hub of the world ocean, the Antarctic circumpolar sea exerts an influence on each of the three arms. The melting ice of Antarctica and excess of precipitation over evaporation in summer tend to freshen the Antarctic coastal waters, but at the same time the climate is so frigid that these waters are still dense enough to sink and produce a layer of cold, relatively fresh water overlying the bottom of most ocean areas in the Southern Hemisphere. In winter the production of new sea ice increases the water density even more by slightly enriching its salt content. Through these effects on its density and its sheer abundance, Antarctic circumpolar water extends its influence even into parts of the world ocean in the Northern Hemisphere.

The bergs characteristically calved from the shelf ice in the Weddell and Ross Seas are tabular blocks sometimes tens of miles long, several miles wide, and hundreds of feet thick. These are different from the splintered bergs (Fig. 2–3), usually encountered in the Northern Hemisphere, which are calved from distributary glaciers connected with the Greenland icecap. The few tabular bergs that exist in the Arctic Sea (Fig. 2–4), the "ice islands," appear to be derived from piedmont glaciers and are therefore very small in comparison with the Antarctic variety, although large enough to support geophysical observatories for a period of years.

Peripheral to the world ocean are the *marginal seas*. These bodies of water are characteristically adjacent to continents and enclosed by island arcs. They vary widely in depth, shape, and size. Most of them are found on the east coast of Asia—the South China Sea, East China Sea, Sea of Japan, Sea of Okhotsk, and the Bering Sea—but the Kara and Barents

Fig. 2–4. Tabular berg off Melville Bay. [Courtesy International Ice Patrol.]

Seas on the north of Asia, the Arabian and Andaman Seas on the south of Asia, and the Caribbean Sea on the southeast coast of North America also qualify.

These features have close relatives in the seas less intimately associated with continents but also delineated by island chains, namely, the Coral Sea, Java Sea, Celebes Sea, and Bismark Sea of the southwest Pacific Archipelago. The Tasman Sea east of Australia and the Scotia and Weddell Seas adjacent to the Antarctic continent are varieties of marginal seas which press the definition by being exceptionally well connected with the world ocean.

The *mediterranean seas* of the earth seem epicontinental but are of nearly oceanic depth and connected to the world ocean by shallow straits. The prototype for these is the European Mediterranean, but the Arctic Ocean and the Gulf of Mexico can also be classed as mediterranean seas.

Among the least of the arms of the ocean are the inland seas. Some of these are quite salt if they are fed by rivers and drained mainly by evaporation. In time some reach complete dryness, leaving deposits of evaporite in their basins. Others become more and more salt and may eventually become anaerobic at depth if their connection with the ocean is blocked by a shallow sill. These many, and often ephemeral, features of the hydrosphere are more closely akin to lakes* than to oceans.

---

* The study of lakes has lately become a nearly separate science, limnology. Rather than to deal with them too briefly here, the complexity of lakes and inland seas makes it preferable to refer the reader to *A Treatise on Limnology*, 1957, Vol. I, by G. Evelyn Hutchinson, New York: Wiley; and *Fundamentals of Limnology*, 1953, by Franz Ruttner, translated by D. G. Frey and F. E. J. Fry, University of Toronto Press.

The age of the oceans

The age of the world ocean has not been established, but some limits can be stated. The ocean cannot be older than the earth itself. Measurements of uranium-lead ratios in samples from the pre-Cambrian rocks of Canada and South Africa show that the solid part of the earth may be as old as 2800 million years. The primordial earth may be twice that old, or even more, and still not be older than the universe, whose age is variously estimated at between 5 and 10 billion years. It remains even now to learn whether or not the primordial crust was covered by ocean water to anything like the present extent.

One of the first methods used to determine the age of the ocean was to divide the total salt content of the world ocean by the annual increment of salt discharged into the sea by rivers. This procedure was suggested by Edmund Halley in 1715[#] but not implemented until 1899[#] when Joly made the first estimate using data on the abundance of sea salt obtained by the *Challenger* expedition. According to F. W. Clarke's summary of *The Data of Geochemistry*,[1][*] the result of several such computations of the salt age is somewhat less than 100 million years, depending upon the figures used. This period of existence is clearly too brief, because marine organisms very like present-day species have been found in early Cambrian and some late pre-Cambrian rocks of an age which, measured by radioactive decay methods, turns out to be in the order of 500 million years. Recent studies by Woodcock[2] and others have shown that the salt in rivers can be quantitatively accounted for in terms of "cyclic sea salts" derived from spray and droplets formed by bubbles bursting on the sea surface. Because of their small size these droplets evaporate rapidly into the lower air, and the tiny crystals of salt remaining are carried aloft by the winds to be distributed more or less uniformly over the earth. Most of these minute salt particles serve as nuclei for the condensation of rain and eventually find their way back to the ocean either directly or by way of the watersheds and rivers of the world.

It is generally assumed that the oceans are at least as old as marine fossils, and that they have been salt as far back as upper Silurian time when very thick salt beds were deposited.[†]

---

[*] The superscript numerals in brackets are keyed to the References at the ends of the chapters.

[†] Thus far the age of the oceans has also eluded measurement by radioactive decay methods. Carbon-14 methods have been used to show the time elapsed since a parcel of water was last exposed to the air. In the mid-depths of the North Atlantic the values seem to center around 400 years.

The origins of sea water

Closely associated with the age of the world ocean is the question of the origin of such a vast quantity of water (about 1 billion km^3) and of the salt (about 3%) that it contains. Early speculations were concerned with a deluge. At the beginning of the nineteenth century when Hutton (1785)[#] and Playfair (1802)[#] proposed the uniformitarian doctrine, such ideas as the cataclysmic appearance of the lands and sea were in general currency, partly in response to Biblical legend. But it can be shown that at present temperatures the atmosphere, fully saturated, can hold no more than some 13,000 km^3 of water at any one time. As water vapor over a molten earth at perhaps 1200°C, only 16% of the present ocean volume would have remained in gaseous equilibrium, and the outer part of this atmosphere would have been subjected to photochemical dissociation. Moreover, both the molecular and the dissociated water would have been far enough from the earth for the velocity of escape to be exceeded by a large fraction of the ions and molecules. Therefore it seems probable that the water of the oceans may have been supplied in other ways, perhaps by being released slowly from within the earth and at temperatures lower than that of molten rock.[3]

Studies of volcanic gases show that both the halogens and the water of the oceans might have been exsolved from the rocks forming the earth's crust. Goranson[4] found that between 3 and 8% water will dissolve in molten rocks having a composition range from basalt to granite under temperatures and hydrostatic pressures approaching those supposed to exist in volcanic pipes and intrusive magmas within the crust of the earth.* Studies of the emission of volcanoes by Fenner (1926)[5] and Zies (1929)[6] in the Katmai region of Alaska show that in addition to water, quantities of halogens, especially chlorides and fluorides, are liberated with sulphurous substances from active volcanoes.

The history of the earth has included a succession of widespread outbreaks of volcanic activity. It is possible that the present abundance of chloride ion in the ocean may have been supplied by volcanic emission, provided that the halogens in modern volcanic gases have not been derived from oceanic infiltration (Fig. 2–5).[†] If the halogens were supplied to the sea by volcanoes, it seems probable that sodium and other metallic ions have a different origin because they occur in other than the proportions required for nonexotic chemical combination. It has been calculated that

---

* Recently C. W. Burnham and R. H. Jahns (1958, *Bull. Geol. Soc. Amer.*, 69: 1544–1545) have re-examined this question and find that Goranson's values may be too high by nearly a factor of two.

† Most modern volcanoes are fairly close to the edges of the continents or have formed islands in the sea.

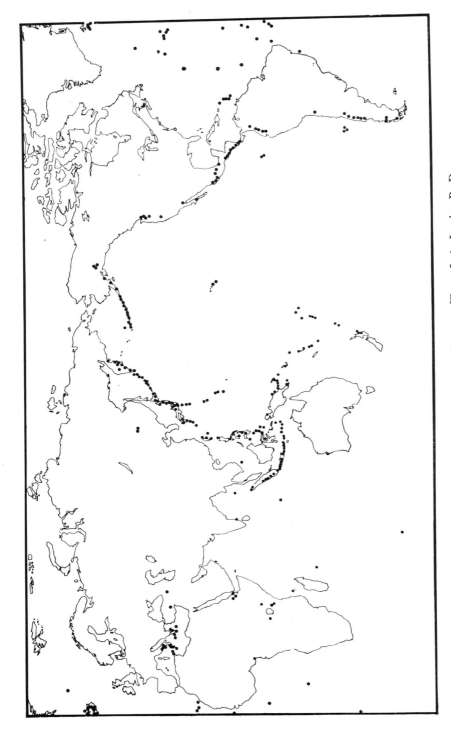

Fig. 2–5. Positions of the world's active volcanoes. [From J. A. Jacobs, R. D. Russell, and J. T. Wilson, 1959, *Physics and Geology*, New York: McGraw-Hill.]

the sodium, magnesium, and other metallic constituents of sea water can be accounted for by the chemical weathering of igneous rocks accompanying rainfall and runoff in the presence of carbonic acid derived from atmospheric carbon dioxide.

In view of the miscibility of water in molten rock under pressure and the fact that this water exsolves as the rock material crystallizes, it is of interest to compute the volume of water that could be derived from the earth's crust and compare this result with the volume of water in the oceans. The land of the earth covers an area of some 150 million km^2. The continental crust averages 33 km in thickness. Therefore the volume of continental crust amounts to about 6 billion km^3. The crust under the oceans is much thinner, about 5 km, but more extensive, covering an area of some 360 million km^2. Therefore the volume of crustal rock in the ocean floor is about 2 billion km^3. The total volume of the earth's crust is about 8 billion km^3. If this rock volume yielded 5% free water upon crystallization, we would expect the ocean volume to be about $\frac{1}{2}$ billion km^3. But the ocean volume exceeds 1 billion km^3, and this result requires either a larger yield of water from the crust than we have assumed or a contribution of water from additional sources such as the upper part of the mantle beneath the crust.

Stability of the ocean basins

The continental blocks of the earth's crust are generally composed of material of lesser density than that of the crust underlying the oceans. Crystal aggregates of the land are rich in aluminum silicates, whereas the denser rocks underlying the ocean tend to contain iron and magnesium in preference to aluminum. The fact that the land is less dense and stands higher than the sea bed, and that the anomalies of the gravitational field of the earth over the land and sea are both relatively small, suggests that the crust of the earth is floating on a still denser mantle.

An increase of density with depth is also required to provide the earth with its proper moment of rotational inertia.

The dissipation of energy in shallow seismic waves rises with the first power of their frequency, which is the case for jointed solids. From this it appears that the outer 1000 km or so of the earth may be jointed but will sustain uncompensated loads up to a point and then fail by elastic or plastic deformation only sufficiently to gain from buoyancy what it lacks in its own strength.

The level at which the crust can be distinguished from the mantle is usually related to the depth of a sharp discontinuity in the velocity of propagation of earthquake waves which has been shown to exist under both the continental and oceanic crustal rocks. This discontinuity was

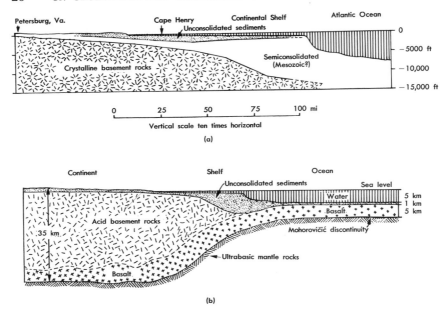

Fig. 2–6. (a) Section across the Atlantic coastal plain, as measured by Ewing. (b) Generalized section across a stable continental margin. [Reprinted from *The Earth as a Planet*, 1954, G. P. Kuiper, editor, by permission of The University of Chicago Press.]

discovered in central European seismograms by A. Mohorovičić in 1909 and independently by several others from observations made elsewhere on the earth. Seismic data taken to establish the depth of the Mohorovičić discontinuity show that the crustal rocks of continents have an average thickness of about 33 km, whereas the crust under the oceans is relatively thin, averaging about 5 km (Fig. 2–6). This, in conjunction with the difference in density of continental and oceanic crustal rock, suggests that they represent different end products of the chemical and physical processes that produced the earth's crust. Moreover, it is observed that the gravitational field of the earth is not markedly different from that which would be expected for a rotating spheroid in hydrostatic equilibrium. The continents rise above the general level of the crust because they extend their bulk deeply enough into the mantle for the buoyant forces in the displaced volume of mantle rock to support their excessive weight in comparison with that of the oceanic crust.

Because of the major chemical and physical reorganization that would be involved in changing the crustal thickness and composition from an oceanic to a continental configuration, it seems improbable that extensive transformations of this type have occurred.

Accumulation of material in sedimentary basins or geosynclines, followed by compression and folding to produce mountains, can increase the area of continental crust and presumably decrease the area allotted to the oceans. To thin the continental crust and restore it to an oceanic thickness and composition requires the invocation of sufficient heat from the mantle to melt away the continental block from below. There is geologic evidence to support the first possibility, but at present there is understandably little to suggest that the second is important. From this we are led to presume that although mountain building has thickened the crust in some areas, such thickening is not easily restored to the oceanic configuration even under the influence of prolonged erosion.

It is thought that as erosion removes the upper surface of a continental block, a continental mass rises step by step, under its own buoyancy in the heavier rocks of the mantle, to replace from below a large fraction of the volume of material removed layer by layer from the surface. Since the rate of subaerial erosion decreases with lessening elevation and ceases altogether when the block is planed off at sea level, it would take an extraordinarily long time, even in the geological sense, for erosion to obliterate any major positive feature on the earth unless sea level were steadily lowered or submarine slumping were commonplace.

The process of crustal equilibration may, at times, become "stuck" by the strength of the crust acting to retard the final stages of isostatic recovery. The fact that there are gravity anomalies also suggests that the earth's crust has strength of its own. Jeffreys estimates this strength to be sufficient to support differential pressures of as much as 100 bars.

Geological studies of the extent of continental glaciation made it seem probable that during the Pleistocene period, when large volumes of water were locked up as ice on the land surface, storm waves beat on what are now deeply submerged shelf regions of the continental margins. In 1936 R. A. Daly[7] suggested that these soft sediments under the pounding of storm waves would produce great quantities of suspended matter which, as a highly dilute mud, would be more dense than sea water and tend to flow down the continental slopes (Fig. 2–7), eroding canyons on their way into the abyss.[8] In 1937 Ph. H. Kuenen[9] developed these ideas experimentally, giving the name *turbidity current* to the muddy slurries and gravity-driven suspension flows so derived. Spectacular evidence of a contemporary suspension flow (Fig. 2–8) was found by Heezen and Ewing (1952)[10] who reasoned, in connection with an earthquake in the Grand Banks region, that a mass of slope material may have slumped onto the Telegraph Plateau southeast of Newfoundland to account for the successive rupture

Fig. 2–7. A relief map by Aero Service Corporation of a topographic survey made by Veatch and Smith of the continental shelf and abyssal slope off New England.

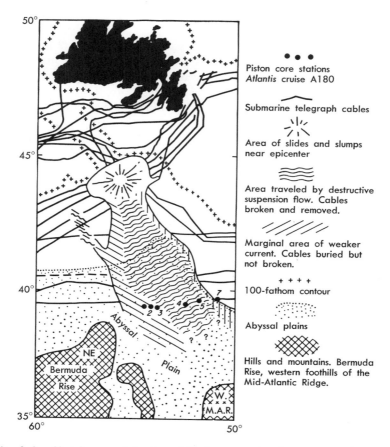

Fig. 2–8. Sketch map of the Grand Banks suspension flow of 1929. [From J. A. Jacobs, R. D. Russell, and J. T. Wilson, 1959, *Physics and Geology*, New York: McGraw-Hill.]

of five transatlantic telegraph cables in the course of a few hours. The positions of these cables were known, as well as the time at which each one parted. This permitted computation of the speed of flow, which turned out to be on the order of 50 knots. The erosive power of suspension flows of clay and silt-laden water is now considered, in addition to rafting by melting icebergs, to be a cause of the anomalous appearance of sand and other rather coarse materials in the deep parts of the ocean basins. Thus it is beginning to appear that glacial conditions are not necessary to the existence of suspension flows and that if submarine slumping is commonplace it may contribute, perhaps significantly, to the erosion of continents below sea level.

Fluctuations of sea level

Within the period of recorded history, sea level has been remarkably stable. At present it is only in areas where the continental ice sheet of Pleistocene time has melted and removed its weight from the land that sea level shows a rapid and persistent local change. In the northern Baltic, the Gulf of Bothnia, the deglaciated land is rising at a rate of approximately one meter per century. Similarly the basins of the Great Lakes of North America are observed to be tilting toward the south as a consequence of the elastic and plastic recoil of the continent relieved some 12,000 years ago of the massive load of the continental ice sheet. In most other parts of the world the average sea level is probably rising slowly because of general melting of land-born ice.

If estimates of the thickness of the ice during the Pleistocene glaciation are reliable, it seems possible that sea level was perhaps 100 meters below the present level in mid-Wisconsin time. This is sufficient to expose most of the present continental shelves. If the continuing amelioration of climate causes the polar glaciers in Greenland and Antarctica to melt completely, sea level may be expected to rise as much as several tens of meters above its present stand.

In the longer view of time, there is evidence of extensive flooding of the continents during the Paleozoic, Mesozoic, and Cenozoic eras. These events took place prior to the spasm of mountain building taken to mark the close of each of these major divisions of geologic time. There is no evidence to indicate the kind of mechanisms that caused the level of the world ocean to change. A variation of the total volume of sea water could be imagined but is hard to accept. A general flooding could result from the coordinated subsidence of the several continental blocks. But it is more plausible to imagine that flooding was caused by a buckling of some part of the sea floor. This possibility requires that the stratographic sequence of marine deposits should tend to follow a similar pattern over the entire land area of the earth. Such is only very roughly the case.

At the present time, the average height of land above sea level is 840 meters, and the average depth of the ocean is 3790 meters. The maximum height of land above sea level is 8840 meters, and the greatest depth in the ocean may exceed 10,860 meters.* The extremes of relief amount to about $1/324$ of the earth's radius, which is only a little less than the earth's rotational ellipticity, $1/297$.

---

* A sounding of 5940 fathoms (10,860 meters) was made in the Marianas trench at latitude 11°20′N, longitude 142°16′E by H.M.S. *Challenger* in 1951. See T. F. Gaskell, J. C. Swallow, and G. S. Ritchie, 1953, *Deep-Sea Res.*, 1: 60–63. Also see J. N. Carruthers and A. L. Lawford, 1952, *Nature, London*, 169: 601–603.

In general, the greatest depths of the ocean are associated with the trenches bordering island arcs. These trenches support negative gravity anomalies, which suggests either that light roots extend deep into the mantle or that the trenches are tension cracks where crustal rock substance is simply absent.

Rigidity of the earth

Those studying the strength and structure of the earth commonly consider that rocks have viscous rather than rigidly solid properties. This apparently strange point of view arises from the fact that rock substance exposed to deforming forces for different lengths of time will behave in ways which through ordinary experience we associate with materials such as wax, tar, or even "silly putty." That is, rock substance will transmit vibration by elastic deformation; sustain substantial temporary loads but break under moderate tension; flow a little in response to prolonged pressure differences; and, under sufficient confining pressure and high enough temperature, move like a liquid without actually melting.

It is often assumed that at depth in the earth the temperature as well as the confining pressure is sufficiently high to permit rock materials to change their properties enough to flow or possibly even to develop cells in which convective overturning is sustained by the evolution of radiogenic heat in the mantle. The crustal rocks floating on top of such convective cells would be exposed to horizontal shearing stresses. There is a question of quantitative sufficiency involved, but many geologic theories of mountain building and continental migration invoke convective motion in the earth's mantle as the prime mover.

Following this line of thought, it is not inconceivable that the continents have drifted over the earth from time to time in the geologic past, thereby changing the shapes but not necessarily the total areas of the oceans (Fig. 2–9).

The quasi-fluid behavior of the earth's interior would lead one to expect the settling of dense materials so as to produce a stable layering of the substances within the earth. This possibility is shown to have a basis in fact through the simple but very exacting Cavendish experiments. From these the total mass of the earth has been found to be 5.976×10^{27} gm. From the known volume of the earth it follows that its mean density is 5.522 gm/cm^3, while that of surface rocks is near 2.7 gm/cm^3.

These results require an increase of density toward the center of the earth (Fig. 2–10) but give no information as to the gradient of increase. The probable distribution of density can be estimated by reconciling the required total mass with a number of other facts and conditions. For instance, the rotational inertia of the hypothetical mass distribution must agree with that obtained from astronomical measurements of the rate of

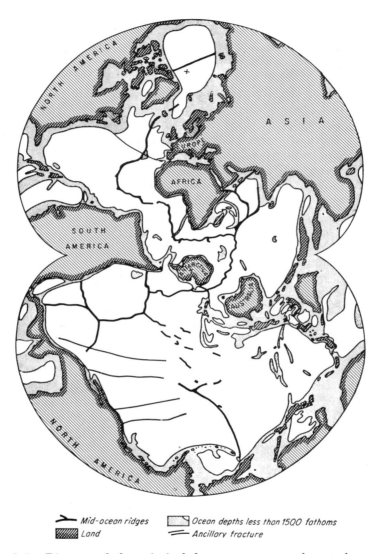

Fig. 2–9. Diagram of the principal fracture systems and central mountain chains of oceanic crust. [From J. A. Jacobs, R. D. Russell, and J. T. Wilson, 1959, *Physics and Geology*, New York: McGraw-Hill.]

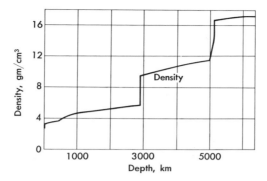

Fig. 2–10. Density of the earth as a function of depth. [Reprinted from *The Earth as a Planet*, 1954, G. P. Kuiper, editor, by permission of The University of Chicago Press.]

equinoxial precession and the observed secular change in the earth's rate of rotation. The ratio of the volume elasticity and density of the material must provide seismic velocities that agree with the observed values and change at the proper levels to account for strong seismic reflections. Since there is, as yet, no direct evidence of the internal composition of the earth, and the effects of high pressure and temperature are known only for surface materials, it is only natural that a variety of earth models have been proposed. Although there is no agreement in detail, the main features of such models are alike in resembling the structure outlined in Table 2–1.

The important fact concerning us here is that the earth seems to behave as a viscous fluid or plastic solid under sustained stress but as an elastic solid when the stress is, geologically speaking, instantaneous. Most of the

TABLE 2–1 *

	Thickness, km	Volume, $\times 10^{27}$ cm³	Mean density, gm/cm³	Mass, $\times 10^{27}$ gm	Mass, %
Atmosphere	. . . .	. . . .	. . . .	0.000006	0.0001
Hydrosphere	3.80	0.00137	1.025	0.00140	0.024
Crust (land)	33.	0.0049	2.7	0.013	0.23
Crust (sea)	5.	0.0018	3.0	0.005	0.08
Mantle	2888.	0.902	4.5	4.059	68.1
Core	3471. (equivalent)	0.175	10.7	1.876	31.5
Whole earth	6371.	1.083	5.52	5.976	99.95

* Adapted from B. Mason, 1954, p. 261, *The Earth as a Planet*, G. P. Kuiper, editor, Chicago: University of Chicago Press.

short-lived stresses on the earth are periodic or quasi-periodic. Frequencies ranging from cycles per day or a week to a full year exist in the tide-producing forces, in the diurnal and longer-period variations in the surface pressure of the atmosphere, and in the weight of water substance precipitated on the crust. To all these forces the earth seems to respond elastically. It is, in part, the difference between the elastic inphase response of the solid earth (some 30 cm) and the lagging fluid response of the oceans that permits the rise and fall of the astronomical tides. But under the sustained centrifugal force of rotation, the solid earth and its fluids respond alike and acquire an ellipsoidal figure.

Figure of the earth

If the earth were a nonrotating homogeneous or concentrically layered sphere, the change in the direction of a plumb line would be uniform for every unit of distance traversed over the earth's surface. This was the assumption made by Eratosthenes when he measured the circumference of the earth from observations of the difference in the slope of the sun's noon rays at two points in different latitudes. But the fact is, of course, that plumb-line verticals taken all over the earth's surface do not converge at a point. The earth has a more nearly ellipsoidal shape produced by the centrifugal reaction to rotation of every part of its substance.

According to the Newtonian concept of gravitation, each particle of the earth is attracted to every other particle with a force that varies directly as the product of the two masses and inversely as the square of the distance between them. A free particle at rest in some latitude on the rotating earth is on the whole attracted toward a point near the earth's geometrical center by Newtonian gravitation and at the same time urged away from the earth's axis by the centrifugal force of rotation. The composition of the centrifugal force of rotation with the force of gravitation, would tend to produce a component of force directed toward the equator were it not that the ellipsoidal figure of the earth provides a "downhill" component of the force of gravitational attraction away from the equator which balances exactly the horizontal component of the centrifugal force directed toward the equator (Fig. 2–11).

If there are no unbalanced horizontal forces acting, the free surface of a dish of water will come to rest parallel with the terrestrial ellipsoid, and a plumb line will hang at right angles to it. This means that owing to rotational ellipticity, the direction of *gravity*—the resultant of Newtonian attraction and centrifugal force—differs from the line directed toward the geometrical center of the earth everywhere except at the poles and on the equator of the earth. This difference in direction is greatest at latitude 45°, where it amounts to about 11′ of arc. It also means that the accelera-

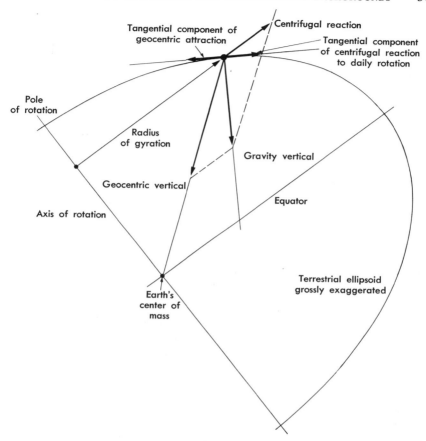

Fig. 2–11.   Relationship of geocentric to plumb-line vertical.

tion of gravity will decrease equatorward because of the combined effects of the increase of distance of the surface of the earth from its center and the increased radius of gyration of points increasingly close to the equator.

Clairault showed in 1743[#] that the acceleration of gravity varies, to a first approximation, as the square of the sine of the latitude.   The average reduction of the acceleration of gravity between the poles and the equator is 3.38 cm/sec², or about 0.3% of g_{90}.   Local anomalies of g have importance in many precise physical measurements.   Occasionally corrections may also be applied for the still smaller but time-dependent forces of lunar and solar gravitational attraction.[11]

Because of the ellipsoidal figure of the earth, distances between points on its surface are not related to the angular separation of points by simple spherical equivalence.   Instead of the shortest distance between any two points on the earth being an arc segment of a great circle, it is a segment of some other curve, a *geodesic*. A geodesic—the shortest distance between two points on a geoid—can range from nearly a circle when measured along the equator to nearly an ellipse when measured along a meridian. Study of the relationships of such curves between points on the earth's surface to their astronomical latitudes and longitudes and horizontal distances actually chained out between them is the concern of *geodesy*.

From geodetic studies of the shape of the earth, it has been learned that the local differences in the density of the rocky materials of the crust cause small anomalies in the acceleration of gravity and of the direction of local vertical. If the earth were flooded with motionless water, the surface of the water would be seen to have minor bulges and depressions. This equilibrium surface would depart from the ideal ellipsoid quite noticeably in some places (about 30 meters in the northern Indian Ocean) and would represent the true figure of the earth—the *geoid*. The geoid is a surface which lies at a variety of distances from the nominal center of the earth. But because of gravity anomalies and the rotational ellipticity of the earth, it would require an exactly equal expenditure of physical work against gravity to reach any point on the geoid from the nominal center of the earth. The geoid is, therefore, a surface which has uniform gravity potential above the geocenter: a surface of constant *geopotential*. Geopotential surfaces are level. There are enveloping the earth as many level surfaces of constant geopotential as one may choose to specify, but they never intersect and are rarely parallel.

A fluid particle on any geopotential surface will show no tendency to move unless acted on by some unbalanced force. Such forces may arise from thermal disturbances of hydrostatic equilibrium which tends to be restored by motions developed under the earth's gravity or, in the case of tides, from the action of the external gravitational fields of the sun and moon. These latter forces take on a variety of angular relationships having periodic changes and limits that are well known in terms of the rotation rates and orbital motions of the earth and moon.

Motions of the earth

The systematic motions of the planets around the sun and of the moon around the earth that are important to us in connection with the tidal problem were reduced to simple terms by Johann Kepler (1571–1630) in the form of three laws.

(1) The orbit of each planet is an ellipse with the sun at one focus.

(2) Each planet revolves at such a rate that the line joining it with the sun sweeps equal areas at equal intervals of time (the law of equal areas) (Fig. 2–12).

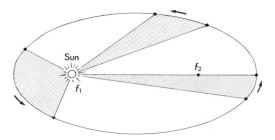

Figure 2–12

(3) The squares of the periods of any two planets are in the same proportion as the cubes of their mean distances from the sun (harmonic law).*

These laws led Newton to postulate a single comprehensive law of universal gravitation:

$$f = G \frac{m_1 m_2}{d^2}, \tag{2–1}$$

where f is the gravitational force, m_1 and m_2 the masses of the respective particles, d their distance apart, and G the gravitational constant. When a spherical assemblage of particles is very distant and its mass is either uniformly distributed or concentrically layered, its gravitational pull can be calculated as though the whole mass were concentrated in a mathematical point at its center. Most celestial bodies are not spherical but ellipsoidal as a consequence of their rotation, but when the distances separating the bodies are large compared with their mean diameters, the point mass assumption works well enough for practical purposes.

From universal gravitation we know the magnitude of the force acting and from Newton's third law of motion we know that the gravitational attraction of the sun for the earth is equal and oppositely directed to the earth's attraction for the sun. The same can be said for the earth-moon system.

In a steadily revolving binary system, where m_e and m_s are the masses of the earth and sun, the force of gravitational attraction f is balanced by the centrifugal force of revolution mv^2/r. If v is the orbital velocity and

---

* Kepler was a student of the Danish astronomer Tycho Brahe (1546–1601), who made the first long series of precise angular observations of the motions of celestial bodies. It was from a study of these observations that Kepler induced the three laws of planetary motion. The first two laws were published in 1609# in a book entitled *Commentaries on the Motion of Mars*. The third, or harmonic law, was published in 1619 in a book called *The Harmony of the World*. They were not subject to physical explanation until Newton (1642–1727) stated the principle of universal gravitation.

r is the orbital radius, we may write for the earth and sun respectively,

$$m_e \frac{v_e^2}{r_e} = m_s \frac{v_s^2}{r_s}. \tag{2-2}$$

Since the sun, s, and the earth, e, revolve about each other in the same time, they must have the same angular velocity, ω; and since $\omega r = v$, we have

$$m_e \omega_e^2 r_e = m_s \omega_s^2 r_s, \tag{2-3}$$

where

$$\omega_e^2 = \omega_s^2, \tag{2-4}$$

so that

$$m_e r_e = m_s r_s. \tag{2-5}$$

The sun's mass is nearly 330,000 earth masses. Thus the earth and sun revolve around each other in orbits of similar shape, but the radius of the earth's orbit measured from the center of mutual revolution of the earth and sun is as much larger than the sun's orbit around this point as the sun's mass is greater than the earth's.

The annual variation in the distance between the earth and the sun's center of mass amounts to a little over 3% of its mean distance, which is close to 149,450,000 km. The orbital period of the earth is 365 days, 5 hours, 48 minutes, 46.0 seconds (365.24220 days) of mean solar time. This interval is the *tropical year* or the year of the seasons to which ordinary calendars are made to conform. The tropical year is the interval between two successive passages of the plane of the earth's equator from north to south through the center of the sun. This annual event, which occurs in the spring and is accompanied by days and nights of equal length, is the *vernal equinox*. The sun's position at the time of the vernal equinox precesses slowly and steadily westward relative to the background stars owing to the conical precession of the earth's axis of rotation through an angular radius of 23°27′. The plane of the earth's orbit extended to the background of fixed stars is the plane of the ecliptic (Fig. 2–13).

The orbit which the earth traces in space has an eccentricity of about 0.017 or 1/60. The *perihelion* point is that at which the earth's orbit most nearly approaches the sun, *aphelion* is the point of most distant separation, and the line joining these two extremities, the major axis of the orbit, is the *line of apsides*. The earth is at perihelion early in January and at aphelion in early July.

The calendar dates vary somewhat because of leap years, but the precession of the equinoxes advances steadily at the rate of 50.26 sec of arc per year, and the line of apsides of the earth's orbit rotates eastward about 11 sec of arc per year.

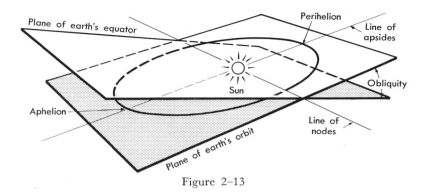

Figure 2–13

Thus the sun is closest to the earth during the Northern Hemisphere winter and farthest away in midsummer. This cyclic change of distance gives rise to a small annual term in the variation of total heat reaching the earth which must be added to the seasonal fluctuations produced by the inclination of the earth's axis of rotation to the plane of its orbit.

The moon is very much smaller than the sun, but it is so very much closer to the earth that its effect on the earth's tides, as will be shown in the next chapter, is considerably greater. The moon's diameter of 2160 miles is little more than one-fourth of the earth's. The moon has 1/81.56 of the earth's mass. The earth-moon system revolves around a point about 2900 miles from the center of the earth, which is therefore below the earth's surface. The mean distance between the center of the earth and moon is 384,400 km (about 240,000 miles) or about 60 times of the earth's equatorial radius. The lunar orbit is noticeably elliptical, which causes the moon's distance from the earth to vary some 13% between perigee and apogee.

The orbit of the moon lies in a plane inclined to the orbit of the earth, hence to the ecliptic, by an angle of 5°09'. This, in addition to the 23°27' obliquity of the ecliptic to the equatorial plane of the earth, permits the moon's path to be inclined to the celestial equator between the limits 28°36' and 18°18'. The variation of the moon's maximum declination has a period of 18.6 years.

The *synodic month* is the period of revolution of the moon relative to the sun or, in other words, the interval between successive new moons—the 29½-day month of phases. Since it is measured relative to the sun and the sun appears to move eastward among the stars, it is longer than the sidereal month of 27⅓ days but varies by more than half a day. The motion of the moon around the earth is influenced by the sun's attraction and several other effects which make its exact prediction a problem of considerable difficulty.

Synodic Month – period of moon's revln in reln to sun

SIDEREAL – 27 ⅓ days
Sun moves eastward rel. to stars

new-Moon – new Moon. 29½ days

The moon revolves around the earth so slowly that its path through space is, surprisingly, always concave to the sun. The attraction of the moon causes the orbital velocity of the earth's center to lag slightly while the moon passes through the full phase and to speed up in times of new moon. It is the center of mutual revolution of the earth-moon system that moves smoothly around the sun in accordance with Kepler's laws.

The angular velocity of the earth's rotation relative to the fixed stars is -0.729211×10^{-4} radian/sec, corresponding to a sidereal interval of $23^h56^m4.091^s$ of mean solar time. The solar day is by definition $24^h00^m00^s$ long and represents the annual mean interval between successive upper transits of the sun. Because the earth moves each day very nearly $1°$ in its annual circuit around the sun, the solar day is longer than the sidereal day by the time required for the earth to rotate the extra degree (about 4 minutes of time, see Fig. 2–14).

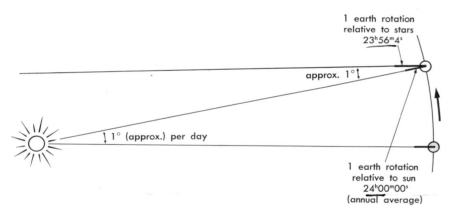

Figure 2–14

The solar day varies in length because the earth rotates more or less steadily on its axis, while the angular motion of the earth in its orbit around the sun varies through the course of a year in accordance with Kepler's law of equal areas. Therefore the length of the apparent solar day tends to be several seconds shorter when the earth is at aphelion in July than when it is at or near perihelion in January. This effect is averaged out in the concept of mean solar time.

STUDY QUESTIONS

1. What evidences suggest that the volume of water and salts in the modern ocean may have been released from the crustal and subcrustal rocks of the earth?

2. Give the evidences which suggest that the continental portions of the earth's crust are not easily transformed into submarine crustal structures but that the oceans may have changed shape during the course of geologic time. Discuss alternative views.

3. Describe the process of subaerial erosion with reference to a recent estimate of the strength of the earth's crust and the principle of isostasy.

4. Discuss the principal mechanisms of submarine erosion and sedimentation.

5. What major astronomical effects are smoothed away in the concept of mean solar time? Distinguish between the sidereal and mean solar days, explaining which is shorter and why.

6. What observed properties of the earth must be reconciled in constructing an idealized model of the interior of the earth?

7. Explain what is meant by the statement that the geoid is a surface of constant geopotential.

8. What minimum diameter hole would be required to see the sky from the center of the earth if the sides of the hole were everywhere vertical and the top of the hole were at latitude 45°?

9. Distinguish between "gravity" and "gravitation."

REFERENCES

1. F. W. Clarke, 1924, *The Data of Geochemistry*, 5th ed., U. S. Geological Survey Bulletin 770, 841 pp.

2. A. H. Woodcock, 1952, *J. Meteorol.*, 9: 200–212.

3. W. W. Rubey, 1951, *Bull. Geol. Soc. Amer.*, 62: 1111–1147. Also R. Revelle, 1955, *J. Mar. Res.*, 14: 446–461.

4. R. W. Goranson, 1931, *Amer. J. Sci.*, 5, 22: 481–502.

5. C. N. Fenner, 1926, *J. Geol.*, 34: 673–772.

6. E. G. Zies, 1929, *Nat. Geog. Soc.*, Contrib. Tech. Papers, Katmai Series, I(4): 61–79.

7. R. A. Daly, 1936, *Amer. J. Sci.*, 5, 31: 401–420.

8. C. A. Veatch and P. A. Smith, 1939, *Spec. Pap. Geol. Soc. Amer.*, 7, 101 pp.

9. Ph. H. Kuenen, 1938, *Geol. Mag.*, 75: 241–249.

10. B. C. Heezen and M. Ewing, 1952, *Amer. J. Sci.*, 250: 849–873.

11. A. B. Malone, 1952, *Geophysics*, 17: 615–619.

SUPPLEMENTARY READING

BAKER, R. H., 1959, *Astronomy*, 7th ed., New York: D. Van Nostrand.

CLARKE, F. W., 1924, *The Data of Geochemistry*, 5th ed., U.S. Geological Survey Bulletin 770, 841 pp.

DALY, R. A., 1940, *Strength and Structure of the Earth*, New York: Prentice-Hall.

FLÜGGE, S., editor, 1956, *Handbuch der Physik* Bd. XLVII, Geophysik I, Berlin: Springer-Verlag. G. D. GARLAND, "Gravity and Isostasy," pp. 202–245; W. M. EWING and F. PRESS, "Structure of the Earth's Crust," pp. 246–257; J. A. JACOBS, "The Earth's Interior," pp. 364–406; K. JUNG, "Figur der Erde," pp. 534–639.

HESS, H. H., 1955, "The Oceanic Crust," *J. Mar. Res.*, 14: 423–439.

JACOBS, J. A., R. D. RUSSELL, and J. TUZO WILSON, 1959, *Physics and Geology*, New York: McGraw-Hill.

KENNEDY, G. C., 1959, "The Origin of Continents, Mountain Ranges and Ocean Basins," *Amer. Scient.*, 47: 491–504.

KOPAL, Z., 1960, *Figures of Equilibrium of Celestial Bodies*, Univ. of Wisconsin Press, Madison.

KUIPER, G. P., editor, 1954, *The Earth as a Planet* (*The Solar System*, Vol. II), Chicago: University of Chicago Press.

REVELLE, R., 1955, "On the History of the Oceans," *J. Mar. Res.*, 14: 446–461.

RUBEY, W. W., 1951, "Geologic History of Sea Water: An Attempt to State the Problem," *Bull. Geol. Soc. Amer.*, 62: 1111–1147.

RUBEY, W. W., 1955, "Development of the Hydrosphere and Atmosphere, with Special Reference to Probable Composition of the Early Atmosphere," pp. 631–650, "Crust of the Earth," Arie Poldervaart, editor, *Spec. Pap. Geol. Soc. Amer.*, 62.

UREY, H. C., 1952, *The Planets*, New Haven: Yale University Press.

CHAPTER 3

Tides and Other Waves

The periodic rise and fall of the sea surface that coordinates itself with the angular position of the sun and moon is referred to as the *astronomical tide*. (Winds, earthquakes, and other terrestrial forces can produce long waves and periodic changes of sea level which may resemble astronomical tides but are not to be confused with them.)

The periodic rise and fall of the tide was first attributed to the motions of the moon by Pytheas in the fourth century B.C. as a consequence of his journey from Massilia in the Mediterranean northward into the English Channel and southern part of the North Sea. To a native of the Mediterranean coast, where the tides are almost unnoticeable, the great tides on the French coast of the English Channel must have been impressive.

Despite this early and promising beginning, no adequate physical theory of the tide-generating forces produced by the sun and moon was proposed until Newton laid the conceptual foundation of universal gravitation in his *Principia Mathematica* of 1687.# Using Newtonian concepts the equilibrium tide was first studied by Daniel Bernoulli in 1740# and the constituent species by Laplace in 1775.# In subsequent work the tidal problem has been generally divided into the shallow-water and ocean-tide cases with notable contributions being made by G. B. Airy (1845), William Thomson (Lord Kelvin) (1867), G. H. Darwin (1879), Horace Lamb (1879), S. S. Hough (1897), Harold Jeffreys (1920), Albert Defant (1924), G. R. Goldsbrough (1928, 1929, 1933), and several others. And yet, for all the genius and dedicated effort of these distinguished men, the tidal problem remains to be solved from first principles, that is, without reference to tide gauge observations.* A historical review of certain efforts to solve the tidal problem has been prepared by S. F. Grace (1931).[1]

---

* C. L. Pekeris of the Weizmann Institute, Rehovot, Israel, reported, at the 1960 meeting of the I.U.G.G. in Helsinki, a successful attempt to predict the real tide by solving Laplace's equations for the tidal potential. The only marine observations employed were those describing the topography of the ocean bottom.

The differential force of attraction

The gravitational attraction of the moon and sun can be regarded as balancing exactly the centrifugal force of the earth's motion around their respective centers of mutual revolution, but there are small differences in this balance from point to point over the earth's surface. The side of the earth nearest the moon, for example, is attracted to the moon more strongly than the side of the earth farthest from the moon. Since the centrifugal force of revolution is in balance with the gravitational pull of the moon at the center of the earth, it follows that the net force on the side away from the moon is directed away from the moon and the net force on the side toward the moon is directed toward the moon. These differential forces tend to elongate the solid earth along a line joining the center of the earth with the center of the moon. But since the solid earth is more rigid than the oceans, there is a net yield of the sea surface which causes the tides to rise and fall relative to the solid crust. The lunar tide in the solid earth, which amounts to some 30 cm, was first detected by Michelson and Gale in 1914.# *

The magnitude of the differential forces can be computed from Newton's law of universal gravitation. The moon is distant from the earth's center nearly 60 earth radii. Since the attraction of gravity varies inversely as the square of the distance, the attractive forces on the near and far sides of the earth are as $1/(59)^2$ is to $1/(61)^2$. If we subtract from both these figures the average force of the moon on the earth's center of mass, $1/(60)^2$, we find that the force acting on the near side of the earth is about 10^{-5} greater than that acting at the center of the earth, while that on the far side is nearly but not quite an equal amount less than that at the center of the earth. The differential force of attraction varies as the *inverse cube* of the distance of the attracting body, so that the tidal ellipsoid tends to be slightly egg-shaped. The amplitude of the tidal bulge varies directly with the mass of the attracting body. The maximum differential force ΔF on the earth is proportional to the gravitational force on the earth's center of mass, plus and minus the force of attraction at the points nearest and farthest from the attracting body respectively:

$$\Delta F \propto \left[\frac{m}{(R \pm r)^2} - \frac{m}{R^2} \right] \propto \pm 2m \frac{r}{R^3}, \tag{3-1}$$

where m is the mass of the attracting body, R its distance from the earth's center of mass, and r is the radius of the earth.

---

* For a discussion of these and more recent results see P. Melchior, 1966, *The Earth Tides*, London: Pergamon Press.

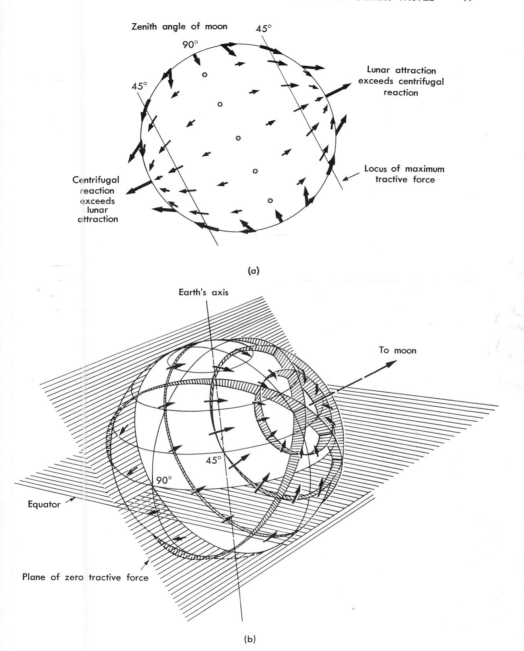

Fig. 3–1.   (a) Configuration of the tractive forces, and (b) the figure of the equilibrium tide.

From this it can be seen that the tidal bulge raised by the sun, with its mass of 330,000 earths, is less than that of the moon (1/80 earth mass) because of the even larger difference in the ratio of the cubes of their mean distances. The average solar tide is somewhat less than half (46%) as great as the lunar tide.

To look at this problem more closely, it is helpful to consider that the earth and moon revolve mutually around a common point approximately 2900 miles from the center of the earth. From this it will be seen that the centrifugal force accompanying the mutual revolution of the earth and moon is parallel with the plane of the moon's orbit and is always directed away from the center of revolution. This, in composition with the vector of lunar gravitational attraction, yields a pattern of resultant forces directed toward the moon and away from the moon on the near and far sides of the earth respectively, tangent to the earth's surface near 45° and directed toward the earth's center at 90° from the earth-moon line as shown in Fig. 3–1. Inasmuch as these forces are on the order of only 10^{-7} of the earth's gravitational pull, the vertical component can exert little influence on the elevation of the sea surface. It is the horizontal component of these forces together with the mobility of water that causes a flow toward the moon on the near side and away from the moon on the far side of the earth. Thus the tractive components of the differential force of attraction are the significant ones from the standpoint of tide generation.

The distribution of tractive forces is not exactly symmetrical, being some 4% stronger on the hemisphere facing the moon when the moon is at its mean distance from the earth. Moreover, the moon is not infinitely far from the earth. The direction of its gravitational attraction differs by about 1° from one side of the earth to the other. For this reason, the great circle through the earth at right angles to the line joining the earth and moon is not precisely the locus of zero tractive force. This discrepancy is ordinarily neglected in the first-order theory of the tidal ellipsoid.

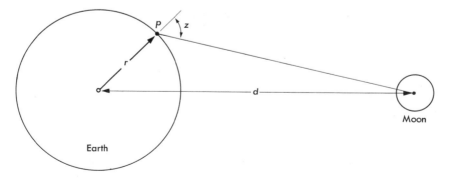

Figure 3–2

To a first approximation, the magnitude of the horizontal component of the lunar tractive force F per unit mass at any point p (Fig. 3–2) is

$$F_p = \frac{3}{2}\, G\, \frac{m_m}{m_e}\, \frac{r^3}{d^3} \sin 2\zeta, \qquad (3\text{–}2)$$

where G is the constant of gravitation, m_e is the mass of the earth, m_m the mass of the moon, r the radius of the (spherical) earth, d the geocentric distance of the moon, and ζ the zenith angle of the moon measured from p. The solar tractive forces can be found by substituting values for the sun's mass and distance for those of the moon.

The equilibrium tide

Were the ocean surface to be everywhere perpendicular to the resultant of the force of gravity and the attractive forces of the sun and moon, and the condition of hydrostatic equilibrium fulfilled, we would have an *equilibrium tide*. Under equilibrium conditions, the greatest rise of water should occur on the line joining the center of mass of the disturbing body with the center of the earth, and thus be inclined to the plane of the earth's equator by an angle equal to the declination of the disturbing body.

In order for the tidal bulge excited by the sun and moon to keep step with the apparent diurnal motion of these bodies, it must move at a rate which would carry it completely around the earth in about 24 hours. When, for example, the moon is at the celestial equator, the wave of the astronomical tide must advance at a rate approaching 1000 mi/hr near the equator and diminish speed with increasing latitude to zero at the poles. The celerity c of a surface wave which is long in comparison with the depth of the water, the so-called *shallow-water wave*, is $c = \sqrt{gd}$, where g is the acceleration of gravity and d is the depth of the water. The depth of water required at the equator to permit the equilibrium tide to remain stationary under the sun as the earth rotates is 22.5 km. The equatorial depth required for the lunar tide is somewhat greater owing to the eastward motion of the moon during the course of each day.

In view of the relative shallowness of the world ocean, 4 km, it is understandable that a free wave cannot keep step with the apparent diurnal motions of the sun and moon except in very high latitudes—poleward of 66°—where there is either no ocean at all, as around the South Pole, or the rather small Arctic basin hemmed in by broad shallows. These geographic effects prevent the equilibrium tide from standing continuously beneath the sun and moon anywhere on earth.

It is found, instead, that the astronomical tides consist of forced oscillations which have periods commensurate with the tide-generating forces but amplitudes and phases that are constrained both by friction and the depth of water. The latter factors cause the crest of the astronomical tide to pass sometimes many hours after the sun or moon has crossed the local meridian.

The equilibrium theory of the astronomical tide fails to explain these daily lags but clarifies, nevertheless, the changes of tidal amplitude that occur during each lunar month and during each year. The sun changes its declination by ±23.5° in the course of a year, and the moon's declination may change by as much as ±28.5° during each month. Because of the comparative slowness of these changes, the equilibrium wave crests can reach maximum amplitudes in the corresponding latitudes on the earth. When the sun and/or moon are not on the celestial equator, the tide-producing forces (and the idealized tidal ellipsoids) are no longer symmetrically arranged with respect to the earth's equator. If, for example, the moon is in southern celestial latitudes, the greatest height of the tidal ellipsoid should be under the moon in the Southern Hemisphere, and the antilunar bulge should be on the opposite side of the earth in the Northern Hemisphere. With daily rotation, the high tide at some stations in the Northern Hemisphere will, therefore, be expected to be higher when the moon is below the horizon than when the moon is visible in the sky, and show a *diurnal inequality* of semidiurnal tidal amplitudes. Generally speaking, the equilibrium theory is borne out by these occurrences. It also helps to explain the spring and neap tides associated with the monthly sequence of positions assumed by the sun and moon relative to the earth, and the long-term variations of tidal amplitude related to the changing distances of the sun and moon.

Spring tides are those of greatest amplitude occurring in the course of each half lunation at or near the time when the sun and moon are in syzygy, their differential forces of attraction being in phase. Neap tides (those of least amplitude in the course of each lunation) occur when the sun and moon are at or near quadrature with respect to the earth, their tractive forces being 90° out of phase. The annual terms associated with the earth's changing distance from the sun at perihelion and aphelion are also reflected in tidal ranges, as are those associated with the monthly change in the moon's distance. Beyond these effects, however, the equilibrium theory does not provide a basis for explaining the observed tide because, in general, the oceans are too shallow and too irregularly shaped for the ideal wave pattern to remain intact.

The astronomical tide is produced by sets of tangential forces which, because of the earth's eastward rotation, tend to change direction in the clockwise sense around the horizon in the Northern Hemisphere and in a counterclockwise sense in the Southern Hemisphere. It was suggested by Harris in 1904[#] that the crest of the actual tide can be considered to move as a forced wave in a circular mode not unlike the rotary sloshing or *amphidromic system* of surface elevations that can be induced when rinsing a tea

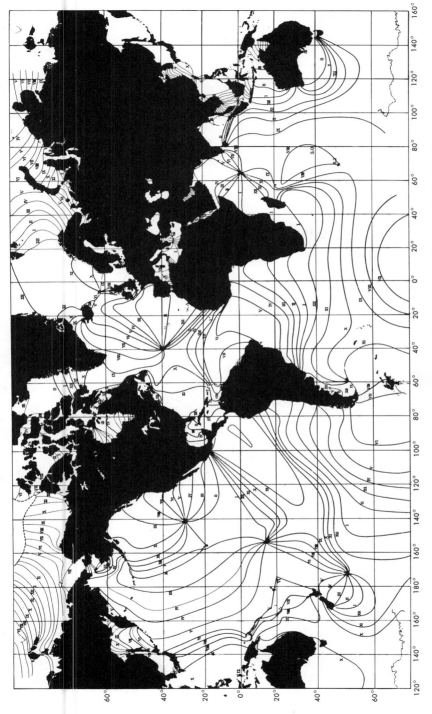

Fig. 3–3. Semidiurnal cotidal lines of the world ocean, according to R. A. Harris.
[From H. Poincaré, 1910, Leçons de Mécanique Céleste, Vol. 3, Paris: Gauthier-Crofts.]

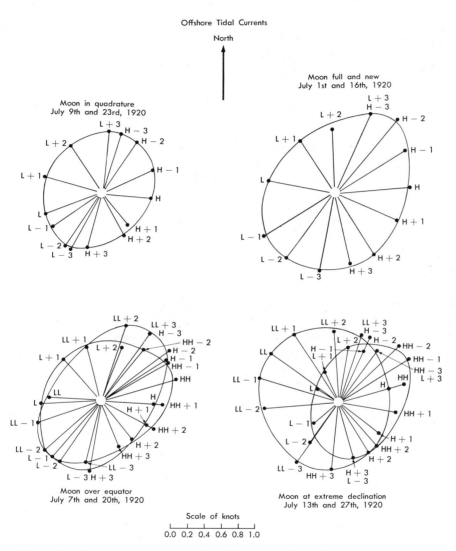

Fig. 3–4. Current curves observed at Nantucket Shoals light vessel for various phases of the moon. [From H. A. Marmer, 1926, *The Tide*, New York: Appleton-Century-Crofts.]

cup. According to this model, the ocean can be presumed to be divided into compartments that permit wave systems of a rotary character to close one complete cycle in a time which is commensurate with the periods of the astronomical tide-generating forces. It is found, however, that neither the *cotidal lines* of equal height nor the currents that are set up by the tide-generating forces necessarily follow the directional regime of the tractive forces (Figs. 3–3 and 3–4). These effects are especially conspicuous in shallow water, perhaps because of their significance to navigation and the abundance of tide-gauging stations. Particularly in bays and channels, the tide tends to develop a transverse motion which has a period of oscillation related to the period of rotation of the plane of a swinging pendulum —the pendulum day—which is an inertial consequence of the earth's daily rotation.

The pendulum day

If, in imagination, we consider the behavior of a swinging pendulum suspended on a starlit night exactly over a pole of rotation of the earth, we would observe that the plane of oscillation of the bob would remain fixed relative to the stars while we who stand on the ground would rotate with the earth. In that the direction of vibration of a pendulum appears to reverse each half cycle, a half turn of the plane of vibration owing to the earth's rotation would to all appearances restore the experiment to the starting point. This interval is called the half-pendulum day, which for a pendulum mounted at either pole would require half a sidereal day or approximately 11 hours and 58 minutes of mean solar time.

If next we imagine the same experiment to be performed on the equator, with the pendulum swinging in an exactly east-west plane, we will recognize that the plane of vibration will remain fixed in the equatorial plane, since the rotation of the earth does not alter its direction relative to the fixed stars. Under these conditions the length of the half-pendulum day is infinite in time. From these limiting cases and noting that the effect of earth rotation around the vertical at any latitude ϕ varies as $1/(\sin \phi)$, we may anticipate that the length of the half-pendulum day in sidereal time is $T = 12 \text{ hr}/(\sin \phi)$.

Foucault actually performed the first of these experiments in 1851[#] in latitude 49°N using as a pendulum a massive ball of copper hung by a wire more than 50 meters long from the dome of the Pantheon in Paris. He started the oscillation by burning away a silk thread that held the pendulum bob in a deflected position. The rate of rotation of the plane of vibration was one-half turn in about 16 hours. This and many subsequent experiments in as many different latitudes not only confirms physical expectations but is one of the proofs of the absolute rotation of the earth.

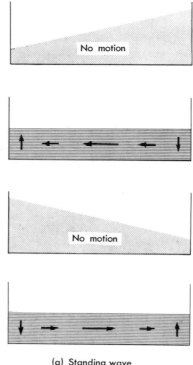

(a) Standing wave

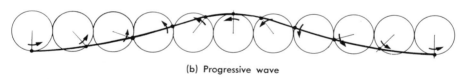

(b) Progressive wave

Fig. 3–5. In standing waves (a) maximum horizontal motion occurs at mid-water. In progressive waves (b) maximum horizontal motion occurs at highest and lowest water.

Particle motions in simple waves

In *standing* waves the particle motions are similar to those of a pendulum, with the greatest horizontal motion occurring at mid-tide and the least horizontal motion at high and low water. The pattern is similar to the simple half-wavelength oscillation of water in a tank (Fig. 3–5a). At the instant when the water level is highest on the right and lowest on the left, there is no motion anywhere. But at the instant when the water surface is level, the horizontal motion in the middle of the tank is maximum and directed toward the position of subsequent high water at the left end of the tank. When high water is established at the left end of the tank, motion again ceases everywhere for an instant and then reaches a maximum as the attitude of the water surface passes through the horizontal.

The particle motions in *progressive* waves are related to the local water level in a quite different way. The elevations of the surface can be considered to be represented in a crude way by the distribution of points on a closely spaced succession of discs turning in a vertical plane with the wave period, as shown in Fig. 3–5(b). If the discs are one-tenth of a wavelength apart and the points on their rims are situated successively 36° farther from the vertical around the circumference, a wave can be drawn through the points. As the discs rotate steadily and in the same direction, the wave through the points will move in the same direction as the tops of the discs at the rate of one wavelength per revolution. Although this analogy is a caricature of real waves, it serves to point out an important fact—the particle motion, represented by the points on the discs, is greatest in the direction of propagation where the water level is highest, greatest in the direction opposite that of wave propagation where the water level is lowest, and there is no horizontal component of motion where the water surface is passing through the rest position.

Both progressive and standing wave oscillations have counterparts in natural embayments and canals. Usually there is some attenuation of the waves as they move shoreward from the open sea, and the reflections that occur are both oblique and imperfect so that real cases contain a mixture of standing and progressive wave phenomena with one type or the other predominating.

The Kelvin wave

Consider a canal in the Northern Hemisphere through which a progressive wave is free to pass with each tide. With the progress of the wave through the canal the associated currents will tend always to flow in the direction they had as they entered the canal. As time passes, however, the canal itself is rotated counterclockwise by the earth's daily rotation. This causes the water current apparently to turn toward the right-hand side of the canal. Since in a progressive wave the maximum flood current of a tide is associated with the crest of the wave, the result will be that high tide will be somewhat higher on the right of flood current in the channel than it is on the left (Fig. 3–6). With the ebb current associated with low tide, the deflection will again be to the right of the current direction, which will cause the low tide to be somewhat lower on the same side where the flood crest was highest. In general, in the Northern Hemisphere a progressive wave tide may produce larger tidal amplitudes on the side which is the

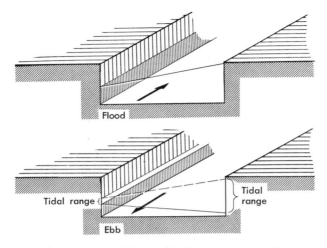

Fig. 3–6.   Inequality of tidal amplitudes on opposite sides of a canal.

right bank during flood.   For example, at Cherbourg the spring range is
18 feet, whereas the amplitude of the same wave at Ventnor on the Isle of
Wight has a spring range of only 9.5 feet.   In the Southern Hemisphere
the side which is on the left of flood may have the larger tidal amplitude.

Embayment tides

Progressive tidal waves entering embayments often produce standing
waves owing to reflections at the head of the bay.   Because of the tendency
for tides in bays to have standing-wave characteristics and because of the
similarity of the currents in standing waves to the oscillations of a pendulum, it is to be expected that effects similar to the motions of a Foucault
pendulum may occur.   Indeed, the parallel is especially close when the
period of tidal oscillation matches that of the half-pendulum day.

Consider that a bay with a standing-wave tide is situated in some high
latitude of the Northern Hemisphere where the length of the half-pendulum
day is nearly the same as the period of the semidiurnal tide (Fig. 3–7).   One
can expect that the standing-wave tide may rise at the head of the bay
which is, let us say, at the north end, and begin to ebb.   With the passage
of time the water particles of the ebb current tend to flow toward their
starting point which, through the eastward rotation of the earth, is now
nearer the western margin of the bay than its mouth.   This will cause the
water surface to slope upward toward the west, and when full ebb is in
progress, it will appear as though the high stand of tide has moved from
the north side of the bay to the west side while low tide has moved from the
south side of the bay to the east side.   Continuing this process through the

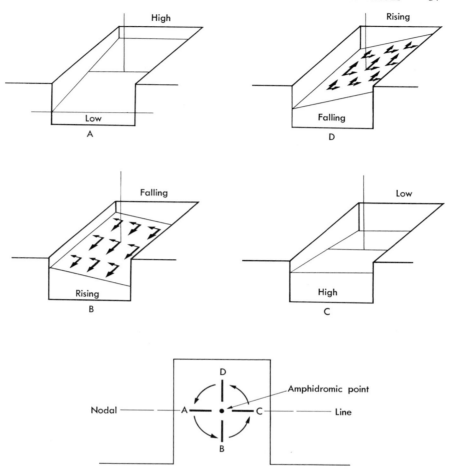

Fig. 3–7. Amphidromic motion in an embayment.

completion of ebb and through the next flood, it will be seen that the peak of high tide will rotate in a counterclockwise direction around the bay once each half-pendulum or semi-diurnal period. The result of this is that the nodal line of the standing-wave tide which would normally lie in an east-west direction across the bay has itself rotated and thus degenerated to a point somewhere near the center. This is the *amphidromic point*, believed to exist in many closed basins in which the period of standing co-oscillation of the tide and Kelvin wave components can remain 90° out of phase. In latitudes near 30° where the length of the half-pendulum day is increased to 24 hours, one would expect that the amphidromic system would be best developed in connection with tides of diurnal frequency.

Owing to the imperfect reflections produced at the heads of bays, an ideal standing wave is hardly ever developed. There is instead, in natural situations, a mixture of progressive and standing-wave types of tidal propagation. This complicates matters, as does the fact that the depth of the bay may be irregular and cause the crest line of the tide to be bent by being retarded around shoals and accelerated in deep spots. Because of these effects and others, it is possible for the phasing of the Kelvin wave and the tide to be such as to produce an amphidromic system in an anticyclonic as well as the normal cyclonic sense.*

ie clockwise in N.H rather than a. clockwise.

Prediction of the real tide

The interaction of amphidromic cells with one another, and the fact that the sun and moon may stand at different angles to the earth's equator depending upon the time of month and the season, can cause the motion of the tides to be very complicated. More than this, the oceans have irregular shapes and depths, and there are centers of frictional retardation in shallow water which are strong enough to damp the tides and may also have some cumulative effect on the rotation rate of the earth. Meteorological tides of a nonperiodic character are also superimposed on all the foregoing effects by the variations in atmospheric pressure and by the stress of the winds on the sea surface in and around storm centers.[2]

Because the real tide is such a complicated response of the ocean volume to known astronomical and chance meteorological forces, it has been the practice to observe the tide at many different points on the earth for long periods of time so as to average out the meteorological influences and accumulate a significant number of the longer period changes. Following this, predictions of the astronomical components of the tide are made by extrapolation. Because of the number of configurations of the sun and moon and axis of rotation of the earth that are possible in the course of

---

* The words "cyclonic" and "anticyclonic" are borrowed from the terminology of meteorology. A cyclone is a major center of low atmospheric pressure toward which winds tend to blow inward from all directions in an area perhaps 1000 miles across. Owing to the earth's rotation, these winds do not trace straight paths but spiral into the low pressure center. The winds spiraling into cyclonic centers form a regime of counterclockwise atmospheric motion in the Northern Hemisphere. In the Southern Hemisphere, the deflecting force of earth rotation is toward the left so that cylonic inflow produces regional air motion in a clockwise sense. Because of this reversal of rotation for the same physical phenomenon in the two hemispheres, it is convenient to refer to *cyclonic rotation*—meaning counterclockwise rotation in the Northern Hemisphere and clockwise rotation in the Southern Hemisphere—whenever fluid rotation is dependent upon the effects of earth rotation. *Anticyclonic rotation* is simply the reverse situation accompanying the outflow of air from centers of high atmospheric pressure.

time, the tides repeat themselves only roughly during successive lunar months. For this reason the observed tide at any station is subjected to a kind of harmonic analysis in which the amplitudes are determined for each of the several astronomically identifiable frequencies. These frequencies, usually called *tidal species*, are surprisingly numerous, as shown in Table 3–1, where the angular speed of each of the tidal species listed is related to the motions of the sun, moon, and earth.

When the tidal signature at a given station on the earth has been analyzed for the amplitudes of each of these tidal species and the phase angles for each of the harmonic terms are known for a given initial time, it is possible to reconstruct the tide for any time in the past or future relative to the initial moment. In general, the predictions are good to within 0.1 foot over the period of the ensuing year, neglecting of course the aperiodic influences of wind and atmospheric pressure on local sea level. The values of the tidal coefficients for each station are constantly under review as the record available for analysis becomes longer with each passing year. When the station is tied in with a precise leveling survey, the secular change of sea level at each station is a by-product of the tidal harmonic analysis.

TABLE 3–1

TIDAL HARMONICS*

Species	Symbol	Period, in hr	Angular speed†
Some important semidiurnal terms			
Principal lunar	M_2	12.42	$2(g - s)$
Principal solar	S_2	12.00	$2(g - e)$
Larger lunar elliptic	N_2	12.66	$2g - 3s + p$
Luni-solar	K_2	11.97	$2g$
Some important diurnal terms			
Luni-solar	K_1	23.93	g
Principal lunar	O_1	26.87	$g - 2s$
Principal solar	P_1	24.07	$g - 2e$
Significant long-period terms			
Lunar fortnightly	M_f	327.86	$2s$
Lunar monthly	M_m	661.30	$s - p$
Solar semiannual	S_{sa}	2191.43	$2e$

* Adapted from Paul Schureman, 1940, *Manual of Harmonic Analysis and Prediction of Tides*, revised ed., U.S. Coast & Geodetic Survey Special Publication 98.

† e = mean motion of sun; s = mean motion of moon; g = rotation of earth (sidereal); p = motion of lunar perigee.

Fig. 3–8.   The large Kelvin machine for tidal predictions at the Hydrographic Institute in Hamburg, Germany.   [From G. Dietrich and K. Kalle, 1957, *Allgemeine Meereskünde*, Berlin-Nikolassee: Verlag Gebrüder Borntraeger.]

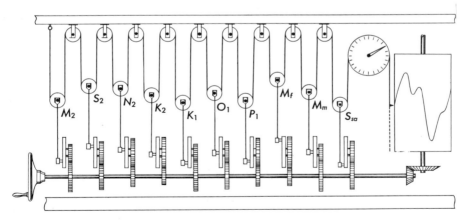

Fig. 3–9.   Arrangement of the working parts in a Kelvin tide-prediction machine.

Predictions based on the tidal harmonics at a given station have been made with a machine of the type devised by Lord Kelvin in 1872[#] (Fig. 3–8). As shown in Fig. 3–9 the machine consists of a succession of separate discs coupled together mechanically so each turns at a rate appropriate to each successive tidal species. It remains for the operator to feed into this machine the amplitude and phase angle of each species for a given origin of time. The contributions of all these species are added by means of an inelastic string threaded alternately over pulleys attached to each of the component generators and the bed of the machine, so that in the end the length of the string emerging from the train of pulleys is proportional to the sum of all the constituents acting together. A pencil on this string will draw on a sheet of moving paper the tide prediction for the station as a function of future time. The tide tables published each year are obtained by reading the time of high and low water shown on the resulting graph for each successive day.

The tidal bore

Because of their very great wavelength, tidal wave crests travel at the shallow-water rate, $c = \sqrt{gd}$. From this it is to be expected that as a tidal crest approaches a coast, it will be slowed down in proportion to the square root of the depth and tend to steepen. Such steepening may be quite pronounced as the crest enters a very shallow estuary. If the entry is also impeded by the discharge of the river, it is possible for the rate of river flow to be even greater than the rate of advance of the tidal wave in shallow water. Under these circumstances, the wave is held back in deeper water where v, the velocity of the river flow, equals and opposes c, the wave celerity in that depth. Eventually, however, the continued rise of the tide outside the estuary will deepen the water at the entrance sufficiently for the tide to make progress against the river discharge. Once this condition has been reached, the tidal crest sweeps into the estuary as a conspicuous wave or even as a frothing front—the tidal bore—following the passage of which the water level differences inside and outside of the estuary remain more nearly equal. Stoker (1957)[3] suggests that the tidal bore is analogous to a shock wave, since it obeys the same laws of motion and reflection.

During spring tide conditions when the moon and sun exert their tractive influence in phase, bores occur in many river estuaries and in some straits, such as those of Messina, the passage between Scylla and Charybdis. The most famous tidal bore is that in the Tsientang Kiang estuary in China (Fig. 3–10). Here junks tie up securely against the rush of the bore, which on occasion may be as much as 25 feet high, and schedule their trips to ride upstream on the current that follows. There are also bores in the Amazon River, the Severn River in England, the Petitcodiac River at the head of

Fig. 3–10.   The great bore of the Tsientang Kiang. [Photograph by H. Thorade, 1931, from G. Dietrich and K. Kalle, 1957, *Allgemeime Meereskünde*, Berlin-Nikolassee: Verlag Gebrüder Borntraeger.]

the Bay of Fundy, and in the Seine, Orne, and Gironde Rivers of France. In many of these places the bore is emphasized by a funneling effect as the tidal wave passes from the broad reaches of the estuary into the narrower confines of the river channel.   However, the process by which a bore is formed is primarily dependent on diminishing depth.

The tsunami

In addition to the great organized motions of the sea due to the attraction of the sun and moon, there is a nearly complete spectrum of other waves produced by nonastronomical forces.   Among these are the stress of wind on the sea surface and geologic disturbances such as earthquake shocks. The *tsunami*, or so-called "tidal wave" associated with earthquakes and volcanic eruptions, is most like the astronomical tide.   The tsunami is a long, low wave train traveling at the maximum speed for open ocean depths (about 400 nautical miles per hour) from the site of some cataclysmic disturbance such as a slipping fault, submarine landslide, or volcanic explosion. These waves spread outward over the ocean at the shallow-water rate $(c = \sqrt{gd})$ with an energy density that decreases as the inverse first power of the distance traveled.   Near the site of wave generation a tsunami may consist of a single long-crested trough or crest; but by the time it has traveled a significant distance it becomes a train of long, low waves.   Since their wavelength is long compared with the depth of the open ocean, they "feel bottom" continuously.   The wave fronts may therefore be dispersed or focused on a coastline, with a corresponding decrease or increase of wave amplitude.

The tsunami may, but usually does not, come in as a rushing wall of water. Rather it is a long swell which rises and falls several times within a few hours. A vivid description of the Unimak tsunami which occurred on April 1, 1946, is given by Macdonald, Shepard, and Cox.[4] This tsunami was composed of a succession of crests 2 feet high and 122 miles apart in mid-ocean, moving at better than 400 knots so that the crests arrived at intervals of about 15 minutes. In Hawaii these waves crested to 55 feet at Pololu Valley. Farther down the windward coast the crest was half as high but caused heavy loss of property and some life in the city of Hilo. The master of a ship lying offshore near Hilo reported feeling no unusual waves pass under his ship, although he could see great waves breaking on the shore. This same wave train was observed at Bikini Atoll in the Marshall Islands to have an amplitude of 1.5 feet. It passed unnoticed except by those who watched for it very carefully.

Wave concepts

The characteristics of natural waves, which (in contrast with tides) involve appreciable curvatures of the free surface, are not susceptible to accurate representation in simple terms. For this reason much of the description of ordinary waves is referred to ideal wave forms which are even simpler than the most elementary waves that can be produced in laboratory tanks. These ideal waves are minute, long-crested, sinusoidal undulations on the surface of a perfect liquid. The waves are considered to occur in indefinitely long trains of equal amplitude that move without loss of energy for an indefinite time, over an infinite distance.

Trains of sinusoidal waves can be said to have a wavelength, L, defined as the horizontal distance separating successive points in phase, measured in the direction of wave travel, which is also the direction of the normal to the crestline. The period, t, of a train of identical waves is the time elapsed between the passage of a given phase on one wave and the arrival of the same phase on the next succeeding wave, as observed from a fixed station. The velocity of a wave is the distance traversed per unit time by any wave phase measured in the direction normal to the crestline. Since the velocity of a wave relative to a stationary observer can be influenced by currents in the liquid, the wave velocity measured relative to the liquid is given a special name, celerity, symbolized by c. The wavelength, period, and celerity of an ideal wave are related to each other by the wave equation

$$c = \frac{L}{t} \quad \text{or} \quad c = nL, \tag{3-3}$$

where $n = 1/t$, the wave frequency or number of waves passing a point in the liquid per unit time.

When two identical wave trains move in exactly opposite directions, standing waves are formed which have the same frequency and wavelength as the parent trains but a greater range of amplitudes. When two identical trains of sinusoidal waves move in not quite the same direction, the amplitude range is again increased and the waves become short-crested.

As a further step toward the natural situation we may consider two sinusoidal wave trains of equal amplitude but very slightly different frequency moving in exactly the same direction. This produces a succession of beats as the phases of the two long-crested wave trains interact constructively and destructively to change the local amplitudes. Such beating produces a succession of uniformly spaced wave groups separated by bands of nearly smooth water. The complexity of a natural water surface begins to be approached when the two wave trains of slightly different frequency are moving in slightly different directions. These conditions make it difficult to use with exactitude the concepts of wavelength, wave direction, period, and celerity. Nevertheless, it is fruitful to consider the properties of natural gravity waves in terms of a train of long-crested sinusoidal or trochoidal waves of small amplitude, moving on an ideal fluid of uniform depth, d. For sinusoidal waves the celerity of phase propagation, c, is

$$c^2 = \frac{gL}{2\pi} \tanh 2\pi \frac{d}{L},$$ (3–4)

where L is the wavelength and g is the acceleration of gravity. For *deep-water waves*, $d/L > 1/2$, the general formula reduces to $c^2 = gL/2\pi$, and the group speed is half the phase speed. For *shallow-water waves*, $d/L < 1/20$, the phase speed is given by the already familiar formula $c^2 = gd$. The group speed of shallow-water waves of small amplitude is independent of wavelength and equals the phase speed.*

Nontidal waves

Nontidal surface waves come in a variety of shapes, and in lengths which range from those of ripples to longer than tides. As already indicated, when a wave has a length that is short compared with the depth of water, it obeys quite different laws of propagation from those which apply to the shallow-water variety of which tides and tsunamis are examples. The characteristics of real waves have not been ascertained from basic physical principles.

---

* For exhaustive treatments of various theoretical and practical aspects of these phenomena, see C. A. Coulson, *Waves*, 7th ed., Edinburgh: Oliver & Boyd; J. J. Stoker, 1957, *Water Waves* (Pure and Applied Mathematics, R. Courant, L. Bers and J. J. Stoker, editors, Vol. IV), New York: Interscience; H. Lamb, 1945, *Hydrodynamics*, 6th ed. revised, New York: Dover, and London: Cambridge University Press.

The shape of deep-water waves depends to a large extent on their length. Waves longer than 1.73 cm, *gravity waves*, are controlled by gravitational and inertial forces. Their shapes range from a nearly sinusoidal form to a sharply peaked crest and rounded trough, which can be approximated by a trochoid for mathematical purposes. Waves shorter than 1.73 cm in length are generally controlled by forces associated with surface tension, causing the short *capillary wave* to have rounded crests and V-shaped troughs. Gravity waves are *normally dispersive*, that is, travel faster as their wavelengths increase, as shown in Fig. 3–11. The celerity of capillary waves, however, increases with decreasing wavelength, which is the case of *anomalous dispersion*. The minimum wave speed is 23 cm/sec, corresponding to a wavelength of 1.73 cm.

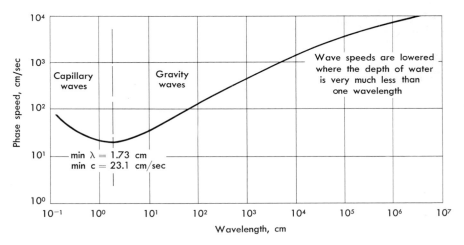

Fig. 3–11.   Celerity of gravity waves and capillary waves as a function of their length.

In ordinary practice it is a commonplace to distinguish waves by their periods rather than their lengths because wave period is an easier quantity to measure.

The period of gravity waves ranges from approximately one-half second through several hours. Gravity waves with a period of less than one second have been designated by Munk[5] as *ultra-gravity waves;* those with periods between 1 and 30 seconds as *ordinary gravity waves;* those with periods from 30 seconds to 5 minutes as *infra-gravity waves;* those with periods longer than 5 minutes as *long-period* waves, which are distinct from tides; those with very long periods, greater than 12 or 24 hours, as *trans-tidal waves.*[6]

Fig. 3–12.   Capillary wave groups superimposed on gravity waves.

When the wind is fresh locally, one can see on the water surface or on the backs of ordinary gravity waves a fine structure of small ripples of nearly capillary dimensions (Fig. 3–12).   It is thought that much of the grip that the wind has on the water to produce waves and currents is provided through these very small wavelets, since they are very numerous and move slowly before the wind.[7]   (Gravity waves speed up with growth to the point where they can eventually move at the same speed as the wind and can extract no further energy from it.)

The "sea clutter" that is observed on radar PPI scopes is also more readily associated with the smaller ripples and capillary waves than with larger waves.   Existing evidence suggests that it is the asymmetry of small waves that causes the sea clutter on a radar screen to extend farther upwind than in any other direction from the ship.   Because of this asymmetry, the path of glittering reflection on water surfaces ruffled by cross winds may also be very noticeably asymmetrical in width or quite out of the vertical plane which contains a bright source, such as the moon or a lighthouse beacon, and the observer's eye.

The ordinary gravity waves normally observed at sea are composed of two varieties: the *swell*, which is a long and relatively symmetrical wave having a period in the order of 10 seconds produced by winds and storms at some distance from the point of observation, and *sea*, which is the local

swell – ~10s period

wind-generated wave motion usually of shorter period with unsymmetrical slopes and steep or white-capped crests depending upon the wind speed and _fetch_—the length of sea surface over which the wind blows without appreciable change of direction. A classic study of the empirical relationship between wind, sea, and swell was made by Sverdrup and Munk (1947)[#] in an effort to provide forecast criteria.

As a purely practical approach which recognizes the interdependence of wind force and sea state, wave conditions are often described in relation to the scale of wind speeds introduced in 1806[#] by Admiral Sir Francis Beaufort and given, with modern amendments, in Table 3–2. Under most circumstances sea and swell can be distinguished by a practiced eye.

A mixture of locally generated wind waves with swell from more distant centers of disturbance is the result of the normally dispersive propagation of ordinary gravity waves. Since there are always one or more centers of wave generation over an ocean and the longer waves, especially the longest, travel with surprisingly little attenuation over hundreds if not thousands of miles of sea surface, it is difficult to describe the surface motion, composed as it so often is of many different wave trains running in as many different directions. Within the past decade a new approach has been developed in which the _wave spectrum_ (Figs. 3–13 and 3–14), a concept introduced by G. E. R. Deacon, N. F. Barber, M. S. Longuet-Higgins, and F. Ursell of the National Institute of Oceanography in England, is used to describe the distribution of energy among waves of different period. Because wave speed increases with wavelength, the wave spectra associated with a distant storm will at first show the bulk of the energy in long-period waves. Later on the successively shorter periods will arrive and dominate the spectrum. From a knowledge of the relationship of wave period to velocity of propagation some estimate can be made of the distance of the storm center. In this way, observations of wave spectra made on the southwest coast of England have revealed the passage of storm centers across the South Atlantic. Similar observations on the California coast have shown the motion of storm centers across the Antarctic Ocean southeast of New Zealand.[8]

When the wind rises around a storm center, the sea also rises more or less steadily. Present practices permit the wave spectrum associated with a given wind speed to be predicted within given areas of a storm if the wind speeds are known. The sea is said to be fully developed when all of the possible wave frequencies possess energies appropriate to the spectrum for the prevailing wind speed. With increasing wind speed the area under the curve increases rapidly, but the frequency of maximum energy shifts toward lower wave frequencies. Successful predictions of the sea state have been based on the prediction of wave spectra from the initial assumption that the sea can be considered to be the sum of the contributions of many simple

TABLE 3–2*†

MODERN BEAUFORT SCALE

Beau-fort number	Wind Speed				Nautical term	U.S. Weather Bureau term	Hydrographic Office	
	knots	mph	meters per second	km per hour			Term and height of waves, in feet	Code
0	under 1	under 1	0.0–0.2	under 1	Calm	Light	Calm, 0	0
1	1–3	1–3	0.3–1.5	1–5	Light air		Smooth, less than 1	1
2	4–6	4–7	1.6–3.3	6–11	Light breeze		Slight, 1–3	2
3	7–10	8–12	3.4–5.4	12–19	Gentle breeze	Gentle	Moderate, 3–5	3
4	11–16	13–18	5.5–7.9	20–28	Moderate breeze	Moderate	Rough, 5–8	4
5	17–21	19–24	8.0–10.7	29–38	Fresh breeze	Fresh		
6	22–27	25–31	10.8–13.8	39–49	Strong breeze	Strong	Very rough, 8–12	5
7	28–33	32–38	13.9–17.1	50–61	Moderate gale			
8	34–40	39–46	17.2–20.7	62–74	Fresh gale	Gale		
9	41–47	47–54	20.8–24.4	75–88	Strong gale		High, 12–20	6
10	48–55	55–63	24.5–28.4	89–102	Whole gale	Whole gale	Very high, 20–40	7
11	56–63	64–72	28.5–32.6	103–117	Storm		Mountainous, 40 and higher	8
12	64–71	73–82	32.7–36.9	118–133	Hurricane	Hurricane	Confused	9
13	72–80	83–92	37.0–41.4	134–149				
14	81–89	93–103	41.5–46.1	150–166				
15	90–99	104–114	46.2–50.9	167–183				
16	100–108	115–125	51.0–56.0	184–201				
17	109–118	126–136	56.1–61.2	202–220				

* Adapted from N. Bowditch (1958 edition), *American Practical Navigator*, U.S. Navy Hydrographic Office Publication No. 9, p. 1059.

TABLE 3–2

MODERN BEAUFORT SCALE (*Continued*)

International		Estimating wind speed	
Term and height of waves, in feet	Code	Effects observed at sea	Effects observed on land
Calm, glassy, 0	0	Sea like mirror.	Calm; smoke rises vertically.
		Ripples with appearance of scales; no foam crests.	Smoke drift indicates wind direction; vanes do not move.
Rippled, 0–1	1	Small wavelets; crests of glassy appearance, not breaking.	Wind felt on face; leaves rustle; vanes begin to move.
Smooth, 1–2	2	Large wavelets; crests begin to break; scattered whitecaps.	Leaves, small twigs in constant motion; light flags extended.
Slight, 2–4	3	Small waves, becoming longer; numerous whitecaps.	Dust, leaves, and loose paper raised up; small branches move
Moderate, 4–8	4	Moderate waves, taking longer form; many whitecaps; some spray.	Small trees in leaf begin to sway.
Rough, 8–13	5	Larger waves forming; whitecaps everywhere; more spray.	Larger branches of trees in motion; whistling heard in wires.
Very rough, 13–20	6	Sea heaps up; white foam from breaking waves begins to be blown in streaks.	Whole trees in motion; resistance felt in walking against wind.
		Moderately high waves of greater length; edges of crests begin to break into spindrift; foam is blown in well-marked streaks.	Twigs and small branches broken off trees; progress generally impeded.
		High waves; sea begins to roll; dense streaks of foam; spray may reduce visibility.	Slight structural damage occurs; slate blown from roofs.
High, 20–30	7	Very high waves with overhanging crests; sea takes white appearance as foam is blown in very dense streaks; rolling is heavy and visibility reduced.	Seldom experienced on land; trees broken or uprooted; considerable structural damage occurs.
Very high, 30–45	8	Exceptionally high waves; sea covered with white foam patches; visibility still more reduced.	
Phenomenal, over 45	9	Air filled with foam; sea completely white with driving spray; visibility greatly reduced.	Very rarely experienced on land; usually accompanied by widespread damage.

† Since January 1, 1955, weather-map symbols have been based upon wind speed in knots, at five-knot intervals, rather than upon Beaufort number.

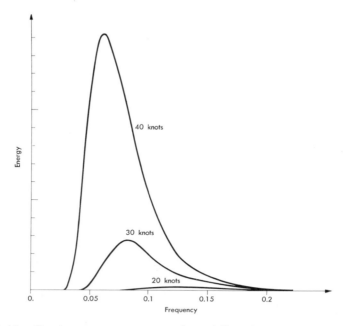

Fig. 3–13. Continuous wave spectrum for a fully arisen sea at wind speeds of 20, 30, and 40 knots. Note the displacement of the optimum band (maximum of spectral energy) from higher to lower frequencies with increasing wind speed. [From W. J. Pierson, Jr., G. Neumann, and R. W. James, 1955, *Practical Methods for Observing and Forecasting Ocean Waves By Means of Wave Spectra and Statistics*, U.S. Navy Hydrographic Office Publication 603.]

sine waves moving in many different directions. These basic wave trains have different wavelengths and periods as well as directions, so that by considering their mutual interferences it is possible to describe very complicated surface configurations (Figs. 3–15 and 16) and their changes in statistical terms.*

Beyond the phenomena of ordinary seas and swell are those of the infra-gravity waves, which have periods of from 0.5 to 5 minutes. These seem to be occasioned by the interference of two different wind wave trains arriving on the same shore line at the same time. The beat frequency between these two wave trains may produce a periodic undulation in the sea level, or *surf beat*, which has an amplitude of a few centimeters. Longer period waves generally arise from the wind shifts and pressure changes accompanying meteorological disturbances. They may appear as standing-wave oscilla-

---

* An illuminating discussion of these methods will be found in *Practical Methods for Observing and Forecasting Ocean Waves by Means of Wave Spectra and Statistics*, 1955, by W. J. Pierson, Jr., G. Neumann, and R. W. James, U.S. Navy Hydrographic Office Publication No. 603.

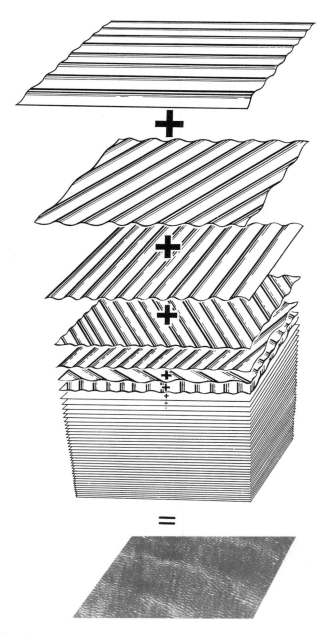

Fig. 3–14. A sea can be represented by the sum of many simple wave trains moving in different directions. [After W. J. Pierson, Jr., G. Neumann, and R. W. James, 1955, *Practical Methods for Observing and Forecasting Ocean Waves by Means of Wave Spectra and Statistics*, U.S. Navy Hydrographic Office Publication 603.]

Fig. 3–15. *Albatross III* encountering a force-7 sea in the North Atlantic drift. [Photograph by D. M. Owen.]

Fig. 3–16.   In strong winds, capillary waves cover the backs of gravity waves and may represent the scale on which the kinetic energy of air is transferred most efficiently to the sea.

tions across the width of a continental shelf and have periods amounting to many minutes or a few hours. Examples of these waves have been studied by Redfield and Miller (1957)[9] in connection with Atlantic hurricanes. Disturbances of a similar nature have been examined by Munk (1949)[10] and by Munk, Snodgrass, and Carrier (1956)[11] as examples of Lamb's *edge waves:* widely spaced wave crests arranged at right angles to the shore line. Waves of this kind can be excited by a wind shift associated with a passing front.

Waves in the trans-tidal portion of the spectrum may arise from the motion of ordinary cyclonic storms at sea which, because of their low pressure, tend to carry a compensating hill of water beneath them that moves with the speed of the storm if the depth permits. When the pressure wave accompanying cyclonic storms runs over shallow water, the portion in which the wind blows with the direction of storm motion may produce a *storm surge* of wind-driven water that often inundates lowlands. There are also occasions when strong squall winds blowing across a shallow arm of the sea may pile water on the farther shore.

Storm surges can be destructive. In recent years there have been major losses of life and property associated with winter gales over the North Sea, with hurricanes in the North Atlantic, and with typhoons in the North and more rarely in the South Pacific.

Once the initial progressive wave or storm surge has been excited by the wind and/or reduced barometric pressure and runs ashore, its reflection may produce a standing-wave oscillation or *seiche* which may contain one or more nodes. Seiching is a phenomenon that can be detected quite frequently as a damped undulation on the trace of the astronomical tide recorded at most tide gauge stations in sheltered waters. It is also known in the Great Lakes of North America, and has been studied intensively in Lake Okeechobee in Florida, Lake Windermere in England, several Scottish lochs, Lake Geneva in Switzerland, and in Lake Baikal in the Far East.

Neglecting the effects of earth rotation, the period, T, of a seiche having n nodes is given by

$$T = (1/n)(2L/\sqrt{gd}), \tag{3-5}$$

where L is the horizontal dimension of the basin measured in the direction of wave motion. This formula was given by J. R. Merian in 1828[#] and is known as Merian's formula.

Both seiches and progressive waves can be greatly influenced by the constraints of their environment. They are obliterated by gently sloping shore lines and reflected almost without modification by steep cliffs. In addition to this they can be diffracted around obstacles and refracted by running across currents or into depths appreciably less than one-half their wavelength.

Refraction

Surface waves entering shoaling water at an angle tend to be refracted as they "feel bottom." The reduction of velocity with decreasing depth (Fig. 3–17) causes the part of the wave front in shallow water to advance more slowly than the part offshore. This, as shown in Fig. 3–18, rotates and bends each wave front so that it tends more nearly to fit the shore line and break upon it almost all at once.[12]

Waves in deep water are also subject to refraction when crossing currents. There is a tendency for waves having a component of motion against the current to be refracted into a stream, while those having a component of motion with the current tend to be turned aside. These effects can produce the marked contrast of sea state that sometimes exists between the Gulf Stream and its surrounding waters. The refraction of swell into and out of a strong current can sometimes be observed from the air.

Diffraction

As in the case of light, water waves can be diffracted into the shadow zone behind steep-sided obstacles, provided these have horizontal dimensions at least in the order of several wavelengths. Diffraction grating patterns may also develop behind a chain of regularly spaced obstacles more than a wavelength apart. It is difficult, however, to be sure that purely diffractive effects are being observed, since obstacles such as islands are surrounded by more or less gently sloping bottom topography that tends to bend wave fronts into the shadow zone by refraction.

In the atolls of the Pacific, some of the fringing reefs of coral rise from the drowned island peak as nearly vertical walls above the submerged talus slopes. In favorable cases the submerged portion of the coralline or algal reefs are about one-half wavelength high, so that diffractive effects may tend to predominate. From the air a characteristic pattern of choppy sea may often be seen to extend some distance downwind from each atoll. It has been suggested that the Polynesians used these choppy areas as navigational guides, for from a small boat or canoe both the observer and the atoll are so near sea level that they are hidden from each other at ranges beyond ten miles by intervening waves and the curvature of the earth.

In the trade-wind zone of the North Atlantic the swell tends to be regular and long-crested by the time it reaches the Caribbean Islands. Diffraction grating effects seem to develop in the Caribbean when the swell passes through the gaps between the islands. However, it should be emphasized that diffraction effects predominate only where the obstacles are defined by nearly vertical slopes for a depth at least as great as one-half wavelength.

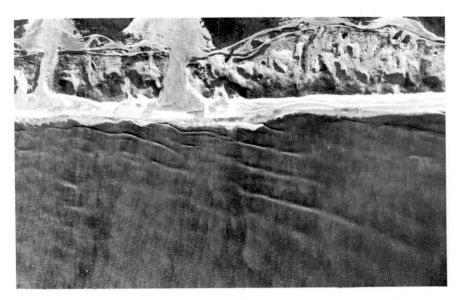

Fig. 3–17. The wavelength of a swell tends to shorten as it enters shallow water near a beach. [Photograph by R. Brigham.]

Fig. 3–18. Approaching a shore, waves are refracted by the decrease of depth, making the crest line lie more nearly parallel to the shore line. [Photograph by R. Brigham.]

Internal waves

In addition to surface waves, there is a class of wavelike oscillations within the volume of the sea which are most easily detected from the rise and fall of isotherms when a body of water is thermally stratified. Under certain circumstances it has seemed necessary to assume that internal waves break as they travel into shallow water, but this interpretation is open to question. Internal waves move most often as smoothly undulating shallow water waves whose behavior can be explained by taking into account the reduced acceleration of gravity arising from buoyant forces. Neglecting the effects of earth rotation, the equation for the celerity of a shallow internal wave is

$$c = \sqrt{g\left(\frac{\Delta\rho}{\rho}\right)\left(\frac{dd'}{D}\right)}, \qquad (3\text{–}6)$$

where $\Delta\rho$ is the difference in density of the lighter and heavier layers in the stratified water, g the acceleration of gravity, d the depth below the density interface, d' the height of the free surface above it, and D the total depth $d + d'$.

Because of the buoyant forces acting, internal waves may have a much greater amplitude than surface waves of corresponding energy density. Internal waves also progress quite slowly and have periods of many minutes. Fundamental studies of these processes have been made by J. E. Fjeldstad (1933,[#] 1937[131]) and others. When internal waves occur beneath a thin surface layer containing silt or plankton, they can be seen from the air as long parallel bands moving with majestic slowness (Fig. 3–19). Gifford C. Ewing (1950)[14] has related the presence of the troughs of internal waves to the visible confluence of the surface contamination. In calm weather such concentrations of surface contamination inhibit the formation of ripples under light winds, so that slicks appear.

A submarine hovering in near-perfect trim beneath the surface can be observed to rise and fall with internal waves in waters where the waves are moving on a strong density interface. Because of the accompanying changes in buoyancy of a cruising submarine as it cuts through internal wave crests and troughs, it is sometimes difficult to maintain a given depth of submergence solely by adjusting the angle of the diving planes.

Very strong density interfaces can be found near the mouths of large rivers where a fresh-water layer may extend far out over the salt water of the sea. These fresh layers may be so thin that a moving surface ship in addition to producing a surface wake may generate an internal wake with a consequent loss of propulsive efficiency. This phenomenon, known as *dead water*, was explained by V. W. Ekman in 1904.[#]

Fig. 3–19. Slicks, such as these on deep water over the abyssal slope off New England, may be associated with internal waves.

In the foregoing discussions of the different kinds of tides and waves, no particular attention has been given to the influence of the deflecting force of the earth's rotation—the Coriolis force. That this neglect is not always justified in the case of seiches in lakes is shown by the recent researches by Mortimer (1952, 1953, 1955).[15] Moreover, in the open sea where flow is sustained enough to be called a current and especially when the flow is nearly free of frictional coupling with the earth, the Coriolis force plays an important, if not a dominant, role. The nature of this force and its effects on fluid motion are discussed in the next chapter.

STUDY QUESTIONS

1. What maximum rate of earth rotation would permit the wave of the equilibrium tide to stay directly under the sun and moon if there were no continents and the ocean were everywhere 3.8 km deep?

2. Calculate the ratio of the solar and lunar tide-producing forces. What four alternative adjustments of the dimensions of the solar system could make the lunar and solar tides equal?

3. Were you to land on a strange and perpetually cloud-covered island (or planet), how could you find your latitude and hemisphere with a wrist watch, a rock, and a long piece of string?

4. Why is the diurnal inequality of the astronomical tide almost nil in the tropics? At what latitude should it be greatest according to the equilibrium theory?

5. Explain why the range of the semidiurnal tide on the French coast of the English Channel is so much larger than the corresponding tidal range on the English coast.

6. Show in a succession of drawings how the water surface elevation and associated currents differ in the case of the standing-wave and progressive-wave tides.

7. Since the tide-producing forces are only in the order of 10^{-7} g, how do you account for semidiurnal tidal currents on the continental shelves having characteristic velocities of 1 knot or more?

8. Distinguish between the tidal bore and the so-called "tidal wave" or tsunami. What conditions favor the occurrence of each phenomenon?

9. Distinguish between deep- and shallow-water waves and give the ratio of the phase velocity to the group velocity in each case.

10. Under what circumstances can gravity-wave amplitudes be increased by environmental influences? What factors can cause wave amplitudes to decrease?

11. What environmental influences can cause the direction of propagation of surface gravity waves to change with the distance traveled?

12. Distinguish between the forms and phase velocities of capillary and gravity waves. Describe as a function of time the surface disturbance near the edge of a pond following the splash of a rock thrown into the middle of the pond. How would events differ if the rock were thrown into a deserted swimming pool?

13. Distinguish between *sea* and *swell*. What property of surface waves makes it possible to determine the distance of a storm at sea from certain observations? Describe the method and explain what factors can alter the apparent direction of the storm center.

14. Under what circumstances can surface waves be diffracted? How could this effect be distinguished from wave refraction?

15. During World War II, the depths of water near invasion beaches were plotted from photographs of the approaching waves and breakers. If you were given this assignment, what observations would you require and what solution of the wave equation would you use?

REFERENCES

1. S. F. Grace, 1931, *Un. géod. géophys. int., Océanogr. phys. Publ. sci.*, 1, 26 pp.
2. A. T. Doodson, 1924, *Mon. Not. R. astr. Soc. geophys.* Suppl. 1: 124–147.
3. J. J. Stoker, 1957, *Water Waves*, New York: Interscience.
4. G. A. Macdonald, F. P. Shepard, and D. C. Cox, 1947, *Pacific Science*, 1: 21–37. See also C. K. Green, 1946, *Trans. Amer. geophys. Un.* 27: 490–500.
5. W. H. Munk, 1951, Chap. 1, pp. 1–4, *Proceedings of First Conference on Coastal Engineering*, J. W. Johnson, editor, Council of Wave Research, The Engineering Foundation.
6. J. Pattullo, *et al.*, 1955, *J. Mar. Res.*, 14: 88–155; J. G. Pattullo, 1959, pp. 778–779, *International Oceanographic Congress Preprints*, Mary Sears, editor, Washington, D. C.: American Association for the Advancement of Science.
7. W. G. van Dorn, 1953, *J. Mar. Res.*, 12: 249–276.
8. W. H. Munk, 1947, *J. Met.*, 4:45–57; N. F. Barber and F. Ursell, 1948, *Phil. Trans.* A, 240: 527–560; G. E. R. Deacon, 1949, *Ann. N. Y. Acad. Sci.*, 51: 475–482; Munk, W. H., G. R. Miller, F. E. Snodgrass, and N. F. Barber, 1963, *Phil. Trans. Roy. Soc.*, London, Ser. A, 255(1062): 505–584.
9. A. C. Redfield and A. R. Miller, 1957, *Met. Monogr.*, 2(10): 1–23.
10. W. H. Munk, 1949, *Trans. Amer. geophys. Un.*, 30: 849–854.
11. W. Munk, F. Snodgrass, and G. Carrier, 1956, *Science*, 123: 127–132.
12. J. W. Johnson, M. P. O'Brien, and J. D. Isaacs, 1948, *Graphical Construction of Wave Refraction Diagrams*, U.S. Navy Hydrographic Office Publication No. 605; R. S. Arthur, W. H. Munk, and J. D. Isaacs, 1952, *Trans. Amer. geophys. Un.*, 33: 855–865.
13. J. E. Fjeldstad, 1937, *Un. géod. géophys. int., Ass. Océanogr. phys. P. V.* 2: 141–142.
14. G. Ewing, 1950, *J. Mar. Res.*, 9: 161–187.
15. C. H. Mortimer, 1952, *Phil. Trans.* B, 236: 355–404; 1953, *Schweiz. Z. Hydrol.*, 15: 95–151; 1955, *Verh. int. Ver. Limnol.*, 12: 66–67.

SUPPLEMENTARY READING

Breakers and Surf, 1944, U.S. Navy Hydrographic Office Publication No. 234.

CORNISH, V., 1934, *Ocean Waves*, London: Cambridge University Press.

DEFANT, A., 1958, *Ebb and Flow*, translated by A. J. Pomerans, Ann Arbor: University of Michigan Press.

DEFANT, A., 1961, *Physical Oceanography*, Vol. II, London: Pergamon Press.

DOODSON, A. T., and H. D. WARBURG, 1941, *Admiralty Manual of Tides*, London: His Majesty's Stationery Office.

LOVE, A. E. H., 1911, *Some Problems of Geodynamics*, London: Cambridge University Press.

PIERSON, W. J. JR., G. NEUMANN, and R. W. JAMES, 1955, *Practical Methods for Observing and Forecasting Ocean Waves by Means of Wave Spectra and Statistics*, U.S. Navy Hydrographic Office Publication No. 603.

TOMASCHEK, R., 1957, "Tides of the Solid Earth," pp. 775–845, *Handbuch der Physik*, S. Flügge, editor, Bd. XLVIII, Geophysik II, Berlin: Springer-Verlag.

CHAPTER 4

Fluid Mechanics*

Fluid motion can be initiated by very small forces. It has been shown that the tides are propelled by the relatively weak differential attraction of the sun and moon. In the waters of the ocean and the gases of the atmosphere it is usually a small component of the earth's gravitational field that provides the motive force whenever static equilibrium is disturbed directly or indirectly by solar heating. Ordinarily the basic cause of the disturbance is overlooked to simplify the question from a problem of considering a totality of effects and interactions to one of measuring pressure differences from place to place on the earth and deducing from Newtonian principles the kinds of motion that should result.

Pressure

When there are no horizontal forces acting, the weight of the oceans and atmosphere above any chosen level tends to be balanced by the pressure at that level. Under ideal hydrostatic conditions, the pressure at every level is assumed to be exactly equal to the total weight of the fluid overburden per unit area. Departures from hydrostatic balance in the oceans are almost vanishingly small, but even so the same pressure is not found everywhere at the same depth. This variation of pressure from place to

---

* The substance of this chapter is largely descriptive, having been included to give those without prior acquaintance with fluid mechanics some knowledge of the elementary concepts required as background for the chapters that follow. The classic work in this essentially mathematical field is that by Sir Horace Lamb, first published by the Cambridge University Press in 1879# under the title *A Treatise on the Mathematical Theory of the Motion of Fluids* but more simply identified in the five subsequent editions as *Hydrodynamics*. In meteorological and oceanographic investigations the principles of fluid mechanics of special interest are those relating to large-scale processes in which the effects of the earth's rotation play an important role. These principles are emphasized in *Physikalische Hydrodynamik*, 1933, by V. Bjerknes, J. Bjerknes, H. Solberg, and T. Bergeron, Berlin: J. Springer. There is also a growing modern literature on this subject, which may be called geophysical fluid mechanics.

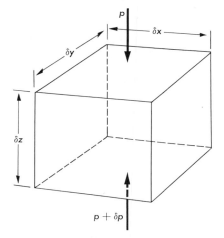

Figure 4–1

place at uniform depth in the ocean is mainly a consequence of the non-uniform density of sea water.

The increase of pressure with depth in the ocean is associated with a vertical *pressure gradient force* which can be described by considering a small cubical volume of dimensions δx, δy, δz (Fig. 4–1). On the upper surface of the cube the pressure is p, and on the bottom surface it is $p + \delta p$. The force per unit area on each of these two surfaces is $p(\delta x\ \delta y)$ and $(p + \delta p)(\delta x\ \delta y)$ respectively. Therefore the net force on the small cubical volume is upward, in the z direction, and amounts to $\delta p(\delta x\ \delta y)$. This is the vertical component of the pressure gradient force. In the limit, as the volume becomes vanishingly small, the total pressure difference between the top and bottom of the cube will approach the local vertical gradient of pressure. Dividing by the mass of fluid in the cube, we find that the vertical component of the specific pressure gradient force is $(-1/\rho)(\partial p/\partial z)$. If this pressure gradient force balances gravity, g, we can assume that $(-1/\rho)(\partial p/\partial z) = -g$. Since no other vertical forces are acting on the oceans which even approach the magnitude of gravity, it is probable that the *hydrostatic equation* is nearly satisfied everywhere.

The increase of pressure δp that accompanies every small increase of depth δz is

$$\delta p = \rho g \delta z, \tag{4-1}$$

where ρ is the average fluid density. The pressure increase from one depth to a new depth in a fluid of varying density is

$$p_z - p_0 = g \int_0^z \rho\ dz, \tag{4-2}$$

where the weights of all members of a vertical stack of unit volumes are summed between the two levels. This form of the hydrostatic equation is ordinarily used to compute the pressure difference between any two levels in the sea.

The pressure of sea water increases by one atmosphere (atm) over approximately 10 meters or 33 feet of depth, the exact value depending upon the water density and local acceleration of gravity. (It is also useful to remember that the pressure below the sea surface increases at the rate of 0.5 lb/in^2/ft.) The pressure on the sea floor at an average ocean depth of 4 km is about 400 atm or 6000 lb/in^2 and can be more than twice as great as this in the deepest parts of the oceans.

In the absence of motion, the pressure at any level below the sea surface is given by the weight of the fluid column between the selected level and the sea surface plus the weight of the atmosphere above. Thus at a depth of 10 meters the absolute pressure is equal to 2 atm. When measurements are made with reference to the sea surface, the pressure at any depth in the sea can be computed by assuming zero pressure at the sea surface; in this case the pressure at depth is read as a "gauge pressure."

The units of pressure usually employed in oceanography and meteorology are related and are generally derived from the metric standard of 1 dyne/cm^2, sometimes called the *barye*. Because the average surface pressure of the atmosphere is 1.01325 $\times$ 10^6 dynes/cm^2, it has become commonplace to consider atmospheric pressures in terms of the *bar* and its decimal parts, which are defined as follows:

$$10^6 \text{ dynes/cm}^2 = 1 \text{ bar} = 10 \text{ decibars} = 10^3 \text{ millibars.}$$

It is convenient to measure ocean pressure in *decibars*, the unit of pressure difference amounting to approximately one-tenth of an atmosphere or (by definition) to exactly 100,000 dynes/cm^2. This corresponds to a pressure change occurring with a change in depth of about 3.3 feet or roughly 1 meter. Therefore when depths in the ocean are measured in meters, the approximate pressure in decibars is also known by numerical equivalence.

Owing to the variation of the force of gravity from place to place on the earth and to the variation of density within the ocean, the exact linear distance through which a change of pressure of 1 decibar will occur varies from place to place and from level to level. To account for these effects, it is customary to refer to the depth interval associated with a given pressure difference in terms of work per unit mass of 10^5 dyne-cm, the *dynamic meter*. More will be said in Chapter 9 about this unit and its relationships to pressure.

Continuity

Wherever there are nonhydrostatic gradients of pressure in the sea there tends to be motion. These motions can be of the simple "downhill" variety or can be complicated by effects of the earth's rotation. In either case it is assumed that where there is no change in state, sea water is essentially incompressible and that the liquid matter involved, when moved, undergoes neither creation nor destruction. These assumptions require that in any vertical contraction of a fluid volume there be a corresponding horizontal extension so that the volume of the original parcel remains constant (Fig. 4–2). This idea can be expressed in the *equation of continuity*.

If we fix in the earth three cartesian coordinate directions x, y, and z and designate the components of motion in these three directions by u, v, and w, respectively, the equation of continuity for an incompressible fluid becomes

$$\frac{\partial u}{\partial x} + \frac{\partial v}{\partial y} + \frac{\partial w}{\partial z} = 0. \tag{4–3}$$

Equation (4–3) implies, for example, that if the vertical velocity w increases with depth z so that the fluid volume stretches, then there must be compensatory shrinking in either or both of the horizontal directions x and y. This form of the equation of continuity is valid for the interior of a fluid but not for a free surface, where motion must always be directed parallel to the surface. In sea water the effects of fluid compressibility on the validity of the equation of continuity are generally small enough to be neglected.

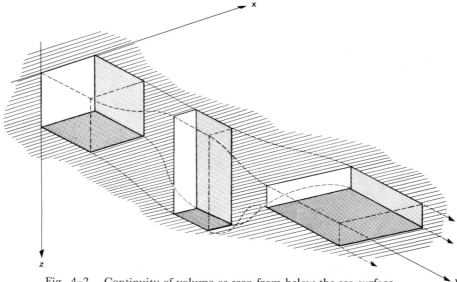

Fig. 4–2. Continuity of volume as seen from below the sea surface.

Inertial space

The generalized concepts of fluid motion are based upon the principles of Newtonian mechanics, particularly the second law of motion. When all the important forces that produce fluid motion are written together within the framework of the second law of motion, we have a statement of the *equation of motion*. This equation treats the motion of particles which can be essentially free of the earth. In preparation for an understandable statement of the equation of motion, it is necessary to examine the properties of inertial space and the concept of the Coriolis force, which has been invented to reduce the problem to convenient terms.

In the parlance of physics it is said that the earth is not a primary inertial system. That is, if a coordinate system of three mutually perpendicular axes is embedded in the earth so as to turn with it, the motions of a free particle are not described in their simplest terms with reference to those axes but rather with reference to coordinate axes at rest with respect to the fixed stars (Fig. 4–3). The latter is, so far as we know, the primary inertial system. Particles of water or air often move with only very slight frictional influence from the rotating earth, and therefore tend to follow paths which are most simply described by reference to the primary inertial system. One exception is the constraint of gravity, which bends particle motions to conform with the earth's curvature.

To those in northern latitudes who watch the stars circle the sky during the night, it is apparent, when they face the east, that everything on the

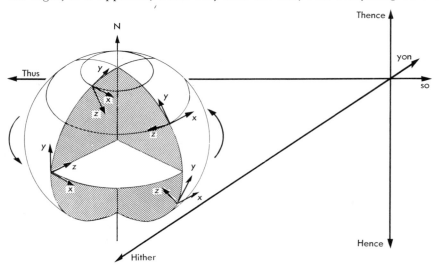

Fig. 4–3. Relation of terrestrial coordinate systems to that of the primary inertial system of the fixed stars.

Fig. 4–4. Gaspard Gustave de Coriolis. [From H. Rouse and S. Ince, 1957, *History of Hydraulics*, Iowa City: Iowa Institute of Hydraulic Research, State University of Iowa.]

earth is being carried by the earth's rotation through a left turn around the earth's axis. Therefore when a body is projected from the earth in northern latitudes, it tends to move toward the star at which it was first headed and, in time, the course of flight will appear to bend toward the right as the observer and all the local terrestrial scenery turn to the left. In southern latitudes the apparent deflection is reversed, because the observer facing east is being carried through a right-hand turn by the earth's daily rotation. The effects of rotational deflection vanish only when objects are moving horizontally and exactly in the plane of the earth's equator. Ordinarily the effects of earth rotation are not noticeable, because the rate of earth rotation is too slow and free-flying objects usually fall back to earth too soon for the rotational deflection to become appreciable. Nevertheless artillerymen are well aware of the necessity to correct their fire slightly to compensate for earth rotation. So too are bombardiers, because the paths of objects moving up or down are also deflected. In the light of knowledge that rotational deflections become noticeable in the flights of missiles, it becomes understandable that the paths of winds and ocean currents, which move for hours and days without appreciable frictional influence from the solid earth, must be strongly deflected. The effect may indeed be so conspicuous that the paths of particle motion in the atmosphere and oceans may become highly circuitous or even successively looped in the course of several days.

We do not have any physiological sense of the earth's rotation beyond our visual evidence of the apparent rising and setting of the sun, moon, and stars. Therefore, instead of referring the motion of free particles to the coordinates of primary inertial space, it is intuitively more agreeable to consider the apparent curvature of motion of free particles as owing to a (fictitious) force acting at right angles to their direction of motion. Taking this imaginary force into account, it is possible to regard the earth as being at rest, that is, possessed of its own inertial system. The concept of a Coriolis force simplifies mathematical descriptions of nearly all the large-scale fluid phenomena of the earth, as well as the free motions of projectiles. This point of view was taken by Laplace in 1775[#] in a discussion of the tides, and presented in general form in 1835[#] by G.-G. de Coriolis (Fig. 4-4). The Coriolis force is a means for accounting for the behavior of particles moving in the primary inertial system but observed from a rotating earth which is assumed to be at rest.

The Coriolis force

It has been argued that a particle at rest on the earth remains in place because the equatorial bulge of the earth provides a slope inclined to the geocentric vertical by just the right amount to produce a poleward component of gravitational force equal to the horizontal component of the centrifugal force accompanying the earth's rotation. As soon as the particle moves over the earth, however, these forces are no longer in balance. If the particle moves eastward over the earth, it will have a centrifugal reaction toward the equator which is slightly greater than that of the earth beneath, and will tend to move toward the equator. If the particle moves westward over the earth, gravitation will tend to move it poleward, since its centrifugal reaction is less than the reaction of the earth.

To find an expression for this imbalance of forces, let us consider first a particle at rest in latitude ϕ where the tangential speed of rotation of the earth is V. Now $V = \Omega r$, where Ω is the angular velocity of rotation of the earth and r is the normal distance of the particle from the axis of rotation —that is, $r = R \cos \phi$, where R is the radius of the earth. If the particle is given a velocity u toward the east, relative to the earth, then its total tangential velocity will be $(V + u)$ and its centrifugal reaction per unit mass will be $C = (V + u)^2/r$. The direction of the centrifugal reaction vector lies at right angles to the axis of rotation of the earth and therefore has a horizontal component

$$C_h = \frac{(V + u)^2 \sin \phi}{r}. \tag{4-4}$$

If we expand this expression by squaring the velocity term, we obtain

$$C_h = \frac{V^2 \sin \phi}{r} + \frac{2Vu \sin \phi}{r} + \frac{u^2 \sin \phi}{r}. \tag{4-5}$$

a) The first term in the expansion represents the centrifugal force on a unit mass at rest, and therefore one balanced out by the poleward component of gravitation produced by the earth's ellipticity; the second term is the Coriolis force per unit mass; and the third term is the centrifugal force per unit mass on a particle moving as though the earth were at rest. The third term is quite small unless the particle is moving over the earth at a speed commensurate with the tangential speed of the earth itself.

The *Coriolis force* per unit mass can be written in a more familiar form if V is replaced by Ωr. Thus

$$C_h = \frac{2\Omega r u \sin \phi}{r} = (2\Omega \sin \phi)\, u. \qquad (4\text{--}6)$$

Now $2\Omega \sin \phi$ has a characteristic value at each latitude circle on the earth and is therefore a parameter which can be tabulated or otherwise specified once and for all. The *Coriolis parameter*, $2\Omega \sin \phi$, appears so frequently in equations of motion that it is often abbreviated as f. The Coriolis acceleration, or force per unit mass, accompanying east-west motion in any latitude is then fu. We know from the previous discussion that the horizontal direction of the Coriolis acceleration is 90° to the right of the direction of motion in the Northern Hemisphere, that is, toward the equator if the direction of motion is eastward and toward the pole if the motion is toward the west. In the Southern Hemisphere the Coriolis acceleration is 90° to the left of the direction of motion.

Arguments based on the conservation of angular momentum show that the deflecting force of rotation is exactly similar for motions directed northward or southward. If a particle is thrust toward the equator, it will have the same angular momentum as the earth in the latitude where its motion began and less than that of other particles at rest in lower latitudes. Since particles at the lower latitudes are farther from the axis of rotation of the earth, they must have a higher tangential velocity to make a full revolution around the earth's axis in one day's time (Fig. 4–5). Relative to the solid earth in lower latitudes the moving particle will lag westward, or have a westward component of velocity relative to the earth. If the particle were thrust northward, it would soon possess more angular momentum than that required to encircle the earth's axis once in one day, and as a consequence it would have an eastward component of velocity relative to the earth. In both cases the deflection due to the change of radius of gyration with latitude results in a deflection to the right for northward and southward motions in the Northern Hemisphere and to the left in the Southern Hemisphere. It can be shown that the Coriolis acceleration acts at right angles to the instantaneous direction of particle motion and has the value $(2\Omega \sin \phi)v$, where v is the particle velocity in the northward direction.

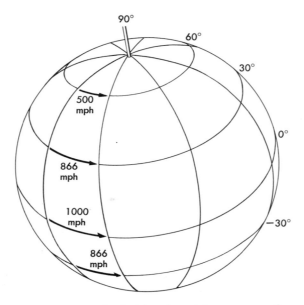

Fig. 4–5. Absolute tangential velocities of particles at rest on the earth's surface at different latitudes.

Thus far we have considered the effects of the earth's rotation only in a horizontal plane. Any change in the distance of a body from the axis of rotation of the earth has inertial consequences of the same character.

A body falling freely in either the Northern or Southern Hemisphere from a point fixed over the earth will be deflected initially toward the east. This is anticipated directly from consideration of the conservation of angular momentum. Similarly, a body initially headed straight up would be deflected toward the west, again in accordance with the intuitive reasoning from the principle of conservation of angular momentum.

The magnitude of the Coriolis acceleration is small, being $2\Omega c$ or approximately $1.5 \times 10^{-4}c$ at the poles and zero at the equator; here c is the horizontal velocity of a fluid parcel in any direction relative to the earth. For average winds, which have a speed of about 1000 cm/sec, the Coriolis acceleration amounts to 0.15 cm/sec^2 or about 0.00015 g. In the case of strong ocean currents which move at about one-tenth the speed of the average wind, that is 100 cm/sec or less, the Coriolis acceleration is one-tenth as great. Small as this acceleration is relative to gravity, it has nevertheless important effects upon the horizontal motions of the air and the water of the open sea. A principal counterbalance of the Coriolis force is the horizontal component of the pressure gradient force.

Pressure gradient forces

In association with the vertical pressure gradient force, which balances the force of gravity in the oceans, there are smaller horizontal pressure gradient forces wherever isobaric surfaces are inclined and intersect local geopotential surfaces. The slope of an isobaric surface of pressure p, relative to a geopotential (level) surface (Fig. 4–6), has components $\partial z/\partial x$ and $\partial z/\partial y$ along which gravity may act to produce force components of

$$g\left(\frac{\partial z}{\partial x}\right)_p \quad \text{and} \quad g\left(\frac{\partial z}{\partial y}\right)_p. \quad (4\text{–}7)$$

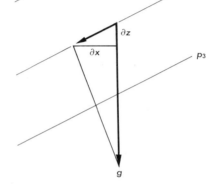

Under hydrostatic conditions these expressions are equivalent to

$$-\frac{1}{\rho}\frac{\partial p}{\partial x} \quad \text{and} \quad -\frac{1}{\rho}\frac{\partial p}{\partial y}, \quad (4\text{–}8)$$

where the minus signs indicate that the horizontal pressure gradient force is directed down the gradient from high to low pressure. In an alternative notation, the specific volume, α, is often used in place of the reciprocal density $1/\rho$.

The total pressure gradient force is a vector, written $-\alpha\nabla p$, which by definition has components

Figure 4–6

$$-\frac{1}{\rho}\frac{\partial p}{\partial x}, \quad -\frac{1}{\rho}\frac{\partial p}{\partial y}, \quad \text{and} \quad -\frac{1}{\rho}\frac{\partial p}{\partial z}. \quad (4\text{–}9)$$

This notation is used and explained in the next section.

Equation of motion

Newton's second law of motion asserts that the product of mass and acceleration equals the vector sum of the forces. This statement is called the *equation of motion*. In atmospheric and oceanic usages it is customary to divide both sides of the equation of motion by the mass so that each term refers to the force per unit mass, or *specific force;* however, the qualifying adjective "specific" normally is only implied.

The important forces which drive the large-scale oceanic motions are the force of gravity, the Coriolis force, pressure gradient force, and frictional forces. The centrifugal force of the earth's rotation is usually included in gravity. The three-dimensional acceleration of a fluid parcel is described most directly by the vector equation of motion, which contains the following terms:

Particle acceleration = Coriolis term
+ pressure gradient term
+ gravity term
+ friction term,

or, in symbols,

$$\frac{d\mathbf{C}}{dt} = (-2\mathbf{\Omega} \times \mathbf{C}) + (-\alpha \nabla p) + \mathbf{g} + \mathbf{F}, \qquad (4\text{-}10)$$

where $d\mathbf{C}/dt$ is the acceleration of a unit mass due to the accumulated effects per unit mass of the Coriolis force $-2\mathbf{\Omega} \times \mathbf{C}$, the pressure gradient force $-\alpha \nabla p$, the force of gravity $\mathbf{g}$, and $\mathbf{F}$, the generalized force due to frictional effects.

The contribution of the Coriolis term, $-2\mathbf{\Omega} \times \mathbf{C}$, is found by taking for $2\mathbf{\Omega}$ a vector parallel with the axis of rotation of the earth, positively directed in the sense of progress of a right-hand screw turning with the earth (that is, northward), and of a length twice that for the spin of the earth. The cross product of $2\mathbf{\Omega}$ with $\mathbf{C}$, the vector representing particle motion relative to the earth, is another vector perpendicular to the plane of $2\mathbf{\Omega}$ and $\mathbf{C}$ and directed in the sense of advance of a right-hand screw turned from the positive direction of $2\mathbf{\Omega}$ toward the positive direction of $\mathbf{C}$ through the smaller of the angles between them. In magnitude the cross product is equal to the arithmetic product of the component vector lengths multiplied by the sine of the angle between them. The direction of the Coriolis force vector can be very much inclined to the plane of the local horizon.

In the pressure gradient term, $-\alpha \nabla p$, the quantity α is the specific volume (volume per unit mass), a scalar. The pressure p is also a scalar, but since it has equal values along certain surfaces embedded in the fluid it is possible to construct normals to these surfaces. The gradient of pressure, ∇p, at any point is a vector normal to the surface of constant pressure and counted positive in the direction in which the pressure increases. Clearly, then, ∇p in most instances is directed nearly parallel with the local direction of gravity but not, as is important here, always exactly so. The horizontal component of the gradient of pressure is directed against the slope of the surfaces of equal pressure. Since a free particle will move down the pressure slope, the specific pressure gradient force is written $-\alpha \nabla p$.

Friction $\mathbf{F}$ is a force per unit mass accompanying real motion which, in certain idealized concepts, arises from the exchange of particles between the slower moving portions of the fluid and other more rapidly moving portions. The particles carry their momentum with them in the exchange process and on arrival share their momentum with their new neighbors. Where rough solid boundaries are involved, the loss of momentum to the earth is so great that the effects of the other accelerating processes can be overwhelmed. But in air and water internal friction is considered to be so slight that it is very often neglected.

Given information as to the magnitudes and directions of all the terms on the right-hand side of Eq. (4–10), it is only necessary to add the terms vectorially to find the change with respect to time of the velocity of a particle of air or water on the earth. This is elementary in principle but difficult in practice. Therefore the general equation of motion is often simplified for practical purposes, and most computation is done with reference to specialized cases.

As a prelude to such simplification and because of the tremendous inequality of vertical and horizontal forces, it is helpful to break up the vector statement of Eq. (4–10) into its components relative to a left-handed coordinate system in which $+y$ is toward the north, $+x$ is toward the east, and $+z$ is directed upward. We will also consider c as being composed of the velocity components u, v, and w along these coordinate axes. Neglecting small terms we can rewrite Eq. (4–10) as

$$\frac{du}{dt} = (2\Omega \sin \phi)v - \alpha \frac{\partial p}{\partial x} + F_x, \qquad (4\text{--}11)$$

$$\frac{dv}{dt} = (-2\Omega \sin \phi)u - \alpha \frac{\partial p}{\partial y} + F_y, \qquad (4\text{--}12)$$

$$\frac{dw}{dt} = +g - \alpha \frac{\partial p}{\partial z} + F_z, \qquad (4\text{--}13)$$

where

$u, v, w \equiv$ velocity components along the coordinate axes x, y, z,
$\Omega \equiv$ angular velocity of the earth's daily rotation, $(0.729 \times 10^{-4} \text{ sec}^{-1})$,
$\phi \equiv$ geographic latitude,
$\alpha \equiv$ specific volume $(\text{cm}^3/\text{gm} = 1/\rho)$,
$p \equiv$ pressure $(\text{dynes}/\text{cm}^2)$,
$F_x, F_y, F_z \equiv$ components of friction force per unit mass or occasionally the tide-producing force per unit mass,
$g \equiv$ acceleration of gravity $(980 \text{ cm/sec}^2$ as an average figure or 10^3 cm/sec^2 for "order of magnitude" computations).

Fig. 4–7. Léonard Euler. [From D. E. Smith, 1923, *History of Mathematics*, Vol. 1, Boston: Ginn and Co.]

These equations define for any particular instant of time the motion of all particles in the space occupied by the fluid. This is the "Eulerian" form of the problem; correspondingly it is possible to inquire into the history of motion of each particle as a function of time. The latter is called the "Lagrangian" statement of the problem, although both were originally discussed by Euler in 1755[#] (Fig. 4–7). We will confine our attention to the Eulerian form of the equations of motion.[*]

---

[*] The Lagrangian equations of motion are given in Vol. 1 of *Treatise on Natural Philosophy*, 1879, by W. Thomson and P. G. Tait, Cambridge University Press. A further description of the Lagrangian method will be found in *Fundamentals of Hydro- and Aeromechanics*, 1934, by L. Prandtl and O. G. Tietjens, translated by L. Rosenhead, New York: Dover Publications, Inc.

Euler's expansion

The acceleration terms du/dt, dv/dt, dw/dt used in the previous section are the components of acceleration of an individual fluid particle moving through a coordinate system (x, y, z) fixed in the earth. The pressure gradient is measured at a fixed point in the coordinate space, while the particles themselves move through it. Therefore the particle accelerations must be resolved into their fixed (local) and moving (advective) parts if the fluid motion is to be studied for intervals longer than an instant.

Euler's expansion permits the transformation of individual derivatives (d/dt) of the behavior of a moving particle with respect to time into local derivatives $(\partial/\partial t)$ plus three advective terms which describe the changing motion of a fluid as it passes through a fixed point:

$$\frac{du}{dt} = \frac{\partial u}{\partial t} + u\frac{\partial u}{\partial x} + v\frac{\partial u}{\partial y} + w\frac{\partial u}{\partial z}, \qquad (4\text{--}14)$$

$$\frac{dv}{dt} = \frac{\partial v}{\partial t} + u\frac{\partial v}{\partial x} + v\frac{\partial v}{\partial y} + w\frac{\partial v}{\partial z}, \qquad (4\text{--}15)$$

$$\frac{dw}{dt} = \frac{\partial w}{\partial t} + u\frac{\partial w}{\partial x} + v\frac{\partial w}{\partial y} + w\frac{\partial w}{\partial z}. \qquad (4\text{--}16)$$

When the local derivatives vanish, the motion is said to be stationary (without change at any given point), and only the advective terms remain to alter the particle motion as it passes into regions of different velocity. The advective terms are nonlinear, so that it becomes difficult to study fluid motions in complete generality.

The steady state

Earth processes such as the circulations of the oceans and atmosphere have been in more or less smooth and continuous operation long before man began to study them. Therefore it is not necessary—although it is undeniably interesting—to inquire how it all started from a state of rest. This point of view has led to a concept of *steady-state* conditions in which the flow at every point exists without time-dependent changes. The assumption is applicable only when the change that actually occurs over a short period of time is small compared with the accelerations required to initiate fluid motion from rest. Where its use can be justified the steady-state assumption simplifies analysis of some otherwise intractable problems.

When local derivatives are expanded, as shown in the previous section, two possible simplifying conditions are presented. If the local derivatives are all equal to zero, that is, if

$$\frac{\partial u}{\partial t} = \frac{\partial v}{\partial t} = \frac{\partial w}{\partial t} = 0, \tag{4-17}$$

the remaining three advective terms equal the sum of the forces and a *steady state* is said to exist. If all the terms on the right are equal to zero or sum to zero, the fluid is in *equilibrium*. Except near boundaries or in small-scale processes, it can sometimes be assumed that the advective terms are so small that it suffices to neglect them entirely. In the open ocean the distinction between steady-state and equilibrium conditions is usually not made. With this limiting condition assumed, and thus ignoring a frequently important part of the problem, we can consider the steady state by setting the individual time derivatives equal to zero, so that

$$\frac{du}{dt} = 0 = (2\Omega \sin \phi)v \quad - \alpha \frac{\partial p}{\partial x} + F_x, \tag{4-18}$$

$$\frac{dv}{dt} = 0 = (-2\Omega \sin \phi)u - \alpha \frac{\partial p}{\partial y} + F_y, \tag{4-19}$$

$$\frac{dw}{dt} = 0 = +g \quad\quad\quad - \alpha \frac{\partial p}{\partial z} + F_z. \tag{4-20}$$

Substituting $1/\rho$ for α in the pressure gradient term and rearranging each statement, we may write the same three equations in the conventional form:

Pressure gradient term	Coriolis term	Friction term	
$\dfrac{1}{\rho}\dfrac{\partial p}{\partial x}$ =	$(2\Omega \sin \phi)v$ +	$F_x,$	(4-21)
$\dfrac{1}{\rho}\dfrac{\partial p}{\partial y}$ =	$(-2\Omega \sin \phi)u$ +	$F_y,$	(4-22)

and

Pressure gradient term	Gravity term	Friction term	
$\dfrac{1}{\rho}\dfrac{\partial p}{\partial z}$ =	$+g$	$+F_z.$	(4-23)

Often, friction too can be shown to be so small compared with the magnitudes of other terms (as in situations far from land and more than 100 meters below the surface or above the bottom) that we may neglect friction without seriously affecting (hopefully) the accuracy of numerical calculations.

Geostrophic motion

When fluid flow is both unaccelerated and frictionless, its motions are said to be *geostrophic*, that is, "earth-turned." In this case the only forces acting are the pressure gradient force, the Coriolis force, and gravity (Fig. 4–8). These must balance one another, so that

$$\frac{1}{\rho}\frac{\partial p}{\partial x} = (2\Omega \sin \phi)v, \tag{4–24}$$

$$\frac{1}{\rho}\frac{\partial p}{\partial y} = (-2\Omega \sin \phi)u, \tag{4–25}$$

$$\frac{1}{\rho}\frac{\partial p}{\partial z} = +g. \tag{4–26}$$

It should be recognized that the third equation of motion is the hydrostatic equation. The other two equations describe geostrophic motion referred to a fixed geographic coordinate system, although one which cannot always be fixed with certainty, especially at sea. For this reason it is customary to make a further simplification: that of transforming coordinates to an arbitrary local system by applying the Pythagorean theorem

$$\frac{\partial p}{\partial n} = \sqrt{\left(\frac{\partial p}{\partial x}\right)^2 + \left(\frac{\partial p}{\partial y}\right)^2} \tag{4–27}$$

and

$$c = \sqrt{u^2 + v^2}. \tag{4–28}$$

As a result we may express Eqs. (4–24) and (4–25) as

$$\frac{1}{\rho}\frac{\partial p}{\partial n} = (2\Omega \sin \phi)c. \tag{4–29}$$

Denoting the Coriolis parameter $2\Omega \sin \phi$ by f, we have

$$\frac{1}{\rho}\frac{\partial p}{\partial n} = fc, \tag{4–30}$$

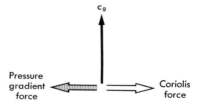

Fig. 4–8.   Geostrophic balance.

where c in this case is the geostrophic velocity at right angles to a measured pressure difference δp along a line segment δn also normal to c. As a reminder of both the geostrophic nature of c and the finite differences actually measured as an approximation to the infinitesimal pressure gradient element $\partial p/\partial n$, this equation is ordinarily written

$$\frac{1}{\rho}\frac{\delta p}{\delta n} = fc_g \qquad (4\text{–}31)$$

or

$$c_g = \frac{1}{\rho f}\frac{\delta p}{\delta n}. \qquad (4\text{–}32)$$

This is the geostrophic equation used in estimating wind and ocean-current velocities from measurements of the horizontal gradients in the field of pressure.

The pressure field in the sea or atmosphere is usually described by a succession of three-dimensional surfaces which may be sloping or buckled but on which the pressure is everywhere the same. Where these *isobaric surfaces* cut an arbitrary plane surface, there is a line of intersection—an isobaric line or simply an *isobar*—along which the pressure is the same at the level of the plane. When a nested sequence of isobaric surfaces is sectioned horizontally, a pattern of isobaric lines is formed which can be interpreted in the same manner as a topographic map. Close spacing of isobars indicates a strong gradient of pressure in the horizontal plane of the section, whereas an open pattern indicates a weak horizontal pressure gradient. Geostrophic flow is aligned parallel to the isobars and varies in speed with the closeness of their spacing. When a number of successive horizontal sections are taken through a three-dimensional field of isobaric surfaces, the corresponding field of geostrophic motion can be mapped. This procedure is common in both oceanography and meteorology.

Although geostrophic motion is, quite strictly, an unaccelerated phenomenon, it is easier to understand how it comes into being than how it is maintained. Consider a particle on a frictionless hillside on the rotating earth (Fig. 4–9). Imagine that as the particle slides freely and gathers speed down the slope it is deflected more and more strongly by the deflect-

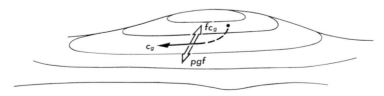

Figure 4–9

ing force of earth rotation so that, in the Northern Hemisphere, its path is turned more and more toward the right. Ultimately, if inertial oscillations are damped out, the particle will run along the contour lines of the hillside at such a speed that the Coriolis acceleration acting at right angles to its direction of motion is directed uphill to oppose and cancel the downhill component of the acceleration of gravity. In a perfectly frictionless world this motion on the hillside might continue forever as geostrophic motion.

But where friction acts ever so slightly to slow the particle's forward speed, the Coriolis force will be reduced and gravity will cause the particle to change its course slightly downhill to acquire the energy it needs to balance friction. Where friction is a steady impediment to free motion, the particle will lose altitude on the hillside at a steady rate and ultimately reach the bottom. Under these circumstances the component of motion down the hillside is called the *ageostrophic* component of motion.

By considering a solid hillside in this example we have made the tacit assumption that the downhill component of gravity remains invariant in space and is independent of the particle's motion. In a fluid, however, the slope of the isobaric surfaces—represented by the hillside—is often influenced by the particle motion or by external forces which, in effect, put changing curvatures and moving waves on the slope of the hillside. These tend to upset the field of geostrophic motion. If the disturbances are small, the field of motion is said to be *quasi-geostrophic*.

The geostrophic approximation provides a basis for computing the vector field of motion from a scalar field of pressure under certain specialized assumptions. These assumptions are more often than not quite unnatural, but the convenience of the transformation is so inviting that geostrophic or quasi-geostrophic motions are often computed to give a useful first approximation of the field of motion.

Margules' equation

Geostrophic currents and winds occur as a dynamical balance between the local gradient of pressure and the deflecting force of the earth's rotation. The horizontal pressure gradient is more often than not associated with a pronounced change in the horizontal gradient of fluid density which can be very marked where two fluid masses are involved. The boundary between the two fluids is a surface of density discontinuity which ordinarily slopes downward under the fluid mass having the lesser density (Fig. 4–10).

It is possible to estimate the slope of the surface density discontinuity associated with geostrophic motions in the sea and of fronts in the atmosphere from a knowledge of the component speeds of geostrophic motion along the interface and the density difference across an interface. An equation relating these variables was derived in 1906[#] by Margules. For a

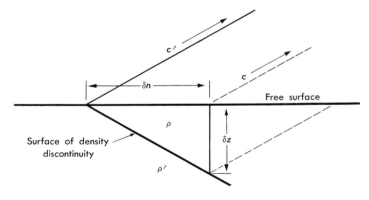

Figure 4–10

stationary front this equation has the form

$$\frac{\delta z}{\delta n} = \frac{f}{g} \frac{\rho' c' - \rho c}{\rho - \rho'}, \qquad (4\text{--}33)$$

where δz is a finite increment of vertical distance, δn a finite increment of horizontal distance along the dip of the interfacial slope, f the Coriolis parameter, g the acceleration due to gravity, ρ the density, and c the geostrophic velocity parallel with one side of the interface, while their primes are the values of corresponding properties on the other side.

Margules' equation is helpful in clarifying the sometimes confusing problem of estimating the change of frontal slope as a function of latitude or, in turn, a change in density contrast across the front at the same latitude but with different velocities of flow.

Meander motion

Where currents are curved or meandering, the centrifugal force accompanying the curvatures of flow must be taken into account. When the motion is anticyclonic, the centrifugal force (CF, Fig. 4–11) augments the horizontal pressure gradient force, $\mathrm{grad}_h\ p$; and when the motion is cyclonic, it augments the Coriolis force, f, in both hemispheres. In these cases

$$\frac{1}{\rho} \frac{\partial p}{\partial n} = f c_m \pm \frac{c_m^2}{r}, \qquad (4\text{--}34)$$

where the minus sign applies to anticyclonic motion, the plus sign to cyclonic motion, r is the radius of the particle trajectory, and c_m is the tangential or meander velocity,[*] which is different from c_g, the geostrophic velocity.

———————

[*] Known in meteorology as the "gradient velocity."

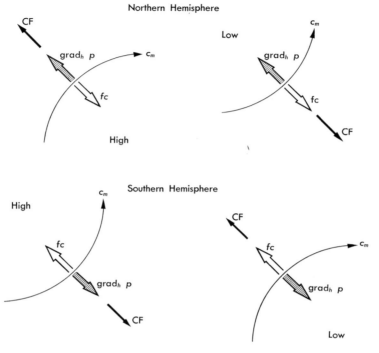

Fig. 4–11. Balance of forces in the four possible cases of meander flow.

The value of c_m is found from measurements of the other factors governing the flow in cases of anticyclonic and cyclonic motion (Fig. 4–12).

In anticyclonic motion the balance of the pressure gradient, Coriolis, and centrifugal forces is as shown in Fig. 4–11. Using $\text{grad}_h\,p$ to denote the horizontal radial component of the total pressure gradient force, and c_m to denote the tangential speed of the particles, we obtain

$$\rho\,\frac{c_m^2}{r} + \text{grad}_h\,p - \rho f c_m = 0 \qquad (4\text{--}35)$$

or, in standard form,

$$c_m^2 - rf c_m + \frac{r}{\rho}\,\text{grad}_h\,p = 0. \qquad (4\text{--}36)$$

Application of the quadratic formula to Eq. (4–36) yields

$$c_m = \frac{rf}{2} \pm \frac{1}{2}\sqrt{\left(rf^2 - \frac{4}{\rho}\,\text{grad}_h\,p\right)r}. \qquad (4\text{--}37)$$

Structure of
warm-core anticyclonic eddies

Structure of
cold-core cyclonic eddies

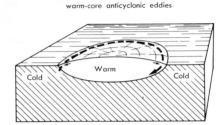
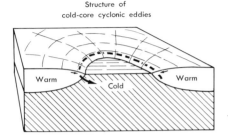

Figure 4–12

The proper sign ($\pm$) is that which makes the flow diminish to zero when the pressure gradient vanishes; that is, ($-$) minus. Equation (4–37) becomes imaginary if $(4/\rho)\,\mathrm{grad}_h\,p \geq rf^2$; therefore the pressure gradient force theoretically cannot exceed the Coriolis force. Indeed, it has been observed in atmospheric anticyclones that weak pressure gradients exist in the central regions where r is small and that calm conditions prevail. Corresponding situations exist in the central anticyclonic eddy of each major ocean.

The corresponding analysis for cyclonic flow leads to the expression

$$\rho\,\frac{c_m^2}{r} + \rho f c_m - \mathrm{grad}_h\,p = 0 \qquad (4\text{–}38)$$

or, in standard form,

$$c_m^2 + rf c_m - \frac{r}{\rho}\,\mathrm{grad}_h\,p = 0. \qquad (4\text{–}39)$$

Application of the quadratic formula to Eq. (4–39) yields

$$c_m = -\frac{rf}{2} \pm \frac{1}{2}\sqrt{\left(rf^2 + \frac{4}{\rho}\,\mathrm{grad}_h\,p\right)r}. \qquad (4\text{–}40)$$

In this case we choose the $+$ sign to allow the current to decrease to zero when the pressure gradient vanishes. But since the radius and pressure-gradient terms are added under the radical, this theory predicts that the gradient of pressure is free to increase without limit as r decreases. Indeed, atmospheric cyclones contain strong winds at their centers, but in hurricanes and tornadoes where the horizontal pressure gradient $\partial p/\partial n$ is much greater than f, the Coriolis parameter, this simple theory breaks down, probably as a result of the vertical motions that occur. There is insufficient data to give examples of cyclonic eddies in the sea, but their existence is suggested in observations by Fuglister and Worthington (1951)[1] and Uda (1951).[2]

The Gulf Stream, Kuroshio, and some other strong currents have been shown to possess meandering courses of flow. From the meander-flow equations one would expect the current to flow more swiftly in anticyclonic curvatures than in cyclonic curvatures having the same radius and transverse pressure gradient. From Margules' equation one would also expect that where the flow is swiftest the slope of the frontal interface should be steepest, provided the latitude and transverse density contrast is the same.

Inertial motion

Inertial motion may occur in nature whenever a particle moves in a curved path in such a way that the centripetal acceleration due to the deflecting force of the earth's rotation is balanced by the centrifugal force of path curvature* (Fig. 4–13). This means that we may write $(2\Omega \sin \phi)c = c^2/r$, where c is the velocity of the particle relative to the earth. Inertial motion on the earth is necessarily anticyclonic, and closes a path having a radius $r = c/f$ in the period of one half-pendulum day. Since the Coriolis parameter f is a function of latitude, the path of a body in inertial motion is circular when it is symmetrical with respect to the pole or when it is projected on the equatorial plane of the earth where f is constant; that is, where f is independent of the radial distance from the axis of rotation.

As a consequence of the variation of the Coriolis parameter with latitude, there is a tendency for inertial motion at a given velocity relative to the earth to be more sharply curved in high latitudes than in low latitudes, resulting in an egg-shaped path. Because of the sharper curvature of inertial trajectories at high latitudes, there is a westward migration of successive inertial evolutions. When a particle is in inertial motion across the equator,

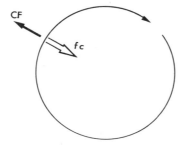

Fig. 4–13.   Balance of forces in inertial motion.

---

* Both the deflecting force of the earth's rotation and the centrifugal force accompanying the curvature of the particle's trajectory are "fictitious" forces in the usage of physics. In this case two fictitious forces are acting to produce an apparent regime of motion relative to the rotating earth.

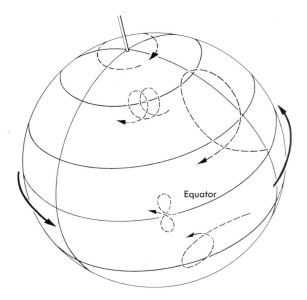

Fig. 4–14.   Various cases of inertial motion.

the sense of curvature is reversed as it passes from one hemisphere to the other (Fig. 4–14). If a particle departs northward from the equator, it will execute a clockwise turn, return to the equator, and cross into the Southern Hemisphere to execute a counterclockwise turn slightly shifted toward the west.

In small-scale processes which do not involve much change of latitude, inertial motion can be considered as a circular path. The radius of an *inertia circle* can be computed from the fact that the path of a particle in motion over a small range of latitudes is determined by an equilibrium between the centripetal acceleration of the Coriolis force and the centrifugal acceleration of its own inertia. Thus if the particle velocity in any horizontal direction is c, then $c^2/R = (2\Omega \sin \phi)c$, where R is the effective radius of the inertia circle, Ω is the angular velocity of the earth, and ϕ is the average latitude. Solving for R we have

$$R = \frac{c^2}{(2\Omega \sin \phi)c} = \frac{c}{f}. \qquad (4\text{--}41)$$

The time, T, for a particle to pass completely around the circumference of an inertia circle is

$$T = \frac{2\pi R}{c} = \frac{2\pi R}{(2\Omega \sin \phi)\, R} \qquad (4\text{--}42)$$

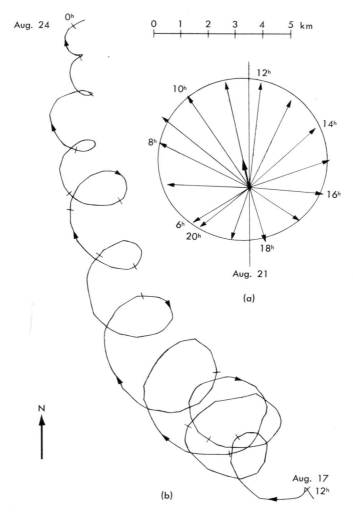

Fig. 4–15. Rotating currents of period one half-pendulum day observed in the Baltic, represented (a) by a progressive vector diagram for the period of 17–24 August 1933, and (b) by a central vector diagram for the period between 6 hr and 20 hr on August 21, according to Gustafson and Kullenberg. [From H. U. Sverdrup, M. W. Johnson, and R. H. Fleming, 1942, *The Oceans, Their Physics, Chemistry, and General Biology*, New York: Prentice-Hall.]

or

$$T = \frac{\pi}{\Omega \sin \phi}. \tag{4-43}$$

Now since the earth requires just twelve sidereal hours to turn through π radians at the rate of rotation Ω, we can simplify Eq. (4-43) to

$$T = \frac{12 \text{ hr}}{\sin \phi} = \tfrac{1}{2} \text{ pendulum day.} \tag{4-44}$$

Motions in an inertia circle have been shown to be quite real in natural situations where there is a sudden impulse which generates fluid motion and allows the system to coast without further interference. Gustafson and Kullenberg (1936)[#] made a classic series of observations of the inertial rotation of currents in the Baltic Sea during the period 17–24 August, 1933, following the passage of a line squall (Fig. 4–15). Their observations revealed a succession of rotary oscillations of the surface water mass through nine full turns, each completed within the period of one half-pendulum day, or 14 hours and 8 minutes at their latitude. This rotary motion was superimposed on a slow northward drift of the water mass as a whole.

Cyclostrophic motion

When the centrifugal force of rotation is exactly balanced by the gradient of pressure surrounding a vortex, another kind of rotary motion develops, which is independent of the rotation of the earth and usually found only in small-scale occurrences such as whirlpools, tornadoes, water spouts, occasional tropical disturbances near the equator, and in household sink drains. Cyclostrophic motion may be given its initial sense of rotation by the larger-scale flow in which it is embedded and upon which earth rotation has an effect, as in the tornado cyclone. But once cyclostrophic rotation is established, the deflecting force of earth rotation becomes relatively too weak to have any decisive influence. For example, it is often argued—quite erroneously—that water spirals down household drains as a result of the influence of the earth's rotation. Were this true the sense of rotation would be predominantly cyclonic, and the tendency to spiral would vanish at the earth's equator. Actually the Coriolis force is so feeble in comparison with the chance rotational impulse provided by pulling the drain plug that the effects of earth rotation are simply overwhelmed.

Cyclostrophic flow is ideally represented (Fig. 4–16) by a balance between the centrifugal force of rotation c^2/r per unit mass and the centripetal gradient of pressure $1/\rho \, (dp/dr)$, where c is the velocity of a fluid parcel relative to the earth, r is the radius of curvature, $1/\rho$ is the specific volume of the fluid, and dp is the change of pressure along a small increment of

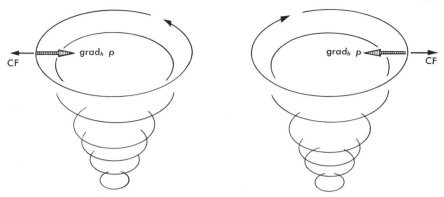

Fig. 4–16.   Balance of forces in cyclostrophic flow.

the radial distance dr from the sink where mass is being drawn off.   Since $c = \omega r$, we may write

$$\frac{1}{\rho} \frac{dp}{dr} = \frac{\omega^2 r^2}{r} = \omega^2 r. \tag{4-45}$$

Cyclostrophic motion is not of major significance in nature, but it illustrates in an exaggerated way the capacity of fluids to rotate around vertical axes and the movement of these rotational systems over the earth.   Fluid columns having some tendency to spin around a vertical axis are said to possess vorticity.   This concept is an important one in much of the discussion in later chapters.

Vorticity

The concept of vorticity as adapted to the problems of meteorology and oceanography refers to any tendency for a horizontal circulation in the motions of particles around a vertical or nearly vertical axis.   The vorticity is the ratio of this circulation to the area enclosed by the chain of particles. A concrete illustration of this idea is to consider the behavior of a pencil suitably weighted to ride vertically in a fluid.   If, as this pencil drifts with the flow, there is some measurable rotation, the fluid motion is said to possess vorticity.   Intuitively it will be seen that the fluid motions do not have to be circular or even curved; the rotation of the pencil can be produced by shear.   In cartesian coordinates the vertical component of vorticity, ζ, in a fluid column can be defined by

$$\zeta = \frac{\partial v}{\partial x} - \frac{\partial u}{\partial y}. \tag{4-46}$$

The convention of signs used in this definition makes counterclockwise

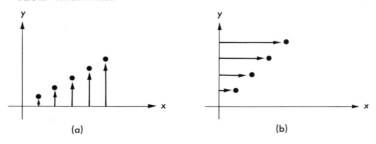

Fig. 4–17. (a) When $\partial v/\partial x$ is positive, the north component of velocity increases with distance to the east. In this shearing flow, a line of particles initially at $y = 0$ will be rotated in a counterclockwise sense. (b) When $\partial u/\partial y$ is positive, the east component of velocity increases with distances toward the north. In this shearing flow, a line of particles initially at $x = 0$ will be rotated in a clockwise sense.

rotational tendencies positive, in agreement with the sense of rotation of the earth in the Northern Hemisphere (Fig. 4–17).

If vorticity is referred to horizontal rectangular coordinates arranged so that one axis is radial and the other tangential to the flow, the definition becomes

$$\zeta = \frac{c}{r} + \frac{\partial c}{\partial r}, \tag{4–47}$$

where c is the scalar speed and r is the distance from the center of curvature of the stream lines; the center of curvature is counted positive when it lies on the left of the flow. The first term in Eq. (4–47) is the *curvature term* and the second is the *shear term*. The curvature term represents the angular velocity of the fluid as it rounds the bend of radius r, considered positive in the cyclonic sense. The shear term represents the horizontal variation of speed across the flow. The two terms are often of opposite sign, in which case the magnitude of the vorticity in curved shearing flow depends on the dominance of one or the other of the two effects. When the sum of the two terms is zero, the flow is said to be *irrotational* (Fig. 4–18).

For solid rotation $c = \omega r$. From this we may write

$$\zeta = \frac{\omega r}{r} + \omega = 2\omega,$$

and conclude that the vorticity in solid rotation is equal to twice the angular velocity.

Fluids at rest on the earth are in fact rotating with the earth at the rate of one revolution per sidereal day. To the extent that there is a component of earth rotation around the local vertical, resting fluids possess *planetary vorticity* relative to the inertial system of the fixed stars. Fluids having some

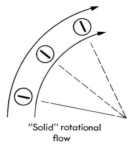

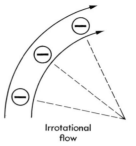

"Solid" rotational
flow

Irrotational
flow

Figure 4–18

spin tendency around a vertical axis beyond that provided by the earth's rotation are said to possess *relative vorticity* with respect to the earth.

Since in solids the vorticity is twice the angular velocity, the Coriolis parameter, $2\Omega \sin \phi$, is a measure of the planetary vorticity at any given latitude. From this we may say that the *absolute vorticity* of a vertical fluid column is given by the algebraic sum of the relative vorticity of the column (ζ) and the planetary vorticity (f) of the earth at the latitude of the column ($\zeta + f$).

The vorticity equation

The vorticity equation is derived by simple operations based on the equations of motion and, as ordinarily used in oceanography and meteorology, is given in a form which applies to cases where fluid flow is frictionless. The equation is

$$\frac{d}{dt}(\zeta + f) = -(\zeta + f)\,\mathrm{div}_h\,c, \qquad (4\text{--}48)$$

where $\mathrm{div}_h\,c$ is an abbreviated notation for the horizontal divergence $(\partial u/\partial x) + (\partial v/\partial y)$, which refers to the horizontal expansion or contraction of a column. Recalling that $(\zeta + f)$ is the absolute vorticity, we may interpret the vorticity equation as stating that absolute vorticity is increased where there is horizontal convergence or decreased in the presence of horizontal divergence. The physical situation is much like that in which angular momentum is conserved as a spinning skater extends or retracts his arms.

But continuity requires that confluence (inward horizontal motion) be accompanied by vertical stretching of an incompressible fluid volume and conversely that diffluence be accompanied by vertical shrinking. This requirement leads to a more interesting and even simpler form of the vorticity equation.

Fig. 4–19. Carl-Gustaf Arvid Rossby. [Photograph by D. Fultz.]

According to the equation of continuity, it is permissible to replace the horizontal divergence by the vertical convergence, $-\partial w/\partial z$. This quantity represents the vertical stretching of a column of water of height D, divided by that height:

$$-\frac{\partial w}{\partial z} = -\frac{1}{D}\frac{dD}{dt}. \tag{4–49}$$

If this expression is substituted for the horizontal divergence in Eq. (4–48), and if the remainder is integrated, we have the result so fruitfully employed by Rossby (1940)[3] (Fig. 4–19):

$$\frac{\zeta + f}{D} = \text{a constant.} \tag{4–50}$$

According to Eq. (4–50) the quantity $(\zeta + f)/D$ is conserved along the fluid trajectory. Thus if, for example, the column stretches vertically, continuity requires that there be compensatory horizontal convergence and therefore increasing absolute vorticity.

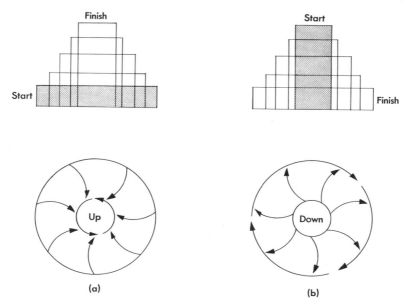

Fig. 4–20. Production of relative vorticity by the stretching and shrinking of fluid columns. (a) Vertical stretching requires horizontal inflow which produces cyclonic rotation $(+\zeta)$. (b) Vertical shrinking requires horizontal outflow which produces anticyclonic rotation$(-\zeta)$.

From these considerations it will be seen that relative vorticity can be changed by (1) vertical stretching or shrinking (Fig. 4–20), and (2) by a change of latitude. This concept has become so basic in modern oceanographic theory that it seems worthwhile to pursue the concept in descriptive terms so as to develop an intuitive grasp of the process.

The change in relative vorticity in the first case arises from the fact that when the column shrinks in height it expands laterally. Lateral expansion produces horizontal motion which is, in turn, deflected by the Coriolis force producing anticyclonic rotation or negative relative vorticity. Stretching produces cyclonic rotation or positive relative vorticity owing to the inflow needed to reduce the diameter of the column as it increases its height.

In the second case, where the height of the column is not changed as it is moved from one latitude to another, anticyclonic rotation or negative relative vorticity is developed when the column moves poleward. Cyclonic rotation or positive relative vorticity is developed as a consequence of equatorward displacements (Fig. 4–21). This effect can be appreciated if we consider a barrel filled with antifreeze to be at rest at the North Pole. The barrel and its contents are both rotating with the earth. Now if the barrel is carried smoothly to the equator, the antifreeze will tend to rotate

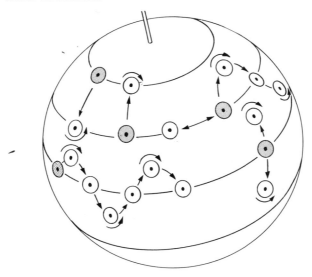

Fig. 4–21. Planetary vorticity: angular momentum tends to be conserved in columns of constant height as they change latitude along different trajectories.

in a counterclockwise sense relative to the earth at the equator, but the barrel will be at rest on the earth owing to its strong frictional coupling with the ground.* If the experiment were initiated at the equator, where the axial component of earth rotation is nil, the barrel would pick up the earth's rate of rotation and the antifreeze would tend to maintain its condition of absolute rest. Since the observer would be rotating with the barrel, the fluid contents would appear to be rotating anticyclonically relative to the earth and could be said to have a relative vorticity of $-f$.

Of course, this hypothetical experiment is an exaggerated distortion of fact. The physical reasoning it is intended to represent is confined to the *tendency* of fluid columns to retain their initial absolute rotation around a vertical axis when they are subjected to *small* displacements in latitude.

Finally we may make the further provision that in large-scale processes in the interior of oceans the changes of relative vorticity are quite small.

---

* This illustrative experiment can be disturbing to those who realize that the axis of rotation of the antifreeze would remain parallel with the earth's axis of rotation since the fluid motion is a real rotation in space. This defect in the barrel experiment can be removed by noting that the fluid rotation would tend to remain in the horizontal if the contents of the barrel were filled with a graded succession of immiscible layers of liquid—dense on the bottom and light on top—which through buoyant effects would tend to preserve maximum rotation around a vertical axis. This is not an unrealistic provision because the oceans possess a similar stratification and stability.

From this we arrive at the simple condition that the ratio f/D must be conserved. In this case, when a frictionless column shrinks vertically, it must move toward the equator, that is, into a region where the Coriolis parameter, f, is decreased.

Frictional effects

Up to this point of the discussion we have considered the motions of the oceanic and atmospheric fluids to be without friction. But where fluids rub against the solid earth or against each other there is frictional coupling that cannot be ignored. Molecular friction is a process in which the momentum of rapidly moving particles of a fluid is exchanged with the momentum of relatively slower particles. The molecules are imagined to collide with one another to produce a diffusive exchange of energies from regions of rapid motion to regions of slower motion. These exchanges produce a *shearing stress.*[5]

Stress, like pressure, is defined as a force per unit area. But whereas pressure is a force perpendicular to the area on which it acts, the stress components of interest here are those which lie parallel to the surface in question.

The components of stress are conveniently denoted by τ with two subscripts: the first describing the direction perpendicular to the plane on which the stress component acts, and the second the direction of stress component in that plane. Thus, τ_{zx} refers to the stress component acting on the xy-plane in the x direction.

Stresses are considered to be produced by molecular or turbulent exchanges of momentum accompanying water motion whenever there is a variation of current speed in a direction at right angles to the flow. In Fig. 4–22 it is apparent that vertical mixing in the region above the parallelepiped will cause some of the fast-moving parcels to descend and produce a stress on the xy-plane at the top of the parallelepiped. This stress component may be labeled τ_{zx}. The stress produced is assumed to be proportional to the vertical shear $\partial u/\partial z$. Thus

$$\tau_{zx} = \mu \frac{\partial u}{\partial z}, \qquad (4\text{--}51)$$

where the coefficient of viscosity μ may be considered to represent the efficiency of exchange, since it determines how much stress will be produced by a given shear. When a fluid is completely at rest or moving only very slowly, the rate of diffusion of heat, momentum, and dissolved solids is determined essentially by molecular motion. However as soon as the fluid is stirred the rates increase considerably. In the oceans and atmosphere, eddies or turbulent motions in the flow can be so effective in moving par-

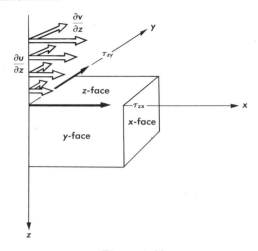

Figure 4–22

ticles among themselves that the effects of molecular diffusion are over-whelmed. For this reason the *molecular coefficients*, μ, of proportionality relating the gradients of heat, momentum, and kinetic energy to the diffusive fluxes of these quantities as measured in the laboratory are replaced by *eddy coefficients*, μ_e, which are several hundred to many thousand times larger. The concept of the eddy coefficient was introduced to oceanography in 1902[#] by V. W. Ekman.

Eddy viscosity is proportional to the velocity of the water within the eddies and to the size of the eddies. Characteristic values for the molecular and eddy coefficients for vertical transfer processes in the oceans are

Kinematic molecular viscosity .... 0.019 cm²/sec,
Kinematic eddy viscosity .... 2–7500 cm²/sec.

Whereas the molecular viscosity is essentially constant with depth, depending only slightly on temperature, the eddy viscosity varies with both the stratification of density and the speed of flow.

Horizontal eddy coefficients vary over an even wider range, from the order of 10 through 10^8 cm²/sec, perhaps because horizontal motion can occur without appreciable work being done against gravity. The determination of horizontal eddy coefficients depends upon the scale of the motion: the smaller values are obtained from the rate of spread of dye spots and clusters of current poles or floats; the larger values are obtained from studies of horizontal motion on an oceanic scale such as the diffusion of mass or momentum associated with the meandering flow of the Gulf Stream.

Through the agency of vertical momentum transfer, much of the energy of the surface layer currents of the oceans is imparted by the wind.[6] The stress of the wind on the sea surface is often considered in terms of an empirical equation of the form

$$\tau = \tfrac{1}{2}\rho C_D U^2, \qquad \qquad (4\text{--}52)$$

where τ is the wind stress, ρ is the density of air, U is the wind speed at some standard height above the sea surface, and C_D is the coefficient of resistance or drag of the sea surface. It is the coefficient C_D that is the subject of much contentious argument. It appears to vary with wind speed from 2×10^{-4} to the order of 10^{-2}, probably because wind speed is the principal factor governing sea surface roughness.[7]

Frictional forces can also be thought of in terms of a *stress gradient force*, by analogy with the pressure gradient force. If we consider a stress to act on the *xy*-plane, the frictional force in the *x* direction can be expressed as

$$F_x = \frac{1}{\rho}\frac{\partial \tau_{zx}}{\partial z}, \qquad \qquad (4\text{--}53)$$

and in the *y* direction as

$$F_y = \frac{1}{\rho}\frac{\partial \tau_{zy}}{\partial z}. \qquad \qquad (4\text{--}54)$$

These forces can be interpreted physically by considering the stress component τ_{zx}, for example, as equivalent to the downward transport of *x* momentum. If the volume gains more momentum than it loses, there is a net gain of momentum which corresponds to the action of a force, according to Newton's second law.

The horizontal equations of motion, including friction terms, can be written

$$\frac{du}{dt} = fv - \alpha\frac{\partial p}{\partial x} + \alpha\frac{\partial}{\partial z}\mu_e\frac{\partial u}{\partial z} \qquad \qquad (4\text{--}55)$$

and

$$\frac{dv}{dt} = -fu - \alpha\frac{\partial p}{\partial y} + \alpha\frac{\partial}{\partial z}\mu_e\frac{\partial v}{\partial z}. \qquad \qquad (4\text{--}56)$$

Ekman solved these equations for the equilibrium case ($du/dt = dv/dt = 0$) and for a homogeneous ocean in which the pressure gradient terms vanish ($\partial p/\partial x = \partial p/\partial y = 0$) to describe the distribution with depth of current speed and direction known as the Ekman spiral.

Ekman spiral

In 1905[#] V. W. Ekman showed theoretically that the effect of wind blowing steadily over an ocean of infinite depth, extent, and uniform eddy viscosity is to drive the surface layer at an angle 45° to the right of the wind direction in the Northern Hemisphere (to the left in the Southern Hemi-

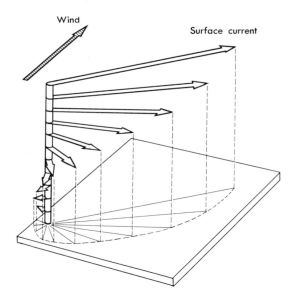

Fig. 4–23.   Schematic representation of a wind-driven current in deep water, showing the decrease in velocity and change of direction at regular intervals of depth (the Ekman spiral). [After H. U. Sverdrup, 1942, *Oceanography for Meteorologists*, New York: Prentice-Hall.]

sphere) and to move the successively deeper layers of water more and more to the right until at a given depth the direction of water motion is opposite that at the surface.   In addition to the motion being directed more and more to the right of the wind, the speed of motion decreases with depth, following the envelope of the equiangular or logarithmic *Ekman spiral* through a succession of revolutions, as shown in Fig. 4–23.

Usually, frictional influence is considered to cease at a depth D where the direction of flow is opposite that of the surface current.   Here the flow has a speed $c_0 e^{-\pi}$, or approximately $1/23$ as great as the speed in the surface skin c_0.   The variation of the Coriolis parameter with latitude causes the depth of reversal in the Ekman spiral to be infinite at the equator but to approach the surface with increasing latitude.   A characteristic value of D is 100 meters.

The Ekman spiral has been observed in the atmosphere and has been produced in the laboratory in rotating tanks and ocean models, but its occurrence in the wind-influenced layer of the ocean has not been demonstrated beyond the tendency for sea ice and some currents to move at some angle to the right of the wind.

Theoretically there is, in association with the rapid decay of velocity with depth in the Ekman spiral, a net horizontal transport of water which

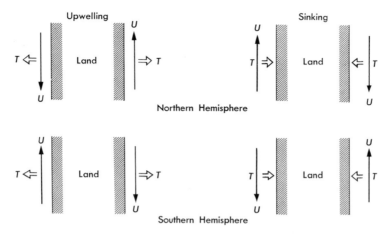

Fig. 4–24.   Ekman upwelling and sinking along coast lines.

is independent of any change of eddy viscosity with depth and is therefore often observed. The mean transport should be developed at right angles to the wind direction, on the right in the Northern Hemisphere and on the left in the Southern Hemisphere. Effects ascribed to the Ekman transport are the upwelling or sinking that sometimes occur when winds blow with a large component of their velocity parallel to a coast line. In the Northern Hemisphere a southerly wind blowing along an east coast should drive surface-layer water seaward. To replace this loss some upwelling will occur. If the deeper water along this coast is colder than the surface layer, the upwelling will cause the water along the coast to be cooler than that some distance farther at sea. The various combinations of wind direction and coastal trend leading to upwelling and sinking in the Northern and Southern Hemispheres are suggested in Fig. 4–24, where U is the surface wind and T is the Ekman transport.

Observations made to demonstrate the existence of the Ekman spiral have not yielded clear confirmation of the theory in oceanic circumstances, but the statistical effects of the Ekman transport on coastal climates and fishing grounds are well known. On the Lofoten fishing grounds off the coast of Norway, for example, the prevailing southwest wind causes the coastal water to sink, bending the isothermal and isohaline surfaces downward to intersect the continental shelf. The cod are found near the bottom and within rather narrow limits of temperature and salinity, so that a regular physical survey of the water structure is made to determine the best distribution of the fishing fleet. The coasts of California and Peru are also influenced by wind-driven upwelling and sinking. When winds blow equatorward on these coasts, upwelling accompanies the flow of the California and Peru currents, increasing their productivity through an increased supply of nutrients from below the euphotic level but at the same time lowering the water temperatures. Poleward motions of air reverse this condition and can produce major changes in the fisheries. Off the coast of Peru, during February and March of certain years, a poleward flow of air produces coastal sinking which permits warm equatorial water to make its way southward along the coast. This current is called El Niño in reference to the child of Bethlehem. When this flow is strong there is a severe reduction in the yield of the fishery owing to a displacement of the fish population, which tends to remain in the colder water.

The oceans, in addition to being responsive to the effects of wind, earth rotation, and gravity, are much influenced by the chemical and physical properties of sea water itself.

STUDY QUESTIONS

1. Neglecting second-order effects, compute the deflection of local apparent vertical to be observed from an aircraft flying eastward at 1000 nautical mi/hr at latitude 60°N. How much does inclusion of the second-order term change this result?

2. Give the logical and symbolic derivation of the oceanographic form of the geostrophic equation from the equation of motion.

3. Distinguish between the steady state and equilibrium state with reference to Euler's expansion.

4. Distinguish between cyclostrophic and inertial motions with reference to the equation of motion.

5. Compute the horizontal pressure gradient across the surface of the Gulf Stream where the current is 100.0 km wide and the water surface is 1.0 meter higher on the right than on the left. What is the corresponding geostrophic velocity of the surface layer in latitude 42.5°N?

6. In 1913 A. H. Compton built a large doughnut-shaped glass tube and filled it with an aqueous suspension of oil droplets to measure the local vorticity of the earth. If the contents of the tube were allowed to come to rest at latitude 30°, what would be the angular velocity of the fluid after the glass doughnut had been very carefully overturned in its mountings? What would be the angular velocity of the fluid be if the tube were quickly and carefully transported from rest at the equator to rest at the North Pole (with proper precautions having been taken to prevent freezing)?

7. For demonstration purposes, the North Pole & Southern Railway Company maintains a frictionless flat car on which is mounted a large and massive frictionless horizontal turntable. The car is frequently left in the United States, at the ninetieth meridian line, for college students to push. Some students push the car northward without touching the turntable; others spin the turntable without disturbing the car. What surprising events ensue in each case?

8. Certain coastal regions of the earth have high biological productivity because of upwelling. If this upwelling is supported by the wind, in what direction must the prevailing winds blow on the east coasts and west coasts of the continents in the Northern and Southern Hemispheres, respectively?

REFERENCES

1. F. C. Fuglister and L. V. Worthington, 1951, *Tellus*, 3: 1–14.

2. M. Uda, 1951, *J. Oceanogr. Soc.*, *Jap.*, 6: 181–189.

3. C.-G. Rossby, 1940, *Quart. J. R. met. Soc.*, Suppl. 66: 68–87.

4. V. P. Starr and M. Neiburger, 1940, *J. Mar. Res.*, 3: 202–210.

5. G. J. Haltiner and F. L. Martin, 1957, *Dynamical and Physical Meteorology*, New York: McGraw-Hill.

6. G. Neumann, 1956, *Bull. Amer. Met. Soc.*, 37: 211–217.

7. J. R. D. Francis, 1954, *Quart. J. R. met Soc.*, 80: 438–443; P. A. Sheppard and M. H. Omar, 1952, *Quart. J. R. met. Soc.*, 78: 583–589.

SUPPLEMENTARY READING

BJERKNES, V., J. BJERKNES, H. SOLBERG, and T. BERGERON, 1933, *Physikalische Hydrodynamik*, Berlin: Julius Springer.

DEFANT, A., 1961, *Physical Oceanography*, Vol. I, London: Pergamon Press.

EKMAN, V. W., 1905, "On the Influence of the Earth's Rotation on Ocean-Currents," *Akr. Mat. Astr. Fys.*, 2(11): 52 pp.

HALTINER, G. J., and F. L. MARTIN, 1957, *Dynamical and Physical Meteorology*, New York: McGraw-Hill.

MUNK, W. H., and G. J. F. MACDONALD, 1962, *The Rotation of the Earth*, Cambridge: Cambridge Univ. Press.

PRANDTL, L., and O. G. TIETJENS, 1934, *Fundamentals of Hydro- and Aeromechanics*, Engineering Societies Monographs, translated by L. Rosenhead, New York: Dover.

PROUDMAN, J., 1953, *Dynamical Oceanography*, New York: Wiley.

ROSSBY, C.-G., and R. B. MONTGOMERY, 1936, "On the Momentum Transfer at the Sea Surface," *Pap. phys. Oceanogr.*, 4(1): 30 pp.

SPENCER-JONES, H., 1956, "The Rotation of the Earth," pp. 1–23, *Handbuch der Physik*, S. Flügge, editor, Bd. XLVII, Geophysik I, Berlin: Springer-Verlag.

SVERDRUP, H. U., M. W. JOHNSON, and R. H. FLEMING, 1942, Ch. 12 and 13, pp. 400–515, *The Oceans*, New York: Prentice-Hall.

CHAPTER 5

Characteristics of Sea Water

Sea water is about 2700 times more abundant on the earth than impounded fresh water. The physical properties of sea water are different from those of fresh water, because they vary not only with temperature and pressure but with the concentration of salt. The temperature and the salt concentration of sea water undergo change mainly at the sea surface where the ocean interacts with the atmosphere. Transpiration of water through the sea surface, radiative exchange with the atmosphere, and mechanisms which mix newly conditioned surface water into the volume of the ocean are fundamental processes that are strongly influenced by the physical nature of sea water itself.

A meaningful description of the physical properties of sea water cannot always be given out of context with the natural situation. For example, where small bubbles, silt, or microorganisms are dispersed in natural waters transparency is altered, and where the sea surface is white-capped (or frozen) the ability of the oceans to exchange water vapor and other gases and to emit and absorb radiant energy is markedly changed. But certain other integral properties can be specified reasonably well for a usefully wide range of conditions.[1] Among the most important of these properties is density.

Density

The density of sea water is determined by its total salt content s, its temperature t, and to a lesser degree by its compression under p, the burden of water and air above it. The average density, ρ, of sea water is near 1.025 gm/cm^3. If a given sample has a density of 1.02523, the significant part of this number is generally in and beyond the third decimal. Therefore the convention has been adopted that in place of the density $\rho_{s,t,p}$, a quantity $\sigma_{s,t,p}$ be used. This is defined as follows:

$$\sigma_{s,t,p} = (\rho_{s,t,p} - 1)1000. \tag{5-1}$$

In this way $\rho_{s,t,p} = 1.02523$ gm/cm^3 becomes $\sigma_{s,t,p} = 25.23$, with the units of measurement given by implication.

The density of sea water *in situ* is usually required to be known to an accuracy of 10^{-5}. At the present time there are no direct means for measuring density *in situ*, although interesting possibilities have been proposed by Richardson (1959),[2] Kanwisher (1959),[3] and others. Deductions of this quantity from the equations of state rely on empirical knowledge of the coefficients of thermal expansion of sea water, as well as on knowledge of saline contraction and isothermal compressibility. Because of nonlinear interactions, each of these is a somewhat imperfectly known function of the ambient temperature, salinity, and pressure.[4]

For simplicity it is often assumed in dynamical oceanography that the buoyancy of a parcel of sea water is unchanged relative to its surroundings when the hydrostatic pressure on both the parcel and its surroundings is changed by the same amount. This is to say that the work done in moving the parcel from one level to another is, to a first approximation, independent of the absolute value of hydrostatic pressure change.

In considering the stability of a water column, it is convenient to be in a position to say whether a displaced parcel will be heavier or lighter (Fig. 5–1)

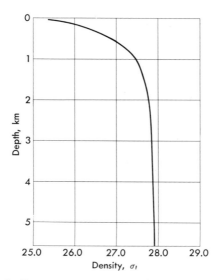

Fig. 5–1.   Schematic diagram of the change in density with depth in the ocean.

than its surroundings from consideration of only its observed temperature and equivalent salt concentration. With this object in view, the density of sea water is often expressed as the quantity $\sigma_{s,t,0}$ (usually abbreviated σ_t and pronounced "sigma-tee"), which is the density of a parcel after the pressure has been reduced to one atmosphere, that is, brought to the sea surface.* This operation neglects adiabatic effects, so that σ_t differs from *potential density* by this very small amount.

Salinity

Sea water is a very complex solution of organic and inorganic salts derived over the course of geologic time from the solution of rocks, the gaseous effusion of volcanoes, biological activity, and probably to a far lesser extent from meteoritic material in the earth's atmosphere. The salinity of sea water has been defined as "the total amount of solid material in grams contained in one kilogram of sea water when all the carbonate has been converted to oxide, the bromine and iodine replaced by chlorine, and all organic matter completely oxidized."† The numerical value of salinity defined in this manner is slightly lower than the amount of dissolved solids in grams per kilogram. The concentration of dissolved materials in sea water is measured most sensitively by its electrical conductivity,[5] but chemical analysis is of both basic and historical interest.

Over the years a large number of samples of ocean water from different parts of the world ocean have been analyzed completely, and these show that regardless of the absolute concentrations of the individual constituents (see Tables 5–1 and 5–2), the relative proportions of the major constituents are nearly constant.‡ Thus it is possible to estimate the total salt content of sea water from measurements of the concentration of only one constituent, usually the chloride ion.

The chloride ion concentration in sea water can be measured by chemical titration against a standard solution of silver nitrate which yields a conspicu-

---

* Tabulated values of this conversion are numerous. Some are to be found in E. C. LaFond, 1951, *Processing Oceanographic Data*, U.S. Navy Hydrographic Office Publication 614, and in greater detail in *Tables for Sea Water Density*, 1952, U.S. Navy Hydrographic Office Publication 615.

† C. Forch, M. Knudsen, and S. P. L. Sørensen, 1902, *K. danske vidensk. Selsk.*, 6, 12: 151 pp. Some other definitions are discussed by D. E. Carritt and J. H. Carpenter, 1959, pp. 67–86, *Physical and Chemical Properties of Sea Water*, Washington, D.C.: National Academy of Sciences—National Research Council Publication 600.

‡ The composition of salts from interior seas, such as the Caspian, Dead, and Salton Seas and the Great Salt Lake, differs from one sea to the other and from the composition of salts in the world oceans.

TABLE 5–1

THE 42 MOST ABUNDANT ELEMENTS IN SEA WATER, IN PERCENT BY WEIGHT*

Oxygen	85.89	Barium	1	$\times 10^{-6}$
Hydrogen	10.82	Zinc	5	$\times 10^{-7}$
Chlorine	1.90	Manganese	5	$\times 10^{-7}$
Sodium	1.06	Lead	4	$\times 10^{-7}$
Magnesium	0.13	Iron	2	$\times 10^{-7}$
Sulphur	0.088	Cesium	2	$\times 10^{-7}$
Calcium	0.040	Uranium	1.5	$\times 10^{-7}$
Potassium	0.038	Selenium	1	$\times 10^{-7}$
Bromine	6.5 $\times 10^{-3}$	Thorium	5	$\times 10^{-8}$
Carbon	2.8 $\times 10^{-3}$	Molybdenum	5	$\times 10^{-8}$
Strontium	1.3 $\times 10^{-3}$	Cerium	4	$\times 10^{-8}$
Boron	4.8 $\times 10^{-4}$	Silver	3	$\times 10^{-8}$
Silicon	2 $\times 10^{-4}$	Vanadium	3	$\times 10^{-8}$
Fluorine	1.4 $\times 10^{-4}$	Lanthanum	3	$\times 10^{-8}$
Nitrogen	0.3–7 $\times 10^{-5}$	Yttrium	3	$\times 10^{-8}$
Rubidium	2 $\times 10^{-5}$	Copper	2	$\times 10^{-8}$
Lithium	1.2 $\times 10^{-5}$	Nickel	1	$\times 10^{-8}$
Aluminum	1 $\times 10^{-5}$	Scandium	4	$\times 10^{-9}$
Phosphorous	5 $\times 10^{-6}$	Mercury	3	$\times 10^{-9}$
Iodine	5 $\times 10^{-6}$	Gold	4	$\times 10^{-10}$
Arsenic	1.5 $\times 10^{-6}$	Radium	7	$\times 10^{-15}$

* V. M. Goldschmidt, 1954, p. 49, *Geochemistry*, Oxford: Clarendon Press.

TABLE 5–2

W. DITTMAR'S (1884)[#] VALUES FOR THE MAJOR CONSTITUENTS OF SEA WATER,* IN PERCENT BY WEIGHT

Cl^-	55.04	Mg^{++}	3.69
Br^-	0.19	Ca^{++}	1.16
SO_4^{--}	7.68	Sr^{++}	0.04
HCO_3^-	0.41	K^+	1.10
F^-	0.00+	Na^+	30.61
H_3BO_3	0.07		

* These values are corrected by Lyman and Fleming for re-evaluation of atomic weight. See J. Lyman and R. H. Fleming, 1940, *J. Mar. Res.*, 3: 134–146; H. U. Sverdrup, *et al.*, 1942, p. 166, *The Oceans*, New York: Prentice-Hall.

ous precipitate of silver chloride.* The silver nitrate solution is first stand-ardized against "normal water" prepared by the Hydrographical Labora-tories in Copenhagen, or alternatively, since 1940, against American stand-ard sea water prepared by F. M. Soule and C. A. Barnes at Woods Hole Oceanographic Institution. This has been carefully compared with the world standard Copenhagen water.

The relation between salinity and chlorinity, according to Knudsen (1901), is

$$\text{Salinity} = 0.03 + 1.805 \times \text{chlorinity.} \qquad (5\text{--}2)$$

Both salinity and chlorinity are expressed in parts per thousand, denoted by $^0/_{00}$. Chlorinity given in parts per thousand expresses the mass of chlorine, bromine, and iodine in one thousand grams of sea-water solution, assuming that the iodine and bromine have been replaced by chlorine.† By chemical means it is possible to determine the concentration of chlorides to an accuracy of ± 0.01 $^0/_{00}$. When the chlorinity is transformed into "salinity," the uncertainty rises to ± 0.02 $^0/_{00}$. Recently, potentiometric titrations of total salts through their effects on the conductivity of sea water have begun to be more widely used. These lead to improved discrimination between samples and reproducibility of measurements on the same sample, but still require translation into salinity units.

As early as 1922 F. Wenner[6] designed a conductivity bridge which can be operated on shipboard and which has been used ever since by the U.S. Coast Guard in the work of the International Ice Patrol. The Wenner bridge is reported to be capable of discriminating salinity with a reproduci-bility of ± 0.005 $^0/_{00}$, a significant improvement over that obtained by chemical titration. A large part of the errors that remain are considered to be due to imperfect temperature control.

The bridge contains six essentially identical conductivity cells. One cell is filled with a reference sample and the remaining five cells are used for unknowns. When the temperature of the reference and unknown samples has been stabilized the relative impedance of each unknown is measured

---

*The procedures used for the chemical titration of the chlorides in sea water generally follow Oxner's directions for implementing the method originally de-scribed by Martin Knudsen. Knudsen's *Hydrographical Tables*, 1901, 2nd ed., 1931, Copenhagen: G.E.C. GAD, London: Williams & Norgate, are still widely used for conversion of the measured chloride into equivalent total salt content.

† Since this definition of chlorinity gives numerical values which change with each refinement of the values of atomic weights, a new definition has been given which is independent of such changes. Chlorinity in grams per kilogram of sea water is identical with the mass in grams of "atomic weight silver" just needed to precipitate the halogens in a 0.3285233-kgm sample of sea water.

(b)

Fig. 5–2.   (a) An early version of the Schleicher-Bradshaw conductivity bridge.   (b) Arrangement of conductivity cells in the Schleicher-Bradshaw bridge. [Photograph by WHOI.]

(a)

against the reference cell. Balance is obtained by adjusting suitable resistors and capacitors to silence a 1000-cps tone. The Wenner design has not been widely used, perhaps because of a reluctance on the part of oceanographers to deal with the difficulties so often encountered at sea in maintaining sensitive electrical equipment in working order. However, in 1956 Schleicher and Bradshaw [7] and shortly thereafter Cox (1957) [8] and Paquette (1958) [9] designed bridges which promise to overcome this prejudice.

The Schleicher-Bradshaw "salinometer" (Fig. 5–2) is like the Wenner bridge in that two cells are used, one containing a standard and the other an unknown sample of sea water; but the instrument is modernized in that the balances are obtained through a phase-sensitive servo-amplifier after rough adjustments have been made manually. The circuit also employs a 1-kc input to the bridge, but the output is fed to a phase detector similar to the chopper amplifiers used in many modern recording potentiometers. The output of the amplifier causes a small motor to turn in one direction or the other to adjust the fine balance of the resistive component of the bridge circuit. Small inequalities in capacitance of the cells containing the standard and unknown sea water samples are nulled by land. The Schleicher-Bradshaw salinometer requires a smaller water sample than the Wenner bridge and therefore comes to temperature equilibrium more quickly but with essentially the same accuracy, $\pm 0.005\ ^0/_{00}$ in salinity being maintained.

In a recent resurvey of the North Atlantic, it was shown that with the increased resolution of conductivity measurements, horizontal gradations of conductivity in the North Atlantic can be distinguished within oceanographic areas that chemical analysis for chlorinity had indicated to be composed of homogeneous layers. This is one of the few instances in geophysics where with increased resolving power the noise-to-signal ratio has been lowered rather than raised. Although the variability of conductivity as a function of position in the open oceans is less than the experimental scatter of chlorinities determined by chemical titration, it remains to be seen where the coherence level of these properties of sea water lies and, moreover, to discover the precise relationship between chlorinity, conductivity, and the total concentration of dissolved solids. The latter property is fundamental to indirect determinations of the density of sea water. Concentration of dissolved solids also influences the freezing point.

The concentration of salt in sea water, as estimated from chlorinity, is related to a depression of the freezing point below that of fresh water by an amount $\Delta T = 0.102710$ Cl, where Cl is the chlorinity per mille.* At

---

* See Y. Miyake, 1939, *Bull. Chem. Soc. Japan*, 14: 58–62. See also M. Knudsen, 1903, *Publ. Circ. Cons. Explor. Mer*, 5: 11–13, for the values of coefficients obtained in earlier determinations.

salinity 30 $^0/_{00}$ the freezing point is $-1.63°C$ and at 35 $^0/_{00}$ sea water freezes at $-1.91°C.$* Distilled water has a maximum density at 4°C, a little above the freezing point, but at a salinity of 24.70 $^0/_{00}$ the temperature of maximum density occurs at the freezing point, $-1.332°C$. Since sea water generally contains more salt than this, freezing usually occurs before the water has cooled sufficiently to attain maximum density.†

Temperature-salinity diagram

With the establishment of systematic, descriptive oceanography since the time of the *Challenger* expedition, numerous separate observations of the temperature and salinity of the oceans at many depths have been accumulated. In the open sea the salinities of water may range from about 33 $^0/_{00}$ to 36 $^0/_{00}$, while the corresponding temperatures may range from as little as $-1°C$ to about 30°C. From tabulations of temperature and salinity data, it can be shown that certain regions of the oceans possess characteristic associations of salinity and temperature both by area and in the vertical, and that in general the density of the sea becomes less with elevation above the bottom. In 1916[#] Bjørn Helland-Hansen proposed that these variables be correlated by means of a *temperature-salinity diagram* (Fig. 5–3). On this the salinity of the sea water sample is plotted with increasing values toward the right, and the temperature of the same sample is plotted on a vertical scale with values increasing upward. When many separate observations of temperature and salinity of various geographical areas are plotted on this kind of diagram, it is not unusual for the points to fall in groups to the extent that oceanic areas and layers can be distinguished from one another as *water masses*. The principal layers of water having characteristic associa-

---

* The vapor pressure of sea water is depressed along with the freezing point as is characteristic of aqueous solutions. Where e is the vapor pressure of sea water at a given salinity and e_0 is the vapor pressure of pure water at the same temperature, then $(e_0 - e)/e_0 = 0.537 \times$ salinity, according to R. Witting, 1908, *Finnl. hydrogr.-biol. Untersuch.*, 2: 173. More recent work by Higashi, *et al.*, (1931), Miyake (1939), Robinson (1954), and Arons and Kientzler (1954) is summarized by F. A. Richards, 1957, pp. 77–128, *Physics and Chemistry of the Earth*, Vol. II, L. H. Ahrens, F. Press, K. Rankama, S. K. Runcorn, editors, London: Pergamon Press.

† The specific heat of sea water is slightly less than that of pure water, but the latent heats of fusion and vaporization are nearly the same because the salts are left behind in the evaporation process and excluded by freezing.

The molecular thermal conductivity of sea water is also only slightly less than that of pure water at the same temperature, but its electrical conductivity is vastly greater. At 0°C sea water of salinity 35 $^0/_{00}$ has an electrical conductivity near 0.029 mho/cm, and at 30°C the same water has a conductivity near 0.058 mho/cm. Electrical conductivity tends to zero as salinity approaches zero. See B. D. Thomas, T. G. Thompson, and C. L. Utterback, 1934, *J. Cons. int. Explor. Mer*, 9: 28–35.

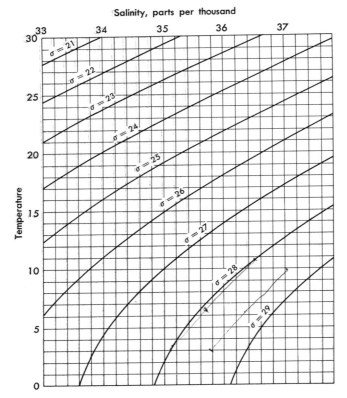

Fig. 5–3.   Temperature-salinity sigma-T grid.

tions of salinity and temperature can be represented on the T-S diagram by a single point. These points define _water types._

The T-S diagram has further uses in that with knowledge of the temperature and salinity of water at atmospheric pressure, σ_t is also specified. At constant pressure the density of sea water increases systematically with an increase of salinity and with a decrease in temperature. Therefore, on a T-S diagram it is possible to rule a grid of lines of constant σ_t. Lines of equal σ_t will tend to slope upward from left to right across the T-S diagram. A change of salinity of 1 $^0/_{00}$ usually produces a larger change in the σ_t of sea water than a change of temperature of 1°C.

The lines of equal σ_t shown in Fig. 5–3 are slightly curved downward, as well as being inclined to the coordinate axes. Thus when two water types of different salinity and temperature lie side by side on the ocean surface and have precisely the same density, their mixture will be represented by a point on the straight line joining their positions in the T-S

diagram. Since this point must lie on the concave side of the σ_t line, the mixture must be slightly more dense than either of the two parent water types and tend to sink. Sinking due to this mixing process is known as *caballing*. Caballing is presumed (but not yet demonstrated) to occur at significant rates mainly in high northern latitudes where there is but little change in water properties with depth, a condition which perhaps bears testimony to the effectiveness of the process.

Elsewhere in the modern oceans both salinity and temperature tend to decrease with increasing depth, so that water of higher σ_t and potential density is found at the lower levels. Because of this circumstance there is a remarkable possibility[10] that sustained motion would occur in a vertical tube embedded in the ocean. If the tube were to extend from the relatively low density surface water to deep water of higher density and if water from the lower layer were pumped to the surface, the flow would continue when the pump was disconnected. Flow would be sustained because the vertical tube would carry the salt without change of concentration from the deeper level to the surface but would permit sufficient heat transfer to enable the water flowing upward in the tube to come to thermal equilibrium with its surroundings. Thus the relatively fresh but cold water from the depths would arrive at the surface as relatively fresh but warm water less dense than the saltier water of the same temperature surrounding the top of the tube. The buoyancy of the warmed deep water would permit the flow to continue for as long as suitably contrasting water types were maintained on the earth or until the tube crumbled with age or became choked with sessile organisms.

If the initial pumping action were directed downward, the salt but warm surface water would be carried to a depth where the surrounding deep water is cooler but fresher. With the attendant loss of heat on the way down, the surface water in the tube would be saltier than the water in its environment, so that it would continue to flow downward without further assistance. In the central North Atlantic a tube 2000 meters long might develop a pressure head of as much as two meters. Flow rates for tubes have been estimated by Groves.[11] Stern has shown that convective filaments actually develop in the absence of any confining tubes because the rate of diffusion of salt in the ocean is about 110 times smaller than that of heat.[12]

A demonstration of the first of these effects has been made on a laboratory scale using a graduated cylinder containing a schematic ocean in which a vertical metal tube is used to define the moving column of water. In this model the flow continues until the conditions of the experiment break down.[13] In Stern's experiment the ocean is simulated in a large aquarium in which hot salt water dyed with fountain-pen ink is made to float over cold, clear fresh water. Vertical filaments develop across the interface within an hour. These suggest a new mechanism for vertical mixing in the oceans which is independent of the large-scale overturning circulations.[14]

Adiabatic effects

As a water parcel is moved upward or downward in the ocean, its temperature varies slightly with the amount of compression. Although the compressibility of water is less than that of steel, increasing hydrostatic pressure causes an increase in the temperature of water. This adiabatic effect of pressure change was shown by Lord Kelvin[15] to be approximately

$$\delta T = \frac{Te}{\mathcal{J}C_p} g \, \delta h, \tag{5-3}$$

where C_p is the specific heat at constant pressure (a function of temperature), T is the absolute temperature, g is the acceleration due to gravity, e is the coefficient of thermal expansion (a function of salinity), $\mathcal{J}$ is the mechanical equivalent of heat, and δh is the vertical displacement of a unit mass. Kelvin's formula for the adiabatic effect of increasing pressure on sea water shows that the temperature rises approximately 0.13°C per 1000 decibars increase of pressure, not a very important amount.

Potential temperature, θ, is defined as $T - \delta T$, where T is the temperature of the water under the hydrostatic pressure prevailing at its place in the water column, and δT is the adiabatic temperature change due to lifting the parcel without exchange of heat with its environment to the pressure at the sea surface. In the very deepest part of the ocean, such as in the trenches flanking island arcs, it can be shown that the potential temperature of the water is virtually uniform from the depth of the surrounding ocean floor to the bottom of the trench, even though the actual temperature increases somewhat with depth. A case of this sort is shown in Fig. 5–4 representing a sounding in the Philippine trench made during the *Galathea* expedition of 1950–52.

One can see in this figure that the actual temperature increases steadily with depth below the 4000-meter level of the surrounding ocean floor, while the salinity remains constant. On the other hand, the fact that the potential temperature between about 4000 meters and about 10,000 meters is effectively constant suggests that there is convective overturning in the trench. Adiabatic warming and cooling due to vertical motions of sea water have been observed only in relatively deep and confined spaces, such as the ocean trenches in the Arctic[16] and European mediterranean basins and parts of the Red Sea, probably because the effect is slight enough to be easily destroyed in the presence of even very weak advection.

Adiabatic effects of a more instantaneous character also play a part in the propagation of compressional (sound) waves in the ocean, but since

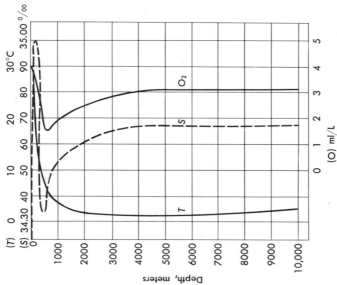

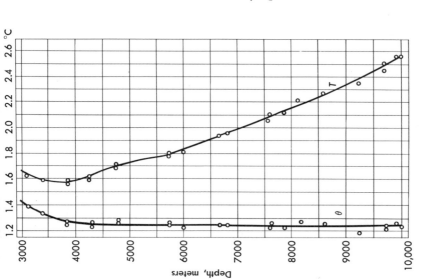

Fig. 5–4.   Temperature, salinity, and oxygen content in the Philippine trench. θ is the potential temperature.   [After A. F. Brunn, *et al.*, editors, 1956, *The Galathea Deep Sea Expedition, 1950-1952*, by permission of Macmillan, New York, and Allen and Unwin, London.]

there is an adiabatic rise and fall of water temperature with the passage of a symmetrical compressional wave, these small effects essentially cancel out and do not influence in important ways the direction of propagation or energy of the wave front. The path of compressional waves in water is therefore governed by Snell's law of refraction.* The speed of sound in sea water at atmospheric pressure has been tabulated by Matthews (1939)[17] and Del Grosso (1952).[18]

Stratification

Owing to the influence of gravity and buoyancy forces, there is a tendency for dense parcels of water to sink, and for less dense parcels to rise toward the surface of the oceans. Ultimately, within each water column, a more or less stable vertical gradient of density is established.[19] While a stably stratified water column may often be cooler at the bottom than at the top, it may also be fresher at the bottom. The balance of the opposing effects of increasing salinity and of increasing temperature on density may cause the separate gradients of salinity and temperature in the vertical to assume apparently unstable configurations in what is actually a stably stratified column from the standpoint of the vertical distribution of density.

The usually gentle vertical gradients of salinity and temperature in the oceans may sometimes be steepened in zones delimited by levels that are closely spaced relative to the total depth. In these layers the vertical gradients of temperature or salinity may be sharply graded from the deep-water regime to that of the surface-layer regime. A steep vertical gradient of temperature in an otherwise gently graded sounding is called a *thermocline*. In the oceans there is a widespread *main* or *permanent thermocline* (Fig. 5–5) which lies deep enough to be almost unaffected by the annual cycle of seasons. In contrast to this there is sometimes a shallower thermocline that comes and goes with the seasons and is therefore referred to as a *seasonal thermocline*. Along with the temperature transition in a thermocline, there may be locally steepened vertical gradients of salinity and density, referred

---

* The acoustic index of refraction of the ocean may be generally lower in the surface layers than at mid-depths owing to the decrease of temperature with depth, so that compressional waves emitted near the surface are on the average bent downward. Below mid-depths the index of refraction is decreased by the increase of pressure by more than the amount it is increased with decreasing temperature. The net effect is to cause compressional waves emitted below mid-depths to be bent upward. The focusing effect of the water column on compressional waves emitted at mid-depths provides a channel or wave guide permitting compressional waves of acoustic frequencies to be transmitted over extraordinarily long ranges, sometimes several thousand of miles. This phenomenon was discovered by W. M. Ewing in 1943. See W. M. Ewing and J. L. Worzel, 1948, *Mem. Geol. Soc. Amer.*, Memoir 27, 35 pp.

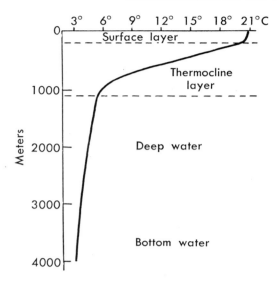

Fig. 5–5.   The vertical subdivisions of the ocean.

to as *haloclines* and *pycnoclines*, respectively. The physical mechanisms which form and maintain these transition layers are thought to be related to a balance of the downward diffusion of heat and salt-enriched water conditioned at the surface against the upward motion of fresher and cooler water from the ocean depths.*

The layering of the water types in the volume of the ocean may take two different forms: barotropic or baroclinic stratification. In a *barotropic* ocean (Fig. 5–6) the surfaces of equal pressure (isobaric surfaces) lie parallel to surfaces of uniform density (isopycnal surfaces), and these in turn may lie parallel to the sea surface. When a barotropic ocean is at rest, the isobaric and isopycnal surfaces are parallel with geopotential surfaces. But when there is geostrophic motion accompanying the barotropic condition, the isobaric and isopycnal surfaces, provided that they remain parallel, may be inclined to geopotential surfaces. Barotropic stratification is not destroyed by regionally uniform fields of parallel force, such as those producing the astronomical tide and on a smaller scale those accompanying a widespread change of barometric pressure. In cases where such force fields are sud-

---

*The processes maintaining the permanent thermocline are intimately connected with the problem of the general circulation of the oceans. Some of the recent work on this question is contained in papers by H. Stommel and G. Veronis, 1957, *Tellus*, 9: 401–407; by A. Robinson and H. Stommel, 1959, *Tellus*, 11: 295–308; and by P. Welander, 1958, *An Advective Model of the Ocean Thermocline*, Tech. Rept. 7, Johns Hopkins University School of Engineering.

BAROTROPIC MODE

Density, ρ, is a function of pressure, p, alone, hence isobaric and isopycnal surfaces do not intersect. The fluid can be motionless when isobaric surfaces coincide with geopotential surfaces.

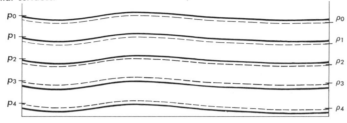

BAROCLINIC MODE

Isobaric and isopycnal surfaces intersect. A baroclinic fluid cannot remain motionless.

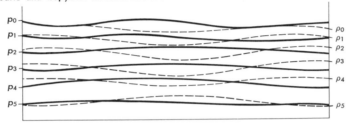

Fig. 5–6.　Ocean stratification.

denly applied and some redistribution of mass is required to achieve equilibrium, the period of readjustment is virtually complete in 12 pendulum hours. Where, however, the field of force establishes steady patterns of current, the period of barotropic equilibration is on the order of 12 to 14 days in middle latitudes.

In contrast to this, the *baroclinic* condition of stratification exists when isobaric surfaces are inclined to isopycnal surfaces. Both may be inclined to geopotential surfaces. (The effects of internal waves are to be overlooked in this connection.) The situation is illustrated by the unequal stands of liquid level in a U-tube when the two arms contain a succession of fluids of differently graduated densities. The baroclinic structure of the oceans is especially pronounced in the upper 500 to 1000 meters of depth. Below these levels the deeper layers approach the barotropic condition.

Where there is sustained baroclinicity there is also water motion, for where isobaric surfaces are inclined there is a horizontal gradient of pressure. Since the regional inequalities of density producing baroclinic structures are usually maintained by persistent external influences, such as climate, the flow tends to be geostrophic. Indeed the geostrophic assumption is often made the basis for interpreting the relationship between the steady component of the primary ocean circulation and the distribution of water characteristics. Because of the great reservoirs of water in each of the oceanic water masses, it takes tens or even hundreds of years for appreciable changes to develop in the baroclinic structure of the major oceanic regions of the earth. This is partly because sea water is not easily penetrated by the sun's light and heat, and yet has an enormous capacity for heat storage.

Heat capacity

Both the near and especially the far infrared bands of solar radiation are rapidly absorbed in the first few millimeters of water depth. Visible light is more gradually transformed into heat, because it penetrates far more deeply into the sea before being absorbed. In the presence of waves and currents, both the shallow and deep radiant heating of the sea is mixed downward within a few hours to a depth which we shall say for convenience is on the order of 100 meters under most circumstances. On land, solar radiation is absorbed by a very thin layer of material and carried downward by relatively slow processes of thermal conduction. It is of interest to compare the temperature changes in the two cases.

With the assumption that solar radiation is absorbed at the same rate on land and sea, the rise in temperature per unit time, dT/dt, of a cubic centimeter of water at the sea surface can be calculated and compared with the rise in temperature of a cubic centimeter of bare rock under like conditions with the aid of the equation

$$\frac{dT}{dt} = \frac{1}{\rho C_p} \frac{dH}{dt}. \tag{5-4}$$

Let dH/dt equal 1 gm·cal/min/cm^2 and assume that ρ, the density of rock, is 2.5 gm/cm^3 while that of sea water is 1 gm/cm^3, that the specific heat C_p of rock is 0.2 cal/gm/°C while that of sea water is 1 cal/gm/°C.* Then we find that if the rock is heated without reflection to a depth of 1 cm, the rate of temperature rise of the rock will be 2°C/min. In the oceans, however, and again neglecting reflection, the rate of temperature rise is 1°C/min, which is not strikingly less until it is considered that the sea mixes its heat downward. The initial centimeter of heated water therefore shares its heat content with a water column perhaps 100 meters deep. Thus the average

---

* Actually the specific heat of sea water is lower than that of fresh water. See J. Thoulet and A. Chevallier, 1889, *C. R. Acad. Sci., Paris*, 108: 794–796. Values of the specific heat of sea water are currently being re-examined by Dr. R. A. Cox in a program jointly sponsored by the National Institute of Oceanography and National Physical Laboratory in England.

temperature rise in the sea surface is 10^{-4} °C/min. From this we are led to conclude that the oceans are warmed only very slowly and, compared with land, can accommodate a very large amount of heat at a given surface temperature because of vertical mixing. Only on the calmest days, when vertical mixing is slight enough to permit a shallow thermocline to develop, will the afternoon surface temperature of the sea rise a degree or two above the temperature of the water a few meters below the surface.

Penetration of visible light

Sunlight incident upon a glassy smooth ocean is partly reflected and the balance is refracted, scattered, or absorbed. The amount of visible light reflected varies with the angle of incidence, as shown in Table 5-3. The variability of the reflection coefficient of smooth water with the angle of incidence is not shared by the rocky, grassy, or forest-covered areas of the earth. However, in the normally ruffled or wavy condition of the sea the angle of incidence is highly varied, and thus absorption continues during the day unless foam is sufficiently abundant to increase the reflection coefficient of the sea surface. It should also be recognized that part of the incoming solar radiation arrives as diffuse energy.[20] On a clear day at noon about 85% of the radiation is direct and 15% is diffuse. When the sun is low in a clear sky, the proportion of scattered radiation is greater, being as much as 40% of the total when the sun's altitude is 10 degrees.

Visible light striking a smooth sea surface is refracted downward toward the normal in accordance with Snell's law, which, it will be recalled, states that the ratio of the sine of the angle of incidence to the sine of the angle of refraction is the same as the ratio of the refractive indices of the two media.* Thus viewed from beneath a glassy smooth sea surface (Fig. 5-7) the sun appears closer to the vertical than it appears when viewed above the sea surface; the horizon is seen at a zenith angle of about 48.6 degrees. The illuminated region overhead is the base of a cone having its apex centered on the observer's eye. This luminous circle can have the appearance of a manhole in perfectly calm weather.† Beyond the edge of the manhole, light from the bottom (and objects in the surrounding water) suffers a single reflection from the sea surface, producing an inverted image. Thus as one scans from the zenith to the horizontal just below the water surface,

---

* The refractive index of sea water is a function of temperature, salinity, and pressure. See C. L. Utterback, T. G. Thompson, and B. D. Thomas, 1934, *J. Cons. int. Explor. Mer*, 9: 35–38.

† For an account of some related physical effects see the "fish eye" camera described by R. W. Wood, 1934, *Physical Optics*, 3rd ed., New York: Macmillan, and various sections of M. Minnaert's (1954) book listed in the Supplementary Reading at the end of this chapter.

TABLE 5–3

Angle of incidence, degrees	Percent reflectivity
0	2.0
10	2.0
20	2.1
30	2.1
40	2.5
50	3.4
60	6.0
70	13.4
80	34.8
90	100.0

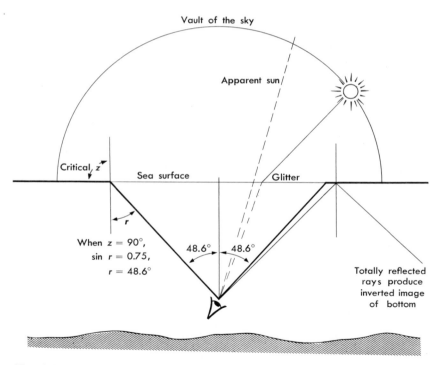

Fig. 5–7.   Factors influencing the radiance of the sky as seen from a point beneath the surface of a glassy sea.

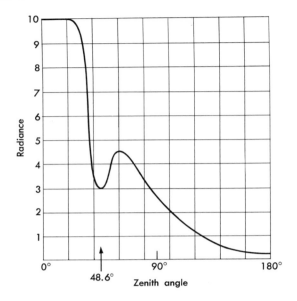

Fig. 5–8. Schematic diagram of the radiance reaching a point in the sea as a function of the zenith angle on an overcast day. [From information provided by J. E. Tyler.]

the light intensity falls off (as the reflection coefficient of the sea increases) to the critical angle where it reaches a minimum; beyond the critical angle the light intensity rises again to a secondary peak before falling off to a second minimum as one looks straight down (Fig. 5–8).

As one goes deeper in calm water the intensity of light reaching the eye diminishes owing to attenuation through scattering and absorption in the water.* The effect of these diffusing processes leads to a change in the direction of maximum radiance: from the direction of the refracted solar beam toward the vertical as the depth is increased. Thus at shallow depths, the sun appears as a bright spot in the manhole; while at greater depths the sun and the manhole become diffuse and the point of maximum radiance shifts more nearly overhead.[21] At the same time there is absorption primarily of red light which causes the visual field to appear more and more blue-green in color. With increasing depth, the scattering process causes the upward flux of light to contribute a large fraction of the total radiance reaching a point, so that when one looks down the water appears to glow. This effect can even be seen from above the surface. In very

---

* Some modern methods for observing these changes are described by G. L. Clarke and G. K. Wertheim, 1956, *Deep-Sea Res.*, 3: 189–205, and others by J. E. Tyler, W. H. Richardson, and R. W. Holmes, 1959, *J. Geophys. Res.*, 64: 667–673.

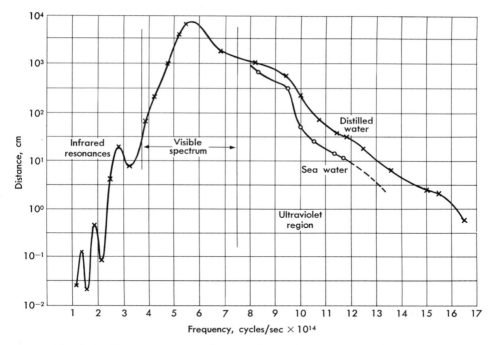

Fig. 5–9. Frequency versus distance for 90% attenuation. [Composite curve drawn by L. A. Jacobsen from material in the *J. Opt. Soc. Am.*, Vol. 24 (1934), pp. 175–177; Vol. 29 (1939), pp. 43–55; Vol. 17 (1928), pp. 15–22.]

clear water in the open ocean the glow has a luminous but velvety cobalt-blue color which makes the phrase "wine-dark sea" take on more than literary significance.

The visible light penetrating the volume of the sea contains only a minor fraction of the total energy of sunlight but this can produce direct heating in the illuminated layer. Long wavelength radiation is strongly absorbed (Fig. 5–9). The very shallowness of long wave heating permits the oceans to distill large amounts of water (about one meter per year) into the lower atmosphere. This evaporative process tends to charge the atmosphere with latent heat which is eventually released in a precipitation process at some other place on the earth.

The transport of fresh water by the atmosphere influences the salinity of the surface layers of the oceans and, through the redistribution of heat, alters the winds which tend to govern at least the surface circulations. These, in turn, produce further modification of the distribution of heat and of the atmospheric circulations, and so it goes.

Because the oceans and atmosphere are so closely interconnected in so many ways and because both are ultimately dependent on solar heating for the energies of their motion, as well as their characteristic properties, it is misleading to discuss either one without the other. In the next few chapters on the circulations and properties of the oceans, frequent reference will be made to the motions and optical properties of the atmosphere as an important part of what really amounts to a single system driven by solar energy.

STUDY QUESTIONS

1. Distinguish between σ_t and potential density.

2. With reference to the temperature-salinity diagram, distinguish between a *water mass* and a *water type;* with reference to lines of constant σ_t on the *T-S* diagram, compare the caballing process with the action of the perpetual salt fountain.

3. Outline Knudsen's method for the titration of chlorinity. What basic assumptions are made in translating chlorinity to salinity?

4. What is the adiabatic process? Where in the oceans is it of demonstrated importance? How can the adiabatic process be discerned in a sounding of salinity and temperature as a function of depth?

5. What factors alter the conductivity of sea water? How are *conductivity* and *salinity* related to each other in practice? Comment on the statement that the conductivity bridge measures salinity to an accuracy of $\pm$ 0.005 $^0/_{00}$.

6. Of what oceanographic significance is it that sea water of salinity 24.70 $^0/_{00}$ reaches maximum density at its freezing point, $-1.332°C$, while fresh water is most dense at 4°C?

7. What is an isopycnal surface? What data are required, at present, to establish the position and shape of an isopycnal surface in the volume of the ocean?

8. Explain how the perpetual salt fountain works and what engineering considerations would optimize its performance.

9. What vertical gradients of sea water properties produce a layer of minimum sound velocity in the deep ocean? What conditions can produce a temporary shallow sound channel?

10. Why does an explosive sound emitted at the axis of the deep sound channel arrive as a low-frequency rumble of rising intensity and with a sharp cutoff?

11. Estimate the variation of sound intensity as a function of range over distances that are large compared with the depth of the ocean when the source and receiver are situated at a level well above the axis of the deep sound channel.

REFERENCES

1. *Physical and Chemical Properties of Sea Water*, 1959, Washington, D.C.: National Academy of Sciences—National Research Council Publication 600.

2. W. S. Richardson, 1959, Ref. 1, pp. 113–117.

3. J. Kanwisher, 1959, Ref. 1, pp. 118–127.

4. C. Eckart, 1958, *Amer. J. Sci.*, 256: 225–240, and Ref. 1.

5. M. J. Pollak, 1954, *J. Mar. Res.*, 13: 228–231.

6. F. Wenner, E. H. Smith, and F. M. Soule, 1930, *Bur. Stand. J. Res.*, Washington, D.C., Paper No. 223, 5: 711–732.

7. K. E. Schleicher and A. Bradshaw, 1956, *J. Cons. int. Explor. Mer*, 22: 9–20.

8. R. A. Cox, 1957, *J. Cons. int. Explor. Mer*, 23: 38–46.

9. R. G. Paquette, 1958, Univ. Wash., Dept. Oceanogr., Tech. Rept. 61, Ref. No. 58–14, unpublished manuscript.

10. H. Stommel, A. B. Arons, and D. Blanchard, 1956, *Deep-Sea Res.*, 3: 152–153.

11. G. W. Groves, 1959, *Deep-Sea Res.*, 5: 209–214.

12. M. E. Stern, 1960, *Tellus*, 12(2): 172–175.

13. H. Stommel, *et al.*, Ref. 10.

14. M. E. Stern, Ref. 12.

15. W. Thomson, 1857, *Proc. Roy. Soc.*, A, 8: 566–569. See also V. W. Ekman, 1914, *Ann. Hydrogr.*, *Berlin*, 42: 340–344.

16. L. V. Worthington, 1953, *Trans. Amer. geophys. Un.*, 34: 543–551.

17. D. J. Matthews, 1939, *Tables of the Velocity of Sound in Pure Water and Sea Water for Use in Echo-sounding and Sound-ranging*, 2nd. ed., H. D. 282, London: Admiralty.

18. V. A. Del Grosso, 1952, *The Velocity of Sound in Sea Water at Zero Depth*, Washington, D.C.: U.S. Naval Research Laboratory Report 4002.

19. M. J. Pollak, 1954, *J. Mar. Res.*, 13: 101–112.

20. W. V. Burt, 1953, *Trans. Amer. geophys. Un.*, 34: 199–200.

21. N. G. Jerlov, 1951, "Reports of the Swedish Deep-Sea Expedition 3," *Physics and Chemistry*, No. 1, 1–59.

SUPPLEMENTARY READING

HARVEY, H. W., 1945, *Recent Advances in the Chemistry and Biology of Sea Water*, London: Cambridge University Press.

MINNAERT, M., 1954, *Nature of Light and Colour in the Open Air*, translated by H. M. Kremer-Priest, revision by K. E. Brian Jay, New York: Dover.

MONTGOMERY, R. B., 1957, "Oceanographic Data," Section 2, pp. 2–115 to 2–124, *American Institute of Physics Handbook*, D. E. Gray, coordinating editor, New York: McGraw-Hill.

Physical and Chemical Properties of Sea Water, 1959, Washington, D. C.: National Academy of Sciences—National Research Council Publication 600.

REDFIELD, A. C., 1948, "Characteristics of Sea Water," pp. 1111–1122, *The Corrosion Handbook*, H. H. Uhlig, editor, New York: Wiley.

CHAPTER 6

Advective Processes

The large-scale advective circulations of the oceans and atmosphere
are dominated by the astronomical circumstances which place the bulk
of solar radiation near the earth's equator and by the rotation of the earth
which distributes this energy zonally. In equatorial regions the radiant
energy received from the sun exceeds the energy radiated by the earth
into space. At higher latitudes the radiation losses from the earth exceed
the solar input. The temperature differences that arise between high and
low latitudes require a poleward flow of heat from the equatorial regions
that is beyond the capabilities of conductive processes to provide. Important
amounts of heat are, therefore, transported by fluid motions. It is known
that above the main thermocline in the oceans and below the tropopause
of the atmosphere, fluid circulations act preferentially in ways which transport heat away from the equator toward the poles.

Heat is supplied to the atmosphere mainly from the sea surface and
land, so that atmospheric circulations are predominantly supported by
heating from below. The oceans, on the other hand, receive their heat
mainly from above, a circumstance that reduces the effectiveness of planetary
temperature contrasts as a primary source of energy for horizontal circulations of the oceans. However, the stress of the wind on the sea surface
produces currents in a pattern which tends to resemble the pattern of the
surface winds. Through this agency of mechanical coupling with the
atmosphere, and perhaps also by regional excesses of evaporation over
precipitation, the advective modes of ocean circulation are made sensitively
responsive to the distribution of solar heat.

Energy units

In discussions of solar and terrestrial radiation it is customary to use
quasi-metric units of the form $cal/cm^2/min$ to express the fluxes of energy.
Much of the original investigation of solar heating of the earth was done by
members of the Smithsonian Institution of Washington under the direction
of Samuel P. Langley (Fig. 6–1). In his honor the unit $1 \ cal/cm^2$ is called
the *langley* (ly). Thus it is appropriate to express $1 \ cal/cm^2/min$ as $1 \ ly/min$,

Fig. 6–1. Samuel Pierpont Langley. [From G. B. Goode, editor, *The Smithsonian Institution, 1846–1896*, Washington, D.C.: The Smithsonian Institution.]

a unit of energy flux roughly characteristic of the solar energy reaching the earth's surface at high noon on a clear day. One ly/min is equivalent to about 700 watts/m^2, so that this unit may also have connotation as a brightness or (since one horsepower is equivalent to 746 watts) as a comprehensible index of power expended over an imaginable area.

In the advection of heat by the currents of the ocean and the winds of the atmosphere, it is usual to refer to horizontal fluxes through vertical surfaces mounted on geographically recognizable lines such as latitude circles. The units often used in this connection are cal/min $\times$ 10^{16} or cal/day $\times$ 10^{19}.

The rate at which energy is supplied to warm the earth and drive the circulations of the oceans and atmosphere is tremendous—something like 2 $\times$ 10^{15} horsepower. But although some of this power goes directly into heat, some into the kinetic energy of winds and ocean currents, some into the mechanical disintegration of rock, some into evaporating sea water, some into plant chemistry, and so on, all of it must eventually be returned to space as radiant energy if the average temperature of the earth is not to rise above its present value.

Insolation

At the outer limit of the earth's atmosphere the intensity of solar energy incident on a surface at right angles to the solar beam is very nearly 2 gm·cal/cm^2/min. Historically, this quantity has been known as the *solar constant*. The earth intercepts a circular sample of this radiation which has

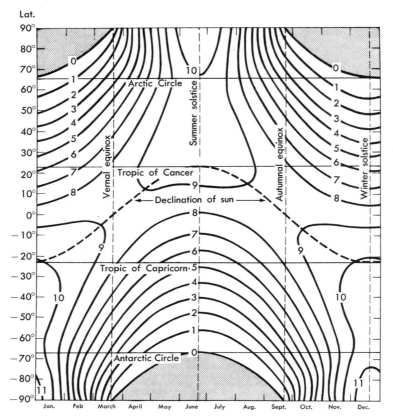

Fig. 6–2.   Undepleted insolation, in hundreds of langleys per day, as a function of latitude and date (after List).   Cross-hatched areas represent latitudes within the earth's shadow. [From Smithsonian Meteorological Tables, 6th edition, Washington, D.C.: The Smithsonian Institution.]

an area of πR^2, where R is the radius of the earth. The earth's daily rotation distributes this sample of energy over the spherical surface of the earth, which has an area of $4\pi R^2$. As a consequence of these geometrical conditions, the average input of solar radiation at the top of the atmosphere is reduced to one-quarter of the maximum, or about 0.5 gm·cal/cm^2/min.

It will be recalled from Chapter 2 that during the course of a year the distance between the earth and the sun varies some 3%, being least at the time of the summer season in the Southern Hemisphere. This causes the earth to intercept a slightly larger fraction of the total solar output at perihelion in early January than it does at aphelion in early July. Figure 6–2, in which is charted the intensity of radiation that reaches the top of the atmosphere at all latitudes and all seasons, shows the effects of the earth's

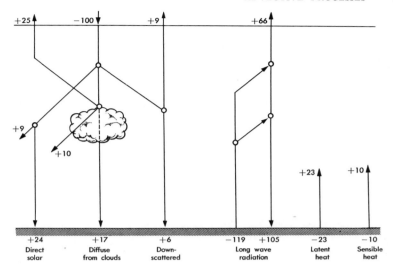

Fig. 6–3.   Schematic representation of heat balance of earth and atmosphere. Solar radiation reflected from the earth's surface is not shown separately, but is included in solar radiation returned to space. [After H. G. Houghton, 1954, *J. Met.*, 11(1).]

orbital eccentricity, as well as the effects resulting from the inclination of its axis of rotation to the plane of its orbit.   A surprisingly large amount of radiation reaches the polar regions during the summer season when the sun is above the horizon twenty-four hours each day.

The energy penetrating the top of the atmosphere is partly refracted, partly reflected, and partly scattered (Fig. 6–3).   As a consequence of these processes, there is a marked change in the intensity of radiant energy that reaches the surface of the earth, as shown in Fig. 6–4.   Because of the intense reflection of sunlight from the snow- and ice-covered polar caps, the heat energy absorbed by the earth's surface in these regions is small. In a similar way, the cloud bands in the equatorial regions reflect sunlight so strongly that the zone of maximum insolation at the earth's surface is somewhat shifted from the equator into each hemisphere.   The inequality of heating of the two hemispheres causes this shift to be asymmetrical, as shown in Figs. 6–2 and 6–4.   The difference between the ratios of land and sea areas for the two hemispheres is also thought to have something to do with the generally northward shift of the meteorological equator.   Because of the asymmetrical heating and the distribution of cloudiness over the earth, the mean surface temperature of the Northern Hemisphere is near 287°K (Kelvin), or about one degree cooler than the surface temperature of 288°K maintained in the Southern Hemisphere.   This small excess of heat apparently requires a transequatorial heat flux into the Northern Hemisphere

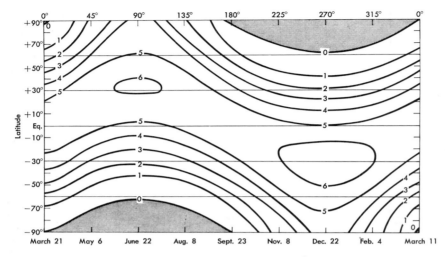

Fig. 6–4. Daily insolation, in hundreds of cal/cm², received at the earth's surface when the atmospheric transmission coefficient is 0.7 (after Milankovitch's computations). [B. Haurwitz, 1941, *Dynamic Meteorology*, New York: McGraw-Hill.]

which, according to present evidence, is maintained by the Atlantic branch of the world ocean.

It can be seen from a comparison of Figs. 6–2 and 6–4 that the heat absorbed at the earth's surface averages near 0.25 cal/cm²/min, or close to one-half of that incident upon the top of the atmosphere. This depletion of energy is caused by direct reflection and scattering of solar energy from the various kinds of land surfaces, the sea surface, and the air and clouds of the atmosphere. The albedo of the earth—its reflection coefficient for the short wavelength emission of the sun—is currently estimated at near 0.34.[1] The radiation which does not reach the earth's surface is absorbed or scattered by the atmosphere or becomes involved in photochemical reactions at high levels. It is mainly the latter process that shields the earth from the ultraviolet portion of the solar emission. The longer wavelength absorptive properties of the atmosphere are associated principally with the presence of water vapor and carbon dioxide. But, these gases, together with ozone, deplete the incoming solar beam only a trifling degree compared with their ability to obstruct the outgoing long wavelength emission of the earth.

Greenhouse effect

Window glass is highly transparent in the visible regions of the spectrum, but to far infrared wavelengths it is increasingly opaque. In a greenhouse the sun's light is readily transmitted through the glass and absorbed by the

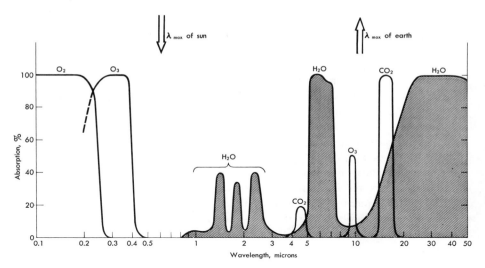

Fig. 6–5. Schematic diagram of the absorptivity of water vapor and other atmospheric gases. [Generalized from R. M. Goody and G. D. Robinson, 1951, *Quart. J. Roy. Met. Soc.*, 75.]

materials inside, which are thereby warmed. Since these warmed substances are usually not made hot enough to emit visible light, the bulk of their radiation lies in the far infrared. The long wavelength radiation is blocked by the glass so effectively that heat accumulates and the temperature in the greenhouse rises. The rise continues until the infrared radiation from the interior of the greenhouse is rich enough in the shorter infrared wavelengths to pass through the glass a flux of energy equal to that entering as visible light. The temperature inside the greenhouse then levels off at some usefully high value. This is called the *greenhouse effect*. Because, in the atmosphere, there exists a similar inequality of transparency to visible and infrared radiation, the mean temperature of the earth is somewhat higher than that warranted by its mean distance from the sun.

The surface of the sun radiates as a nearly black body with a temperature of 5800°K. For a black body at this temperature the wavelength of maximum emission is 0.54μ. Because of the transparency of the atmospheric gases to visible and near-infrared wavelengths, the band of solar energy which penetrates the atmosphere and contributes warmth to the surface of the earth is predominantly in the wavelength band ranging from 0.3μ to 3.0μ. (Fig. 6–5). Wavelengths shorter than 0.3μ are strongly absorbed by the ozone and molecular oxygen in the upper atmosphere. On the long wavelength side of 3μ, particularly in the region 4μ to 8μ and increasingly from 13μ to 30μ and beyond, the absorption due to atmospheric water vapor is also strong. The band from 14μ to 16μ is almost totally absorbed by carbon dioxide. Thus of the total amount of solar radiation that reaches the top of the earth's atmosphere, it is mainly the invisible fraction that is filtered out by atmospheric gases.

The solar radiation that reaches the surface warms the earth so that it, in turn, emits infrared radiation. In the infrared spectrum of the atmosphere, there is a window between the absorption band of water vapor at $8\ \mu$ and the band due to carbon dioxide at $14\ \mu$. In order to return to space an amount of radiation equal to that received by the sun, the temperature of the solid earth and oceans must rise to the level where the flux escaping through the relatively narrow window ($8\ \mu$ to $12\ \mu$) in the lower atmosphere equals the flux received in the short wavelength emission from the sun. This self-adjustment of the earth's temperature whereby the peak of blackbody emission is placed in the window between the infrared absorption bands of atmospheric gases—the greenhouse effect—has the function of a thermal regulator.*

Heat distribution

The distribution of heat derived from sunlight on the earth varies with the season, the time of day, and the local albedo. The earth's albedo is determined primarily by the amount of cloud cover and the condition of the earth's surface. Ice- and snow-covered regions near the poles and clouds everywhere reflect a very large fraction of the light that reaches them through the atmosphere. The oceans, bare earth, and forests reflect much less. The fraction of light returned directly to space by reflection and scattering is best known for the Northern Hemisphere. Owing to variations in cloud and snow cover, the zonal albedo for different latitude belts in the Northern Hemisphere is known to vary considerably. The average values, shown in Table 6–1, tend to be small near the equator where ocean surface predominates and where there is relative scarcity of horizontally layered clouds. In middle latitudes between 30° and 40°, the albedo is somewhat higher, reaching near 38° a maximum of more than 11%. Northward there is again a decrease to the limits of the polar regions where persistent ice and snow cover generally causes a high surface albedo.

Table 6–1 also shows the relationship of the absorbed solar radiation to the long wave re-radiation to space as a function of latitude, and the differences that exist in each 10° latitude zone for the Northern Hemisphere. Smooth curves based on these differences show a net accumulation of heat in the region between the equator and 38° north latitude, and a deficiency between 38° north latitude and the North Pole. From these differences

---

* It is conceivable that the change in the carbon dioxide concentration in the atmosphere produced by the extensive combustion of fossil fuels since the industrial revolution could have some effect on the average temperature of the earth.

TABLE 6-1

Latitude φ, degrees	Visual albedo (surface),* percent	Absorbed solar radiation,* ly/day	Long wave radiation to space,* ly/day	Difference,* ly/day	Poleward oceanic and atmospheric heat flux across latitude circles,* cal/day × 10¹⁹	Average absolute surface temperature†
0	7.1	573	488	85	0.0	299
10	8.0	578	502	76	4.05	300
20	9.8	574	503	71	7.68	298
30	11.0	532	492	40	10.46	293
38						
40	10.2	444	469	−25	11.12	287
50	9.2	352	442	−90	9.61	279
60	9.1	261	419	−158	6.68	272
70	16.8	192	400	−208	3.41	262
80	36	147	385	−238	0.94	254
90	56	117	380	−263	0.0	250
Mean	34	461				287

* After H. G. Houghton, 1954, *J. Met.*, 11: 1–9.
† Adapted from J. v. Hann and R. Süring, 1937, p. 180, *Lehrbuch der Meteorologie*, Leipzig: Verlag Keller.

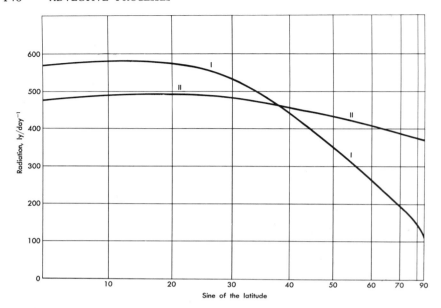

Fig. 6–6. Annual mean solar radiation absorbed by earth and atmosphere (curve I) and outgoing long-wave radiation leaving atmosphere (curve II) as functions of latitude. [From H. G. Houghton, 1954, *J. Met.*, 11(1).]

it is possible to compute the poleward flux of heat across the latitude circles that must be provided by the circulation of the oceans and atmosphere (Fig. 6–6). The next to the last column in Table 6–1 gives the values of the poleward heat flux required to balance the energy budget for the Northern Hemisphere.

The ambient temperature of the earth lies within the narrow range which permits water to remain liquid over most of the earth. That which is frozen is not only a good reflector of sunlight but also floats on the sea surface. This circumstance probably tends to increase the present temperature contrast between equatorial and polar latitudes beyond that which may be characteristic of the predominantly ice-free periods of the earth's geologic history.

Heat transport

Since the energy of black-body radiation to space depends on the fourth power of the absolute temperature of the earth (Stefan-Boltzmann law), the average temperature difference of 50°K between the tropics and North Pole leads to a large decrease in the radiative emissivity of the earth at successively higher latitudes. The heat demand at high latitudes is abated somewhat by the relatively small area between successive latitude circles near the poles compared with that between corresponding intervals of latitude near the equator; nevertheless the deficit must still be compensated by the poleward heat transports provided by advective processes in oceans and atmosphere.

Transports for the Northern Hemisphere are given in Table 6–2. These figures reveal that the atmosphere is the principal agency for poleward heat transport. This is not so much due to the more rapid motions of air than water as to the very efficient means the atmosphere possesses for transporting large amounts of latent heat in admixed water vapor.[2] For example, rainfall at the rate of 1 in./hr releases latent heat in the levels of condensation at the rate of some 25 cal/cm^2/min (or 16,000 watts/m^2), which is 100 times greater than the average solar input to the ground under clear skies. Since the energy of terrestrial radiation to space tends to obey the Stefan-Boltzmann law, the sensible heat of the atmosphere, oceans, and crust must stay near existing values if the heat lost to space at low latitudes is to remain small enough for appreciable surpluses to be exported toward the poles. If water vapor were excluded from the atmosphere, the temperature contrast between the equatorial belt and the poles would be greatly increased. It is quite probable that with sufficient increase of the meridional temperature contrast, the mode of circulation of the atmosphere might change and, in turn, change that of the oceans. Representative effects of this kind have been produced in laboratory models of the atmosphere by Fultz and his colleagues at the University of Chicago.

TABLE 6–2

POLEWARD TRANSPORT OF HEAT BY THE OCEANS AND ATMOSPHERE, IN UNITS OF 10^{19} CAL/DAY

Latitude, degrees N	North Atlantic*	North Pacific*	Northern Hemisphere atmosphere†
0	0.6	0.0	0.0
10	1.2	1.0	1.9
20	1.6	0.9	5.2
30	1.9	0.6	8.0
40	1.1	0.0	10.0
50	0.9	0.0	8.7
60	0.4	0.0	6.3

* Adapted from H. U. Sverdrup, 1957, p. 636, Handbuch der Physik, S. Flügge, editor, Bd. 43, Geophysik II, Berlin: Springer-Verlag.
† Adapted from H. G. Houghton, 1954, J. Met., 11: 8.

TABLE 6–3

Water mass	Toward	Mass transport, $10^6 m^3/sec$	Mean temperature, °C
Upper	North	6	15
Intermediate	North	2	5
Deep	South	9	4
Bottom	North	1	2

Table 6–2 also indicates that there is a transequatorial flux of heat in the Atlantic Ocean. From data obtained during the *Meteor* expedition, Sverdrup (1942)[#] has estimated that the transport of water and heat across the equatorial plane in the Atlantic Ocean must be roughly as shown in Table 6–3. It follows from these data that the rate of advection of heat across the equatorial plane must be the sum of the rate of mass transport times the temperature at each level of the ocean. This gives 0.6×10^{19} cal/day for the northward transequatorial heat flux, assuming that the mass transport is balanced. The value of transequatorial heat flux in the Pacific has been estimated to be southward by Bryan (1962) [2a] but that for the Indian Ocean remains to be determined.

Present evidence suggests that the transequatorial fluxes of heat are primarily an oceanic phenomenon and are minor in comparison with the poleward fluxes within each hemisphere.[*] A major part of the poleward heat flux is maintained by the water vapor transports in the atmosphere, so that the problem of measuring balance of evaporation and precipitation over the oceans is of importance. The winds that transport water vapor also drive some of the major surface currents of the oceans, and are partly responsible for deep-water circulation as well. The nature of these advective processes in the atmosphere and oceans is discussed in succeeding sections.

Structure of the atmosphere

Studies of the structure of the atmosphere reveal a relatively narrow region of strong meridional temperature contrast between the belt of low-

---

[*] The summer hemisphere of the earth sustains a somewhat lower over-all atmospheric pressure than the winter hemisphere as a result of the expanded condition of the atmosphere when it is warm. Over the oceans the atmospheric mass shift is probably compensated by a shift of mass in the oceans, since each millibar of pressure change in the atmosphere is balanced by a change in sea level of nearly one centimeter. To the extent that the oceans respond in this way to atmospheric pressure changes, there must be some slight annual transequatorial shift of the oceanic mass distribution.

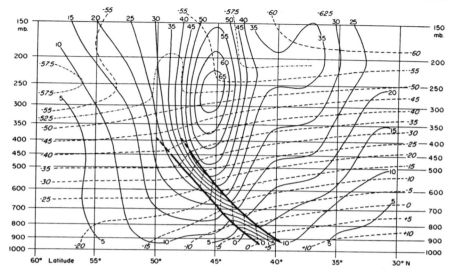

Fig. 6–7. Mean temperature and zonal component of geostrophic wind, computed from 12 cases in December, 1946. The cross section lies along the meridian 80°W. Heavy lines indicate mean positions of frontal boundaries. Thin dashed lines are isotherms (adjacent numbers indicate degrees centigrade). Solid lines are isolines of westerly component of wind (in m/sec). Means were computed with respect to the polar front. [E. Palmén and C. W. Newton, 1948, *J. Met.*, 5(5).]

latitude circulations in the region of heat surplus and the high-latitude circulations in the region of heat deficit. This zone of temperature contrast is associated with the position of the middle-latitude atmospheric "jet stream," a phenomenon named and intensively studied only after it had been penetrated by manned aircraft. The circulation of tropical air seems to make the poleward temperature gradient in the tropical regions of the earth more gentle than might be expected, up to the position of the middle-latitude jet stream. Similar circulations existing in the polar regions blend the temperature of the polar regions as far south as the latitude of the westerly* jet stream, as shown in Fig. 6–7. Since warm air has a greater capacity for water vapor than cold air, it is not surprising that in latitudes where warm, moist air from the tropics consistently meets and overruns colder, drier polar air, there tends to be persistent cloudiness. This belt lies in latitudes 30° to 50°, with a mean near 40°N.

Because of the water vapor in the atmosphere, a great deal of the heat is transported in latent rather than in sensible form. Thus the meridional heat flux is not directly proportional to the product of the northward

---

* In meteorology it is customary to name winds according to the compass bearing *from* which they blow. Ocean currents are designated by the direction *toward* which they flow.

velocity and the temperature of the air. Rather, it is given by the product of the northward velocity and the sum of the sensible heat and the latent heat to be released by the formation of cloud droplets and precipitation.

The existence of a discrete band around the earth where the meridional heat flux, albedo, and atmospheric velocity all reach a maximum is probably related to both the intensity of solar heating and the rate of rotation of the earth. Recent experimental and theoretical studies have dealt with this phenomenon in sufficient detail to permit the character of these motions to be more clearly understood.

The planetary wind field

The circulation of the atmosphere near the earth's surface can be characterized by three principal wind zones in each hemisphere. Between the 30° latitudes on either side of the equator, the winds blow from the east toward the west; between 30° and 60° latitude they blow from west to east; and again in the region poleward of 60° the winds seem to blow more or less from east to west. In both hemispheres there is so much variability in the winds poleward of 30° that the actual wind direction at any instant may differ from the generalized pattern. Only when they are analyzed over a period of months or even years do these zonal motions emerge with statistical significance.

Associated with the zonal winds at the surface are the weaker north-south or meridional motions of the atmosphere (Fig. 6–8). These can be clearly discerned only in the belt of *trade winds*, where the surface level component of meridional motion is toward the equator. These winds show a high degree of directional persistence. North of the equator the ocean trade winds blow from east-northeast, while south of the equator they blow from east-southeast, 80% of the time. The astronomer Hadley, in 1735[#], supposed that convergence of this flow on the meteorological equator results from the conservation of angular momentum as air moves from higher latitudes to supply the upward motion of air in the equatorial regions. Hadley further specified that after the air converges in the equatorial zone it rises to a considerable height and flows poleward, in each hemisphere, as a high-level west or *antitrade* wind, to a latitude near 30°. Here the air sinks to the surface and returns to the equator again as a trade wind. This circulation, known as the *Hadley cell*, has recently been subjected to intensive field study. It is found that the upward motion is not entirely concentrated at the equator and the downward motion is not entirely concentrated in the horse latitudes near 30°. In fact, convective motion in cumulus clouds throughout the trade wind zone seems to account for a large part of the rise of air and water vapor in the tropical atmosphere. Between the cumulus clouds there is some descent of dry air which "short-

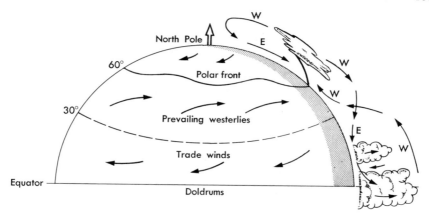

Fig. 6–8. Meridional circulation according to Rossby. [After C.-G. Rossby, 1941, "The Scientific Basis of Modern Meteorology," in *Climate and Man*, Washington, D.C.: U. S. Government Printing Office.]

circuits" the Hadley cell. Nevertheless, the air motion over the equator seems to contain pronounced vertical drafts, and near 30° latitude contains significant downward motions so that, as a first approximation to the actual tropical circulation in the atmosphere, the Hadley cell still serves as a useful model.

Over the poles, circulation is supposed to be characterized by sinking motions as the air is cooled by the frozen surface, and by a subsequent spreading from the poles to a latitude near 60°. In this interpretation, the subsidence of air over the poles is thought to be balanced by air rising near the 60° latitude circle, so that the polar circulation may have a meridional component similar to that in the trade-wind zone. But the observed air motion over the poles resembles more nearly that of the belt of westerlies in its unsteadiness and frontal associations.

Between the Hadley cell and the polar vortex in each hemisphere there is the zone of *westerlies*. Ferrel (Fig. 6–9) described this zone as a meridional cell rotating in a sense opposite to that of the Hadley cell, and supposed it to be driven through frictional interactions with the thermally direct cells at its northern and southern boundaries. The evidence for the meridional motions which compose the *Ferrel cell* is not convincing, and indeed its very existence has often been doubted. According to recent studies it seems more likely that the zone of westerlies lies in the region swept by the writhing band of west winds maintained in the belt of maximum meridional temperature contrast between the tropical and polar atmospheric zones.

This thermal contrast tends to support a horizontal gradient of pressure between low and high latitudes, because pressure falls off more slowly with altitude in warm air than it does in cold air. Therefore the same

Fig. 6–9. William Ferrel. [From John G. Albright, 1939, *Physical Meteorology,* Englewood Cliffs, N. J.: Prentice-Hall.]

pressure in the atmosphere is found at a higher elevation in the tropics than in the polar regions. The flow of air northward in response to this pressure gradient is deflected toward the east by the effects of the earth's rotation to produce west winds, which often extend downward to the earth's surface.

Another way of looking at this system of winds is to consider that the angular momentum of the atmosphere must balance over the earth as a whole in order that the stress of the winds on the earth's surface will tend neither to increase nor to decrease the rate of the earth's rotation. It has been shown that the momentum of east and west winds does tend to balance over the earth at all times. Moreover, if there is an unbalance in one hemisphere, it is corrected by a counterbalance in the other hemisphere.

The form of the balance with reference to observations taken in the Northern Hemisphere is given by Starr and White (1951).[3] The balance over the oceans alone is given by Mintz and Dean (1952).[4] The latter observations are shown in Fig. 6–10. The mechanisms of the momentum exchange are discussed by Palmén (1955).[5]

Coupling

It is well known that winds impart significant amounts of momentum to water if they are reasonably steady in direction and are given enough time. Seasonal variations of winds and the air motions around the semipermanent centers of high and low pressure over the oceans in winter and summer are significantly large, as shown in Figs. 6–11 to 6–14. The responses of

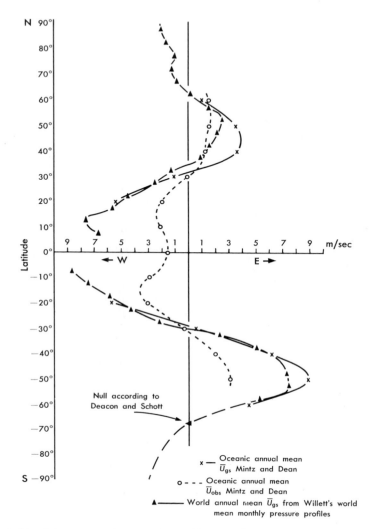

Fig. 6-10. Mean annual zonal wind profiles. [From Chapter 1, *Physics and Chemistry of the Earth*, Vol. 2, 1957, L. H. Ahrens, *et al.*, editors, London: Pergamon.]

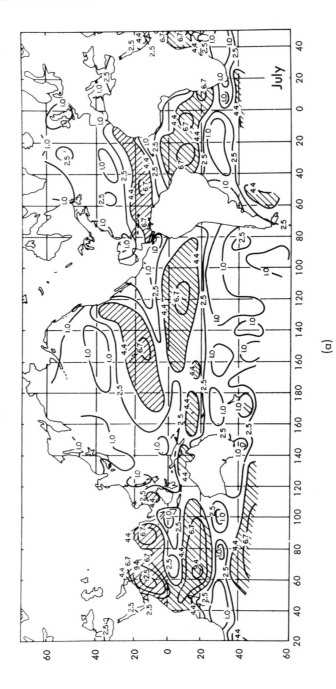

(a)

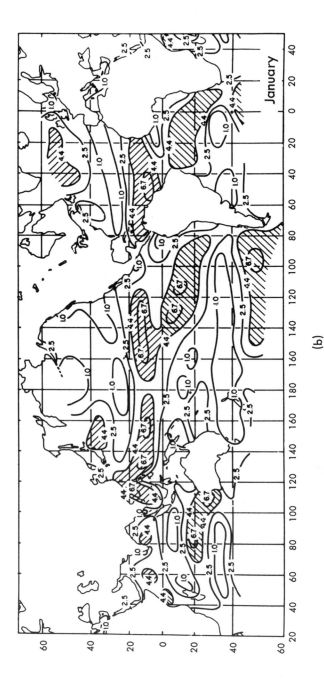

Fig. 6–11. Magnitude of the mean surface wind in (a) July, (b) January. The contours show the magnitude of the mean vector wind at 10 meters anemometer height in m/sec. [From Y. Mintz and G. Dean, 1952, *Observed Mean Field of Motion of the Atmosphere*, Geophysical Research Paper 17, Geophysics Research Directorate, Air Force Cambridge Research Center.]

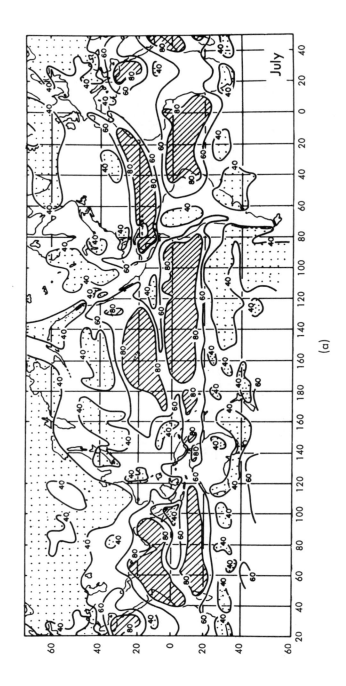

(a)

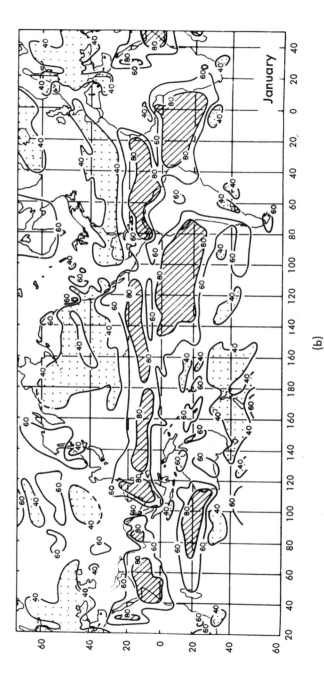

(b)

Fig. 6-12. Constancy of the surface wind direction in (a) July, (b) January. The contours show the percentage frequency with which the wind is within 45° of its modal direction. [From Y. Mintz and G. Dean, 1952, *Observed Mean Field of Motion of the Atmosphere*, Geophysical Research Paper 17, Geophysics Research Directorate, Air Force Cambridge Research Center.]

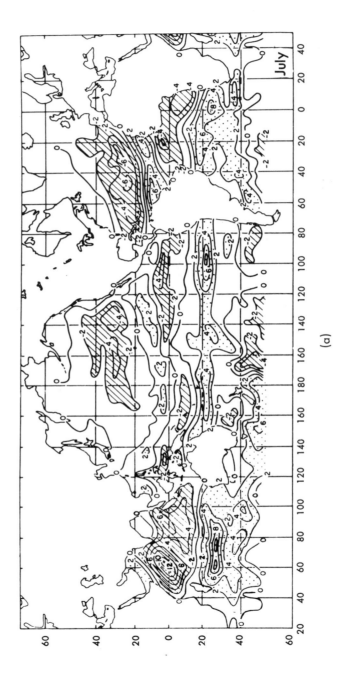

(a)

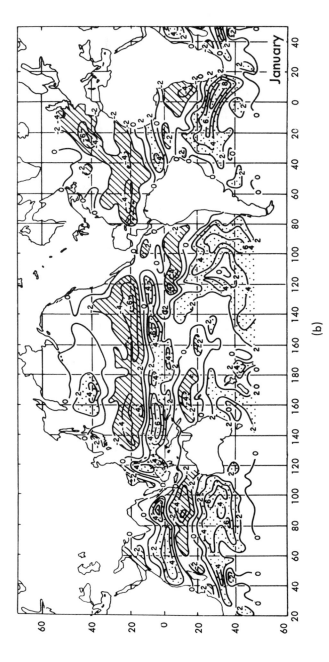

Fig. 6-13.   Vorticity of the mean surface wind in (a) July, (b) January, in 10^{-6} sec^{-1}.   The stippled pattern indicates positive, and the diagonal ruling negative, vertical vorticity greater than 2×10^{-6} sec^{-1}.   [From Y. Mintz and G. Dean, 1952, *Observed Mean Field of Motion of the Atmosphere*, Geophysical Research Paper 17, Geophysics Research Directorate, Air Force Cambridge Research Center.]

(b)

January

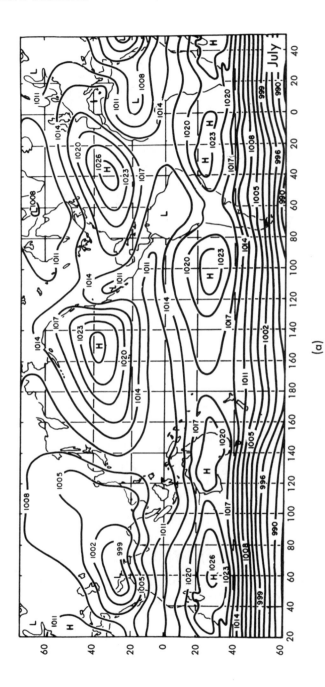

(a)

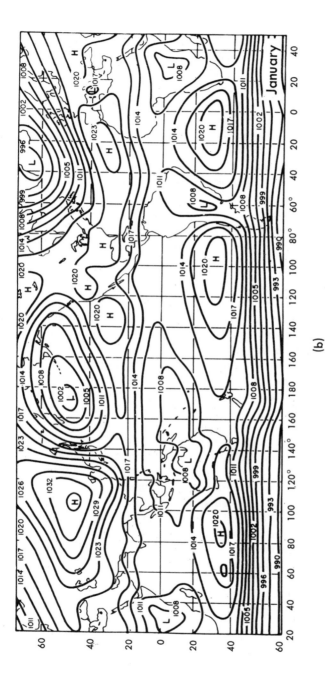

(b)

Fig. 6–14. Mean sea-level pressure in (a) July, (b) January, in millibars. [From Y. Mintz and G. Dean, 1952, *Observed Mean Field of Motion of the Atmosphere*, Geophysical Research Paper 17, Geophysics Research Directorate, Air Force Cambridge Research Center.]

the oceans to these environmental changes can occur in two distinct modes: the barotropic mode, in which the pressure distribution in the sea is altered almost concurrently by a disturbance lasting for days or weeks; and the slower baroclinic mode, which involves the redistribution of density in the sea and therefore requires periods of time which are probably long in comparison with the duration of a year.[6]

The ocean attempts to keep step with the variations of atmospheric pressure and wind stress by revising the barotropic part of its circulation pattern, but because the baroclinic process involves a major reorganization of the temperature and salinity distribution, it does not come to equilibrium with the seasonal wind field before one season has given way to the next. The oceans are constantly in a state of readjustment with respect to the winds, because the baroclinic part of the readjustment process is so sluggish. As a consequence, the amplitudes of seasonal fluctuations in oceanic circulation are relatively slight compared with the changes in atmospheric circulation, except, it is believed, for the circulation of the northern Indian Ocean.

With the more rapid rise of land temperature than sea temperature under the same rate of solar heating and with the greater rate at which land cools off as solar heating diminishes, there is a tendency for the land masses of the earth to become regions of low atmospheric pressure in the summer and high atmospheric pressure in the winter. This alternation produces a contrast of pressure over land in relation to the atmospheric pressure over the oceans which, together with the resulting winds, is referred to as the effect of *continentality*.

The temperature excess on land in the summer hemisphere is associated with lowered atmospheric pressure, which causes air to flow in from the sea. During the winter months the circulation is reversed as the land mass becomes colder than the adjacent ocean. In Asia this effect is so strong that the westerly winds of the northwestern North Pacific are much reduced or sometimes even reversed during the summer months. In the southern part of the Asiatic continent the monsoonal winds are driven northward from the Indian Ocean toward the land in the summertime, and from the land toward the sea in the Northern Hemisphere winter.*

With the reversal of the monsoonal winds over the Indian Ocean, there is a change in the ocean circulation. The flow is predominantly from east to west during the dry monsoon of the Northern Hemisphere winter, and

---

* As warm, moist air from the Indian Ocean blows in toward the Asiatic land mass during the Northern Hemisphere summer, it is lifted and adiabatically cooled over the Deccan Plateau, Burma Highlands, and Himalaya Mountains, to produce some of the heaviest rainfall known on the earth: well over 400 cm per year. Much less conspicuous monsoonal effects exist in the southern United States, in Australia, Central South America, and in south Central Africa.

from west to east during the wet monsoon of the Northern Hemisphere summer. The winds of the Asiatic monsoon are directed essentially north and south, but resulting ocean circulations flow east and west. This is thought to be a consequence of the Ekman deflection of wind-driven water motions and transports.

Only in the northern Indian Ocean does a major feature of the wind-driven ocean circulation reverse direction with the season, and apparently keep in step with atmospheric motions. Although even now little is known of motions below the surface, the seasonal reversal of flow in the surface water of the northern Indian Ocean has been recognized since Roman times. Sustained mechanical coupling between the oceans and atmosphere in the large-scale motions of the surface layers has been acknowledged only in much more recent times.

The wind-driven circulation

It has often been suggested that the surface circulation of the oceans is wind-driven. Edmund Halley made direct reference to this relationship in 1686.[#] Synoptic charts of winds and ocean currents assembled by M. F. Maury in 1855[#] showed the similarity of motion very clearly. In 1905[#] V. W. Ekman advanced his theory of wind-driven boundary-layer motions which Iselin employed in 1936[#] and 1940[#] and which Sverdrup in 1947[#] took into account in describing the circulation of the interior regions of the tropical Pacific. Sverdrup's contribution to a physical explanation of the wind-driven general circulation on a stratified ocean departed from Ekman's barotropic model in that no boundary-layer effects were required next to the ocean floor, but did not explain the asymmetry of surface-water motions in the Northern Hemisphere. A physical explanation of the asymmetry was found by Stommel (1948),[#] and this basic idea was later employed by Munk (1950)[#] and others to account for an important fraction of the observed transport of water by wind-driven currents.

Ekman model of the barotropic ocean circulation

In the presence of frictional interaction between the atmosphere and the free surface of a barotropic ocean, or between the ocean circulation and the sea bottom, there should be a turning and lessening of current with depth according to the form of the Ekman spiral. The effect of wind on a barotropic ocean is to produce two Ekman spirals (Fig. 6–15). One begins at the free surface and is associated with wind stress, and the other is associated with bottom friction. In each friction layer the transport is at right angles to the driving force of the wind, but between the layers of frictional influence the flow is geostrophic.

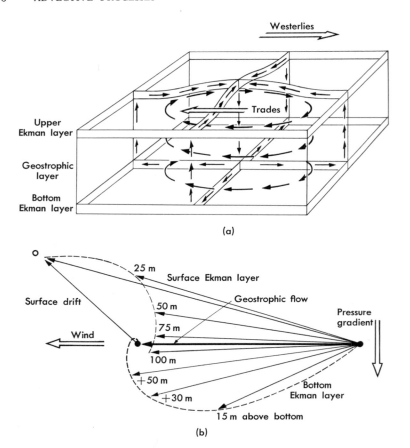

Fig. 6–15.   (a) Ekman's model of the wind-driven ocean circulation.   (b) Motions in the upper and lower Ekman layers.

In the Ekman model an anticyclonic circulation of air over the free surface would drive water in the layer of frictional influence toward the center of high atmospheric pressure, producing a hill of water and a corresponding high-pressure center in the barotropic ocean. Associated with the pressure gradient directed radially outward from the oceanic high-pressure center is a field of anticyclonic geostrophic motion, which is to be expected at all depths save those occupied by the friction layers.

The Ekman transport in the surface friction layer is centripetal, while that in the friction layer over the bottom is centrifugal. Since the centrifugal transport in the bottom friction layer tends to reduce the pressure in the oceanic high-pressure cell by draining away mass, the centripetal transport in the surface friction layer must replace mass in order that a steady geostrophic circulation be maintained in the body of the ocean. The centripetal flow in the surface friction layer is necessarily associated with sinking to supply the diffluent circulation in the bottom friction layer. Thus there is a slow circulation inward at the surface and outward at the bottom. The energy dissipated in friction is assumed to come from the wind stress.

This simple model has no steady-state counterparts in nature, because the solar heating of the surface layer of frictional influence tends to convert a barotropic cell into a baroclinic cell by storing warm, less dense water as a lens under the atmospheric anticyclone. The effect of such baroclinic stratification is to isolate the geostrophic circulation to the upper layer and to remove further interaction with the fluid beneath. In nature the boundary beneath this lens is marked by a relatively strong permanent thermocline and halocline.

Iselin's concept of the North Atlantic circulation

The wind system of the subtropical North Atlantic Ocean is dominated by the Bermuda-Azores cell of more or less permanent high pressure. Associated with this is a persistent anticyclonic circulation of air composed of the trade-wind motions on the south, the westerlies on the north, a less conspicuous but statistically persistent northward flow of air along the east coast of North America, and an even weaker southward flow along the west coast of Euro-Africa. The Ekman transports associated with the Bermuda-Azores high cell seem to produce near 30°N a zone of convergence of warm surface water extending from the Azores to a site somewhat south and east of Bermuda. The subsidence of water beneath this confluence is considered to be responsible for a marked depression of the main thermocline near 30°N and also for the rising of the isotherms both northward and southward of this latitude. (Fig. 6–16) Iselin (1936, 1940)[#] regards the middle of the main thermocline layer in the North Atlantic as being represented by the 10°C isotherm, which is found at a depth of 800 to 900 meters off Bermuda, within 200 meters of the surface near 10°N, and at the surface at 50°N. He also points out that the packing of isotherms in the main thermocline is inversely related to their depth. Owing to the excess of evaporation over precipitation in the horse latitudes, the vertical gradient of salinity is parallel to the thermal gradient, being as fresh as 34.88 $^0/_{00}$ at the bottom of the North Atlantic, 34.98 $^0/_{00}$ under the main thermocline, and rising abruptly through it to a central surface salinity of more than 37.3 $^0/_{00}$. However, the effects of vertically increasing temperature far outweigh this increase of salinity, so that the lens of North Atlantic central water is decidedly less dense than any other major water mass in the North Atlantic basin, and therefore floats in these adjacent and subjacent water masses.

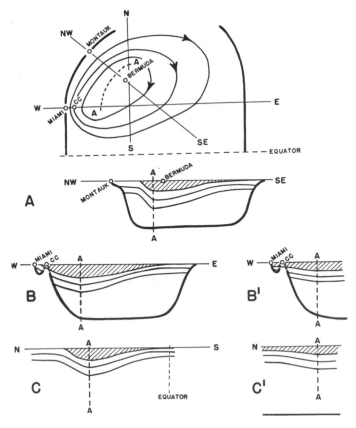

Fig. 6–16.   Schematic representation of density profiles across the North Atlantic in various directions.   For simplicity the sections show a three-layered ocean, the lightest water being shaded.   [After Iselin, 1940.#]

Since the central Atlantic water mass floats hydrostatically, its surface stands higher than that of the surrounding coastal water, but it exerts no extra load on the water beneath. Ideally then, the surfaces of equal pressure beneath the main thermocline should be horizontal, and no geostrophic motion should be expected at these levels.   In the warm lens, however, the vertical change of pressure is more gradual than that in the more dense water masses near the coast.   For this reason the isobaric surfaces should be expected to slope steeply near the edges of the buoyant lens of central water, resulting in strong geostrophic motions. Both expectations are realized along the western margin of the lens in the Gulf Stream system, which forms a kind of dynamical dam that keeps the lenticular central water mass from reaching the coast, and in the Gulf Stream extensions across the North

Atlantic which confine the lens on the north. Moreover, observations by many different methods show that the geostrophic currents do indeed fall off rapidly with depth below the main thermocline, but not to zero.

In this concept of North Atlantic circulation, the Ekman transport is called upon to gather up the surface water heated by the sun, and the wind stress is taken as the driving force needed to overcome friction in the otherwise geostrophic circulation. It is a baroclinic model and has been a useful interpretation of the facts for nearly two decades. To explain the lack of strong currents on the eastern side of the North Atlantic, however, it is necessary to postulate a thickening of the lens on the western side of the ocean. A physical explanation of such westward thickening is not contained in this theory.

Torque of the surface winds

The average velocity of winds blowing east and west across the oceans varies with latitude, as shown in Fig. 6–10. Where the trade winds diminish in strength with increasing latitude and ultimately yield to the flow of the westerlies, there is a stress couple acting on the sea surface. Where the westerlies, in turn, develop horizontal shear between the latitude of their greatest strength and higher latitudes occupied by the polar easterlies, an oppositely directed stress couple acts on the sea surface. In addition to the zonal component of motion in the atmosphere, there are meridional flows of air around the semipermanent and mid-ocean high-pressure cells at latitudes 30° north and south, and low-pressure centers near the sixtieth parallel in both hemispheres. These circulations produce an anticyclonic torque over the tropical and middle-latitude oceans in the Northern and Southern Hemispheres and a cyclonic torque over the polar and temperate seas.

It will be recalled that the meridional flows of air over the earth are less conspicuous than the east-west flows. For this reason there is a tendency for water to pile up against the east coasts of the continents in the trade-wind zone, and on the west coasts of continents subjected to westerly winds. Where water is piled up by the wind, it can flow farther only by turning poleward, equatorward, or by sinking. The stable thermal structure of the tropical and middle-latitude oceans opposes sinking, so that horizontal motion is the preferred course, and this tends to follow the direction of the meridional flow of air.

According to this concept, ocean currents under the steady urging of the trade winds should be strong and flow from east to west in both hemispheres. Similarly, the ocean surface should move with the westerly winds as far poleward as about latitude 60° in both hemispheres. On the eastern sides of the ocean basins, some southward flow should return the water pushed ahead of the westerlies to the tropics. On the western sides of oceans the water piled up by the trade winds should flow poleward to feed the deficit created by the westerlies, and some smaller amount should flow equatorward to drain back toward the east under the weak winds at the meteorological equator.

Most, but not all, of this hypothetical circulation is actually observed. The equatorward flow on the eastern sides of oceans is always weak and often ill defined, while the poleward flow on the western sides tends to be very conspicuous and swift. This asymmetry is most pronounced in the oceans of the Northern Hemisphere. A satisfactory physical explanation of these discrepancies has been advanced only within the past decade.

Vorticity theory of the wind-driven circulation

The vorticity model recognizes first of all that the torque of the planetary wind field on the sea surface generates a field of similar motion in the ocean volume; that is, anticyclonic vorticity is generated in the oceanic cell beneath the trades and westerlies, and cyclonic vorticity in the cell beneath the westerlies and polar easterlies (Fig. 6–17). These motions would be symmetrical around a central point in each cell were it not for the tendency for vertical fluid columns at the eastern and western sides of each cell to conserve their angular momentum as they change latitude. To conserve vorticity in this way it is necessary, in an ocean of uniform depth D, that $f + \zeta =$ a constant, as was shown in Chapter 4.

Under these conditions, we are led to expect a change in the relative vorticity of fluid columns that change latitude which varies inversely with the change of the Coriolis parameter. Fluid columns moving equatorward on the eastern sides of subtropical ocean basins acquire cyclonic vorticity relative to the earth, while columns on the western sides, moving poleward in the subtropic cell, acquire anticyclonic relative vorticity. This effect adds to the existing vorticity of the symmetrical wind-driven motions on the western sides of oceans, and opposes it on the eastern sides. Since to establish a steady state the total vorticity cannot rise without limit, some braking action is required on the lateral boundaries (Munk, 1950[#]) of the ocean cell.

Theoretically, on the eastern sides of the subtropical ocean cells the anticyclonic relative vorticity due to the planetary wind field is nearly balanced by the cyclonic relative vorticity acquired as the water columns move equatorward. This would permit the flow in the eastern parts of the oceans to move as a broad and consequently slow drift, without frictional interaction. The physical existence of this kind of motion is very much in doubt.

On the western sides of ocean basins, however, an intense frictional brake seems to be required, because the anticyclonic relative vorticity of the water provided by the planetary wind field is augmented by the anti-

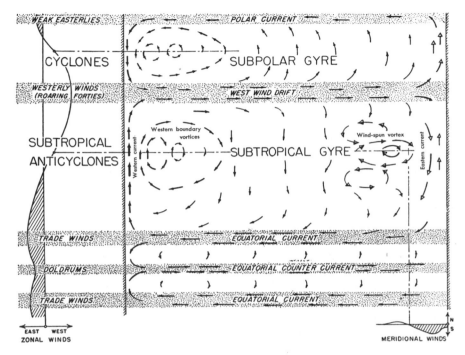

Fig. 6–17. Schematic presentation of circulation in a rectangular ocean resulting from zonal winds (filled arrowheads), meridional winds (open arrowheads), or both (half-filled arrowheads). The width of the arrows is an indication of the strength of the currents. The nomenclature applies to either hemisphere, but in the Southern Hemisphere the subpolar gyre is replaced largely by the Antarctic Circumpolar Current. [From W. H. Munk, 1950, *J. Met.*, 7(2).]

cyclonic spin of vertical water columns moving poleward. On the western side of the ocean the excess of anticyclonic vorticity can be balanced if the poleward flow interacts with the coast to produce a shearing flow with an equal and opposite component of frictionally generated cyclonic vorticity.* This mechanism provides a strong asymmetry of water motion, even though it is driven by a symmetrical wind torque.

From these considerations it becomes plausible that very strong, narrow currents, such as the Gulf Stream in the North Atlantic and Kuroshio in the North Pacific, the Mozambique-Agulhas Currents in the Indian Ocean, and to some lesser extent the East Australian Current in the South Pacific,

---

* Stommel (1948)# has used generalized bottom friction to achieve the same result. It is also conceivable that suitable braking action could be provided by the pressure gradient torque resulting from water being piled by the winds against continents. A head of about 40 cm would be required.

and the Brazil Current in the South Atlantic, should exist and serve as the principal pathways by which tropical water makes its way poleward along the east coasts of the continents. These narrow currents have been called *western boundary currents* by Stommel and his collaborators. As Stommel (1951)[7] has pointed out:

(1) . . . although energy is added to the oceans by the work of the wind over the entire surface, it is dissipated chiefly in the strong western currents; (2) . . . a good representation of the circulation outside of the western currents can be obtained from a knowledge of the field of the wind stress alone, independent of friction. . . .

Stommel's suggestion as to circulation has been illustrated in laboratory model experiments,[8] shown in Fig. 10–10.

The observed circulation

Taken as a whole, the wind-spun vortex of the ocean's circulation is asymmetrical and circulates around a point displaced very considerably to the west of the center of each of the major oceans in the tropical and temperate latitudes of the earth (Figs. 6–18 through 20). On the western side of each ocean the poleward flow tends to become fast and narrow as it moves into close proximity with a solid continental barrier. The arche-type of such boundary currents is the Gulf Stream of the North Atlantic which seems to maintain contact with the coast all the way from the southern tip of Florida to Cape Hatteras in North Carolina. Between Cape Hatteras and the Grand Banks the current is removed from the continent, but develops a strong tendency to meander and cast off eddies every few months. These have been thought to provide frictional braking action by large-scale eddy diffusion.

The transverse profile of velocity in this western boundary current is strongly asymmetrical, being swiftest in a narrow zone somewhat west of the center (Fig. 11–8). On either side of the zone of maximum velocity in a western boundary current, the shear tends to be equal to or greater than the local value of the Coriolis parameter in the cyclonic zone on the left-hand side of the current maximum, but equal to or less than the value of the Coriolis parameter in the anticyclonic zone on the right of the current maximum. This observation is consistent with theoretical expectations.

In the Gulf Stream, the region of cyclonic shear on the western side of the current is approximately 12 to 15 nautical miles wide, while the zone of anticyclonic shear on the right of the current may be in the order of 30 to 40 nautical miles wide, so that the total width of the western boundary current in the North Atlantic may range from 40 to 60 nautical miles, or something like 1/50 of the total width of the ocean.

Studies of the width of the current associated with cyclonic and anti-cyclonic meanders indicate that in cyclonic curvatures the flow tends to

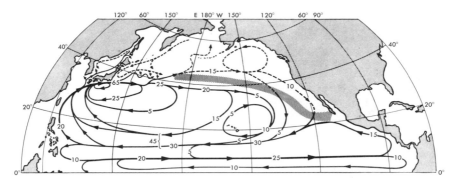

Fig. 6–18. Transport chart of the North Pacific. The lines with arrows indicate the approximate direction of the transport above 1500 meters. The inserted numbers indicate the transported volumes in millions of cubic meters per second. Dashed lines show cold currents; continuous lines show warm currents. [After H. U. Sverdrup, M. W. Johnson, and R. H. Fleming, 1942, *The Oceans, Their Physics, Chemistry, and General Biology*, New York: Prentice-Hall.]

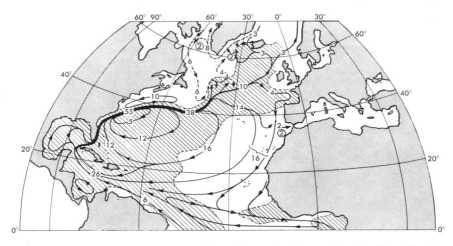

Fig. 6–19. Transports of water above the main thermocline in the Atlantic Ocean. The lines with arrows indicate the direction of the transport. The inserted numbers indicate the transported volumes in millions of cubic meters per second. Dashed lines show cold currents; continuous lines show warm currents. Areas of positive surface temperature anomaly are shaded. [After H. U. Sverdrup, M. W. Johnson, and R. H. Fleming, 1942, *The Oceans, Their Physics, Chemistry, and General Biology*, New York: Prentice-Hall.]

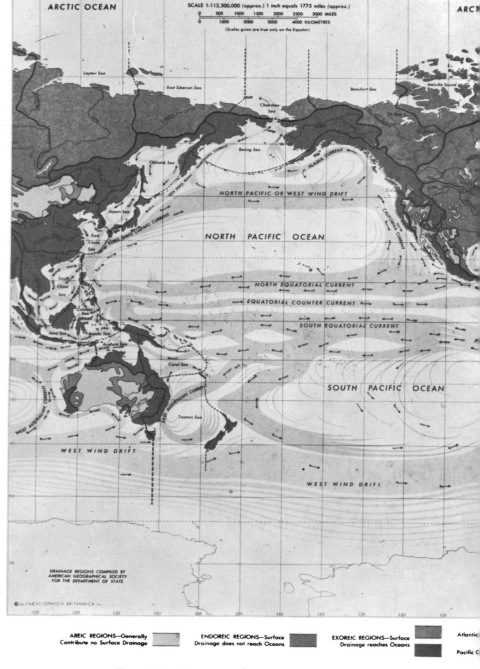

Fig. 6–20. Principal ocean currents and drainage regions of the wo.

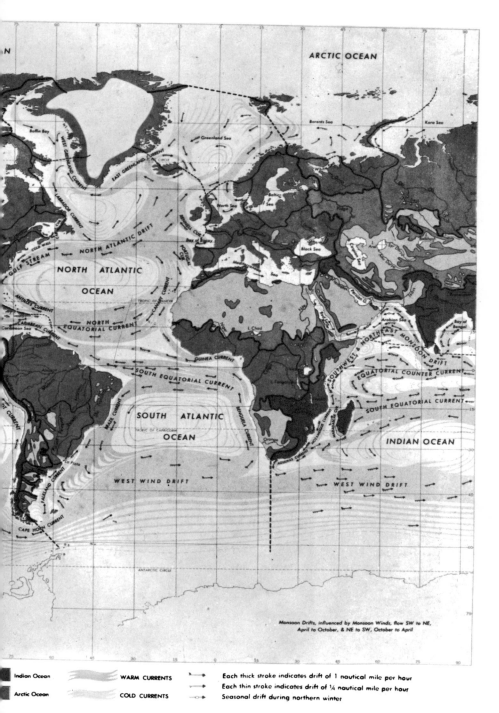

m *Encyclopaedia Britannica World Atlas,* Chicago.]

be narrower and somewhat deeper and faster than in anticyclonic curvatures. This result is different from that to be expected from the meander flow equation when the transverse pressure gradient is assumed to be the same in each sense of curvature.

The geostrophic transport of water by the Gulf Stream is greatest near Cape Hatteras—about 70×10^6 m^3/sec—and ranges from about 20×10^6 to 40×10^6 m^3/sec in the Florida Straits. Rossby's "wake stream" hypothesis (1936)[#] deals with the question of augmented transport, but even he would doubtless agree that a really compelling explanation of this increase in volume transport with distance downstream has still to be proposed.

Successful as the vorticity theory is in accounting for the westward intensification of flow in the major ocean basins, difficulties remain. The westward intensification of surface circulations in the oceans of the Southern Hemisphere is not nearly as conspicuous as that in the North Atlantic and North Pacific. Secondly, there are rather serious oceanographic obstacles to the concept of a slow and broad equatorward drift on the eastern side, especially of the North Atlantic. This simple mechanism for closing the orbit of circulation in the wind-driven layer is incompatible with the observed pattern of water properties along the surface route. Both these dilemmas have been met in the modern conception of thermohaline circulation, which is discussed in Chapter 7.

Equatorial currents

In most places on the earth the wind-driven circulation is influenced by the deflecting force of the earth's rotation. At the equator, however, the horizontal component of the deflecting force vanishes. The sense of the Coriolis deflection reverses upon crossing the equator, and the magnitude of the effect rapidly grows stronger with increasing latitude. These circumstances give rise to two regimes of ocean circulation under the trade winds which are peculiar to the equatorial regions.

In the doldrums between the northeast and southwest trade-wind belts, the wind is predominantly from the east but considerably weaker than either of the trades. Water driven westward by the trades tends to slope up against the east coasts of continents.[*] Most of this water flows westward as equatorial currents and thence poleward into the Northern and Southern Hemispheres,[9] but some drains back toward the east in the doldrums. This drainage forms the equatorial countercurrents, which are well developed some five degrees north of the equator in the Atlantic[10]

---

[*] R. B. Montgomery estimates the slope to be 4 cm/1000 km in the Atlantic Ocean. See. R. B. Montgomery, 1940, *Bull. Amer. Met. Soc.*, 21: 87–94.

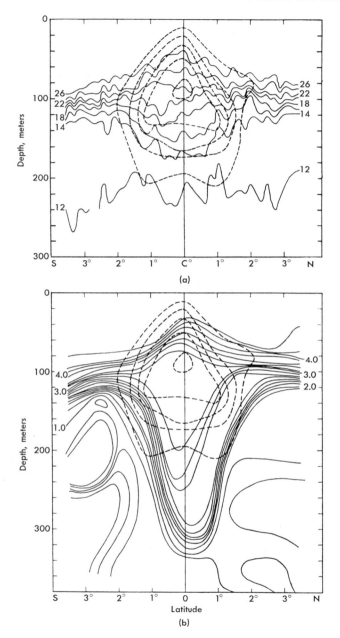

Fig. 6–21.   Velocity cross section (dashed lines) superimposed on (a) temperature and (b) oxygen sections through the Cromwell Current at longitude 140°W. [From J. A. Knauss, 1960, *Deep-Sea Res.*, 6(4).]

and Pacific Oceans[11] and south of the equator in the Indian Ocean. The choice of hemispheres is in response to the asymmetry of the atmospheric circulation. The average position of the meteorological equator is found to be approximately five degrees north of the geographic equator in the Atlantic and Pacific Oceans, but some seven degrees south of the equator in the Indian Ocean sector of the atmosphere.

Studies of the Pacific equatorial fishery resulted in the discovery (August, 1952) of another countercurrent (Fig. 6–21) which lies exactly on the equator, with its core of maximum velocity at a depth of about 100 meters (Cromwell, Montgomery, and Stroup, 1954).# The Cromwell Current— so named in memory of its discoverer, Townsend Cromwell—has been traced several thousand miles eastward from the longitude of the Hawaiian Islands to the vicinity of the Galapagos. It is a surprisingly extensive and important feature of the equatorial circulation. Knauss (1959)[12] found the volume of water transported eastward by the Cromwell Current to be about $40 \times 10^6 \, \mathrm{m}^3/\mathrm{sec}$, which ranks it among the major currents of the world ocean. This discovery suggests that the physical geography of even the upper layers of the oceans is still incompletely known.

STUDY QUESTIONS

1. The short wavelength solar radiation at the distance of the earth's orbit is $2 \, \mathrm{cal/cm}^2/\mathrm{min}$, and the earth emits from its surface about $0.25 \, \mathrm{cal/cm}^2/\mathrm{min}$ as long wavelength radiation. How is it that this inequality of radiant fluxes does not cause the atmosphere to accumulate heat at the rate of $1.75 \, \mathrm{cal/cm}^2/\mathrm{min}$?

2. For the range $0.1 \, \mu$ to $40 \, \mu$, show by an absorption-wavelength diagram the principal wavelengths in which the atmosphere is strongly absorbent. Label the absorption bands according to the composition of the atmospheric gases that produce them.

3. Explain what is meant by *continentality*. Relate the continentality of the Asiatic land mass to the wet and dry monsoons, and give their relationships, in turn, to the coastwise currents of the northern Indian Ocean. How did these currents influence sea traffic in Roman times?

4. Even though the Hadley, Ferrel, and polar cells of the meridional atmospheric circulation are now suspected to be of doubtful significance in nature, they do serve to rationalize the existence of three zonal wind belts in each hemisphere. Show the logical connection between these three cells and the surface wind zones, with reference to the conservation of angular momentum in bands of air parallel to latitude circles.

5. The stress of the wind on the sea surface has been ascribed to the difference in pressure between the upwind and downwind slopes of surface waves. Discuss the effectiveness of this process as a function of the wavelength of deep-water waves.

6. Explain the role of the Ekman transport in the maintenance of a geostrophic circulation in the interior of a subtropical barotropic ocean.

7. In simple terms, explain why the wind-driven circulations of a subtropical ocean tend to produce narrow swift currents along their western boundaries. What influence does this effect have on the planetary heat balance?

8. The equatorial countercurrents flow against the local mean zonal wind direction. How, nevertheless, can these ocean currents be considered to be supported by the wind? Distinguish between the equatorial countercurrents and the Cromwell Current.

REFERENCES

1. H. G. Houghton, 1954, *J. Met.*, 11: 1–9.

2. V. P. Starr and R. M. White, 1955, *J. Mar. Res.*, 14: 217–225.

2a. Bryan, K., 1962, *J. Geophys. Res.*, 67(9): 3403–3413.

3. V. P. Starr and R. M. White, 1951, *Quart. J. R. met Soc.*, 77: 215–225.

4. Y. Mintz and G. Dean, 1952, pp. 12–15, *The Observed Mean Field of Motion of the Atmosphere*, Geophysical Research Paper No. 17, Air Force Cambridge Research Center.

5. E. Palmén, 1955, *J. Mar. Res.*, 14: 465–476.

6. G. Veronis, 1956, *Deep-Sea Res.*, 3: 157–177; G. Veronis and H. Stommel, 1956, *J. Mar. Res.*, 15: 43–75.

7. H. Stommel, 1951, *Bull. Amer. Met. Soc.*, 32: 23.

8. W. S. von Arx, 1952, *Tellus*, 4: 311–318; and 1957, Ch. 1, pp. 1–29, *Physics and Chemistry of the Earth*, Vol. 2, L. H. Ahrens, F. Press, K. Rankama, and S. K. Runcorn, editors, London: Pergamon Press.

9. R. O. Reid, 1948, *J. Mar. Res.*, 7: 74–99.

10. R. B. Montgomery, 1938, *Pap. phys. Oceanogr.*, 6(2): 55 pp.

11. N. G. Jerlov, 1953, *Tellus*, 5: 308–314; R. B. Montgomery and E. Palmén, 1940, *J. Mar. Res.*, 3: 112–133.

12. J. A. Knauss, 1960, *Deep-Sea Res.*, 6(4): 265–286.

SUPPLEMENTARY READING

DEFANT, A., 1961, *Physical Oceanography* Vol. 1, London: Pergamon Press.

HOUGHTON, H. G., 1954, "On the Annual Heat Balance of the Northern Hemisphere," *J. Met.*, 11: 1-9.

KRÜMMEL, O., 1907, 1911, *Handbuch der Ozeanographie*, 2 vols., Stuttgart: J. Engelhorn.

MALONE, T. F., editor, 1951, *Compendium of Meteorology*, Boston: American Meteorological Society.

RIEHL, H., 1954, *Tropical Meteorology*, New York: McGraw-Hill.

SCHOTT, G., 1935, *Geographie des Indischen und Stillen Ozeans*, Hamburg: C. Boysen.

SCHOTT, G., 1942, *Geographie des Atlantischen Ozeans*, Hamburg: C. Boysen.

STOMMEL, H., 1951, "An Elementary Explanation of Why Ocean Currents Are Strongest in the West," *Bull. Amer. Met. Soc.*, 32: 21–23.

STOMMEL, H., 1957, "A Survey of Ocean Current Theory," *Deep-Sea Res.*, 4: 149–184.

CHAPTER 7

Convective Processes

The characteristics of climate that are familiar on land are also apparent in the oceans in terms of the salinity and temperature of the sea surface, the presence or absence of sea ice, and in the balance of evaporation and precipitation. The daily and annual variations of temperature and airborne moisture tend to be smaller at sea. This results from the enormous heat storage capacity of the ocean which tends to stabilize atmospheric circulations and properties and acts as a kind of "flywheel" that holds the extremes of daily weather closer to the mean than is the case on land.

The influences of climate on the sea are found in the surface layers as a gentle gradient of temperature from the equator toward the poles, with an increase of σ_t as the surface temperature decreases. This, added to the requirements of planetary heat balance, suggests that there is thermally direct convection in the oceans which somehow, at depth, carries warm surface water poleward and cold water equatorward. It has been argued that since the heat sources and sinks are all at essentially the same geopotential, the meridional convection process cannot be very efficient. This may be the case if the circulation is likened to a belt conveyor, but in recent studies of energy transformation processes in the atmosphere by Starr and his collaborators and of similar mechanisms in the ocean by Stommel, Worthington, and others, a new conception of the mechanisms of meridional exchanges is emerging.

As in the case of the astronomical tide, it is beyond the capabilities of modern understanding to describe the mechanisms of meridional convection from first principles. Studies are based on observations, and simple physical models are conceived to explain these facts. Among the most important are the distributions of marine climates and the rates of evaporative exchange between the oceans and atmosphere.

Ocean climates

Equatorial ocean climates are dominated by convective rainfall from dense cumuliform cloud masses associated with the convergence of the trade winds of the two hemispheres; the trades on either side of the belt of dol-

180

Fig. 7–1.   Stunted trade-wind cumuli. [Photograph by F. C. Ronne.]

Fig. 7–2.   Giant trade-wind cumuli. [Photograph by F. C. Ronne.]

Fig. 7–3.   Sea smoke in Woods Hole harbor. [Photograph by H. Curl.]

drums carry a steady procession of cumulus clouds which on some days are quite numerous and tall and on others are scattered and stunted (Figs. 7–1 and 7–2). Sunlight penetrating between these clouds to the sea surface is the most intense on earth, warming the seas by direct and indirect processes to a depth approaching 100 meters. Off the coast of West Africa, the land-dried trade winds evaporate water from the sea surface, and their stresses produce upwelling from as deep as the 300-meter level. The nutrients in this deeper water produce a notable increase of biological productivity. The evaporation produced by the dry winds causes the sea to become more salt, and the bright sunshine of the region causes the sea to become warmer to the westward. Similarly, water vapor evaporated from the sea is carried upward in the trade cumuli to increasingly higher levels in the west. This interaction of wind and sea produces a gradual westward increase in the level of the trade-wind inversion in the air[1] and in the depth of the main subequatorial thermocline in the sea.[2]

On the poleward margins of the trade-wind zone in the vicinity of the tropics of Cancer and Capricorn, there is a belt of almost perpetually clear skies associated with the subsidence of air in semipermanent cells of high pressure. This is a desert climate at sea corresponding to the desert belt on land near latitude 30° in each hemisphere. Here evaporation exceeds precipitation and the winds are often light or baffling. These high-pressure cells migrate seasonally with the sun, and are most pronounced in the

Northern Hemisphere summer owing to the tendency for low pressure to develop over the heated continents.

Associated with the poleward limits of the subtropical high-pressure cells is the polar front, the principal interface between cold, dry polar air and warm, moist tropical air.   Along this interface, chains of extratropical cyclones form and migrate eastward to produce strong irregularities in the zone of westerly winds.   Cyclonic storms in this zone develop stratiform clouds which shade the sea and produce heavy precipitation, especially in winter.

In the Northern Hemisphere the polar front tends to bend equatorward over the land and extend poleward over the oceans, so that on the east coasts of continents cyclones encounter the westward intensified ocean currents over which they frequently deepen as a result of the added supply of water vapor from the warm water.   In winter these cyclones produce the worst weather to be encountered at sea, barring hurricanes.   Most storms of the Northern Hemisphere decelerate near mid-ocean in the poleward bend of the polar front, and collect in a semipermanent cell of low pressure near Iceland in the North Atlantic and over the Aleutian Archipelago in the North Pacific.   Were it not for the warmth carried into these regions by the west-wind drift currents and the latent heat of condensing water vapor,* these regions would be totally miserable in winter.

As the North Atlantic drift moves eastward to the vicinity of the British Isles, it, together with the release of latent heat in rainfall, supplies such significant amounts of heat that the winter temperatures of the land in those high latitudes are agreeably higher than those at the same latitude on the western side of the ocean at Labrador.   Similar effects occur in the North Pacific, on the coasts of British Columbia and the northwestern United States.

In the Southern Hemisphere there is less evidence for the existence of semipermanent centers of low pressure.   Cyclones and anticyclones alternate in the zone of high-latitude westerly winds, and migrate eastward to form an endless train of high- and low-pressure centers.   The average winds in this area of the earth are the strongest known.   The character of the circulation in the Antarctic Ocean is probably determined in part by the fact that there are few land masses to retard the winds.   Moreover, the ocean temperature is almost circularly symmetrical around the Antarctic continent, so that there are no heat sources or sinks to lock the planetary waves in the atmospheric circulation in any geographically fixed configuration.

---

* Fog occurs when warm, moisture-laden southerly winds cross the surface boundary between the warm water of the subtropical ocean gyre and the edge of the polar sea. Sea smoke (Fig. 7–3) occurs when polar air moves out over the warmer water. Both can make the visual range from a ship's bridge almost zero, but it is sometimes possible to get a clear view above sea smoke from the masthead.

Hobbs (1926)[3] and others have suggested that in the ice-covered polar regions of the earth there exists a semipermanent anticyclonic circulation. Recent detailed studies of the weather in both the Arctic and Antarctic have shown that this explanation of polar circulation is oversimplified, and that fronts and moving depressions exist in this portion of the earth's atmosphere just as they do in the extratropical zone of west winds. Therefore, the meridional circulation over the poles is probably weak in comparison with the zonal circulation, just as it is elsewhere on the earth. The only region where a meridional cell appears to have synoptic significance is in the trade-wind zone, and even here, as study proceeds, the simplicity of the Hadley cell is gradually being replaced by more complicated interpretations of synoptic motion. These involve fronts and easterly waves, as well as distinct differences between the low-level and 500-millibar patterns of motion, just as is the case in the better-known circulation in the zone of westerlies.

Ocean climates are largely determined by the response of the oceans to atmospheric circulation and, in turn, both ocean climate and atmospheric circulation respond to the distribution of solar heat over the earth's surface.

Heat budget of the oceans

The heat content in the oceans does not vary appreciably from year to year, but maintains a gently shifting balance consistent with external sources and demands. Several processes serve to heat the oceans and others to cool them (Table 7–1).

Of these processes the conductive and radiative exchanges with the atmosphere and space are of greatest importance. The heat of conversion of the mechanical energy of winds and ocean currents is generally less than the heat of incoming radiation from the sun and sky by a factor of 10^{-4},

TABLE 7–1

Heating processes	Cooling processes
Radiation absorbed from the sun and sky	Radiation from the sea surface to space
Condensation of water vapor as dew on the sea surface	Evaporation to the atmosphere
Conduction from the atmosphere	Conduction to the atmosphere
Conduction from the sea floor	Conduction to the sea floor
Conversion of mechanical energy into heat	

but locally may be somewhat greater where tides are accompanied by strong frictional retardation, as in some shallow seas. Conduction to the atmosphere is significant where the ocean temperature is higher than that of the atmosphere but not in the reverse case. This difference in conductive efficiency is due to the strongly developed convection that can develop in cool air over a warm ocean, and the marked stability that develops when warm air is chilled by a cold ocean. Conduction may amount to some 10% of the evaporative heat loss of the oceans. Conductive exchanges with the sea floor are thought to be mainly upward, owing to the evolution of radiogenic heat in the earth's crust. According to recent measurements (Revelle and Maxwell, 1952)[#] the flux of heat through ocean sediments in the Pacific is 1.2×10^{-6} cal/cm^2/sec, which is not significantly different from that in continental rocks. While this contribution of heat is too small to have any marked effect on the heat budget of the oceans, it is put into the bottom water where, in the absence of competing effects, it could conceivably have some influence on the circulation or stability of the water column.

By this process of elimination, the important influences are seen to be associated with radiative exchanges, evaporative processes, and the upward flux of heat in conduction to the atmosphere. These effects are given further consideration in sections that follow.[*]

Flux of sensible heat

The flux of sensible heat is directed mostly from the oceans to the atmosphere, but not exclusively so. In middle latitudes of the Northern Hemisphere, particularly over the eastern parts of the North Atlantic Ocean, the heat flux in summer is directed from air to sea. This effect places the region of maximum atmospheric warming by conduction in the western and northern portions of the Northern Hemisphere oceans, mainly in association with the poleward transport of tropical waters in the Kuroshio of the North Pacific and the Gulf Stream of the North Atlantic Ocean. Over the middle-latitude portions of these currents the annual average upward flux of sensible heat may exceed 90 cal/cm^2/day over the Kuroshio, and 120

---

[*] Direct measurements of the heat balance of the oceans are not easily made. Various indirect methods have been developed which give roughly consistent results, but the problem remains in an unsettled state. Summaries of the theory, observations, and results, by W. C. Jacobs and H. U. Sverdrup, are to be found in articles in *Compendium of Meteorology*, 1951, and again by Sverdrup in *The Oceans*, 1942, *The Earth as a Planet*, 1954, G. P. Kuiper, editor, and in *Handbuch der Physik*, S. Flügge, editor, Bd. XLVIII, Geophysik II, 1957. One of the important early studies of this problem of the influences of climate on the distribution of sea-water properties is that by G. Wüst, 1936, *Länderkundliche Forschung*, Festschrift Norbert Krebs: 347–359, which he later expanded in G. Wüst, 1954, *Arch. Met.*, Wien A, 7: 305–328.

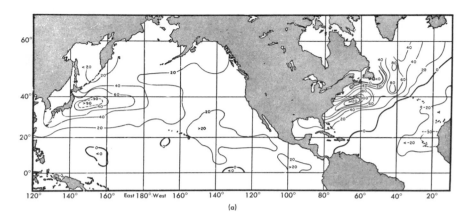

(a)

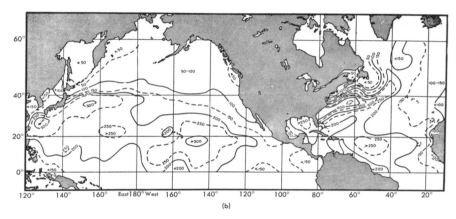

(b)

Fig. 7–4.   (a) The annual values of the rate of exchange of sensible heat between ocean and atmosphere over the North Atlantic and North Pacific, in cal/cm²/day. (b) The annual values of the rate of energy loss from the sea surface through evaporation over the North Atlantic and North Pacific, in cal/cm²/day. [From W. C. Jacobs, 1951, "Large-Scale Aspects of Energy Transformation Over the Oceans," in *Compendium of Meteorology*, T. F. Malone, editor, published by the American Meteorological Society through the support of the Geophysics Research Directorate, Air Force Cambridge Research Center, Air Research Development Command.]

cal/cm²/day over the Gulf Stream (Fig. 7–4). The total energy loss of sensible plus latent heat is about four times as great.

Precipitation, in general, returns sensible but not latent heat to the oceans. Oceanic warming by latent heat can occur only when there is condensation directly on the sea surface as dew, or possibly to some extent in very shallow fogs. In other cases latent heat is released at the level of condensation, and has only the indirect effect of abating radiation losses from the sea surface.

Oceanic radiation

The sea radiates to the atmosphere and space very nearly as a black body and therefore contributes outgoing energy in amounts proportional to the fourth power of its absolute surface temperature. The wavelength of maximum emission for the sea surface is nearly centered on the 10-μ "window" between the absorption bands for atmospheric water vapor and carbon dioxide. Water vapor is, however, such a strong absorber of infrared radiations even in this window that the amount of outgoing radiation from the ocean surface is more closely related to the absolute humidity of the lower air than to the ocean temperature. As the air temperature falls and the absolute humidity of the lower atmosphere is correspondingly decreased, the radiation losses from the ocean tend to increase until a skin of ice is formed. Bubble-free ice is nearly as good a black-body radiator as a free-water surface is, but as soon as bubbles are trapped in sea ice they reflect radiation back into the ocean and reduce the intensity of long wavelength emission to the atmosphere. The latter effect tends to confine heat within the water phase of ice-covered oceans, and indeed explains in part why the ice in the Arctic Ocean is relatively so thin despite the very low air temperatures that sometimes prevail.

The average energy of incoming short-wave radiation from the sun and sky usually exceeds the heat loss through radiative processes from the sea surface. The excess heat in the ocean is first communicated to the atmosphere by the processes of evaporation and conduction, whereupon the atmosphere radiates this energy to space.

Evaporation

Measurements of the amount of water distilled into the atmosphere from the sea surface have been made by direct observations of evaporation from pans of water on shipboard by Mohn (1883),[4] by meteorological measurements of the upward flux of water vapor in the air over the sea by Jacobs (1942),[5] Albrecht (1949),[6] and Sheppard, et al., (1952, 1956),[7] and by evaluating the successive terms in the steady-state heat budget for the oceans proposed by Schmidt (1915).[8] Using Schmidt's method, Mosby (1936)[9] found the average evaporation from the sea surface to be 106 cm/year, a figure that is widely quoted. Wüst's (1936)[10] value, 93 cm/year, from direct measurements in a limited zone of high latitudes, is in substantial agreement considering the degree of uncertainty still attending this kind of measurement.

Schmidt's method is based on the assumption that the annual change of temperature at the sea surface is sufficiently small to consider that all gains and losses of heat must balance from day to day. The important parts of this balance, neglecting advection, can be represented by the equation

$$Q_s = Q_r + Q_e + Q_h, \tag{7-1}$$

where

Q_s = the short-wave radiation received by the oceans
      from the sun and sky,

Q_r = the energy lost by the oceans in long-wave
      radiation,

Q_e = the energy lost by the oceans in evaporation
      at the surface,

Q_h = the heat lost to the atmosphere by conduction.

The quantities Q_s and Q_r are known approximately from measurements of solar radiation at sea and climatological data on cloudiness over the oceans. A further relationship, established by Bowen (1926),[11] permits the *Bowen ratio*, $R = Q_h/Q_e$, of sensible to latent heat lost to the atmosphere, to be determined from meteorological observations provided the sea-surface temperature is also known. Since the amount of water evaporated from the sea surface, E, is defined by Q_e/L, where L is the heat of vaporization of water, it is possible to express the evaporation as

$$E = \frac{Q_s - Q_r}{L(1 + R)}. \tag{7-2}$$

Measurements by this and other methods indicate a greater amount of evaporation in the western reaches of the North Atlantic and North Pacific Oceans, where the Gulf Stream and Kuroshio carry warm water swiftly to high latitudes, than on the eastern sides where the currents are cooler and weak. The zonal maximum of evaporation occurs near 30° north and

TABLE 7–2

EVAPORATION IN MILLIMETERS PER DAY*

Latitude zone, degrees N	Schmidt's method (Mosby)	Meteorological observations (Jacobs)	Ship observations (Wüst)	Area of ocean $\times 10^{16}$ cm^2
50–60	1.03	1.66	1.18	9.3
40–50	1.58	1.81	2.11	13.9
30–40	2.44	3.31	2.90	18.7
20–30	3.69	3.54	3.50	24.1
10–20	3.95	3.68	3.61	26.8
0–10	3.73	3.07	3.20	27.8

* Adapted from H. U. Sverdrup, 1951, p. 1071, *Compendium of Meteorology*, Boston: American Meteorological Society.

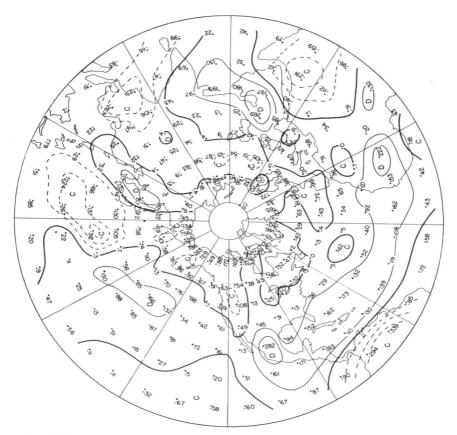

Fig. 7–5. Distribution of the horizontal divergence of the vertically integrated total annual flux of water vapor in the atmosphere during the year 1950. Isopleths are drawn for the equivalents of 100-cm depth of liquid water per year. [From V. P. Starr and J. P. Peixoto, 1958, *Tellus,* 10.]

south latitude, where desert conditions prevail over the sea. From this zone there is a slight decrease of evaporation equatorward and a much greater decrease poleward. Sverdrup (1951) has summarized these zonal measurements, as shown in Table 7–2.

In a fresh approach to this problem, Starr and Peixoto (1958)[12] have computed the vertically integrated divergence of atmospheric water vapor for the year 1950. Since water vapor, like heat, is supplied to the atmosphere from below, the pattern of divergence represents the annual average distribution of sources and sinks of atmospheric water. Figure 7–5 shows the mean annual balance and the effectiveness not only of oceans but of deserts in supplying water to the atmosphere.

Conditioning of surface water

Where the climate is dry, evaporation exceeds precipitation and the salt concentration at the sea surface becomes higher than that in regions of heavy rainfall. Krümmel (1907)[13] and later Wüst (1936, 1954),[14] as well as Wüst, Brogmus, and Noodt (1954),[15] have shown that the salt concentration in the surface layers of the open ocean is, in the mean, directly related to the local balance of evaporation and precipitation, and that this can be regarded mainly as a function of latitude. Figure 7–6 shows some of these results. In a similar way the temperature of the ocean surface is directly related to the amount of sunshine reaching it, but currents carry this warmth to higher latitudes and return colder water in a manner which complicates simple latitudinal dependence. With allowance for the effects of surface motion, the distribution of salt and heat between the equatorial and polar regions of the oceans seems also to be reflected in their vertical structure at low latitudes.

Over the world as a whole, the properties of the surface layer, which has a thickness ranging from 100 meters near the equator to several hundreds of meters in middle latitudes, can be divided into three major subdivisions: the Equatorial water of the mid-tropics, the Central water of middle latitudes, and the subpolar water of high latitudes. In general the coldest climates tend to produce water which is not only cold but somewhat fresher than that from lower latitudes.

At high latitudes, water conditioned at the surface tends to sink as it is cooled and to slide equatorward by some route along or above the ocean floor. Such sinking is only inferred, there being to date no direct measurements of the vertical or horizontal motions involved. It is inferred, however, from the considerable body of biological and chemical information that suggests them, that slow and possibly intermittent vertical motions do actually occur which refresh the deeper layers of the ocean at a density level proportional to the severity of the winter climate,[16] or the excess of surface evaporation over regional precipitation.

Lamination of water masses

In 1814[#] Alexander von Humboldt observed that water is cold at great depths in the tropics, and explained this as a consequence of the sinking of cold water in the polar regions. This possibility has been extended to the point where the distribution of water in the ocean volume has been interpreted in relation to the surface effects of marine climate at all latitudes.

The conditioning of water at the sea surface and its sinking under the influence of climatic stress is thought to be a continuing or repetitive process which can lead to the formation of large masses of each given water type in the volume of the oceans. If renewal of these interior layers of the oceans

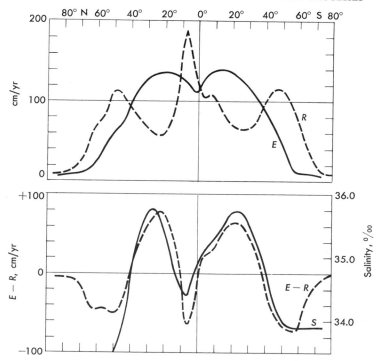

Fig. 7–6. Zonal distribution of rainfall R, evaporation E (above), $E - R$, and salinity S in the surface layer of the ocean (below) (after Wüst). [From G. Dietrich and K. Kalle, 1957, *Allgemeine Meereskünde*, Berlin-Nikolassee: Verlag Gebrüder Borntraeger.]

were not to occur through contact with the atmosphere, the oxygen they contain would eventually become exhausted and, through vertical mixing, their other constituents and properties would become uniformly distributed in the course of time. The fact that there is both a "constancy of composition" and a vertical gradient of salt concentration in sea water offers testimony to the effectiveness of horizontal mixing processes and to the resistance of present oceans to vertical exchanges.

The patterns of T-S correlation can be seen from the temperature-salinity relationships given in Fig. 7–10b for each ocean. The subdivisions have been named partly out of consideration for geographical location and partly to represent the several characteristic and systematic associations of water properties with one another as a function of depth.

It is assumed that all the particles in a given water mass have a similar origin and history. It is also assumed that, except for the mixing of two water types, the density of sea water can be changed only at the surface of the oceans by heating or cooling, by evaporation or precipitation, or by the freezing or melting of ice. Near coasts the influx of river and ground water may have added effects, but these are not far-reaching in terms of the dimensions of the world ocean.

The vertical structure of the ocean, reflecting, as it seemingly does, the climatic influences of each successively higher latitude, results in a layering of water masses which is mainly dependent on temperature. In a meridional section from pole to pole, the structure of the oceans resembles the interlaced fingers of clasped hands.

According to this view it comes as no surprise that sea water of highest density, except for the saline water produced by intense evaporation in the European Mediterranean and Red Seas, is found at the bottom of the Arctic basin. The Arctic mediterranean basin is cut off from the world ocean by the Wyville Thomson and Icelandic Ridges in the North Atlantic and by the shallowness of the Bering Strait at the Pacific entrance. The Arctic basin is subdivided by the Lomonosov Ridge, which extends from Ellsmere Land to the New Siberian Islands. The bottom water of the Beaufort Sea in the Alaskan part of the Arctic basin attains a density of 1.02811 gm/cm^3, while that at the bottom of the part including the Norwegian Sea has an even greater density—near 1.02815 gm/cm^3. These two Arctic basins can communicate above the sill depth, which is at some 2000 meters. The waters below sill depth that remain on the Eurasian side of the Lomonosov Ridge show evidence of adiabatic warming with depth. Waters on the Alaskan side of the Lomonosov Ridge not only show adiabatic gradients but are in general 0.3° to 0.5°C warmer, and correspondingly less dense, than their counterparts at similar depths in the Norwegian Sea.[17] In view of the weak circulation that seems to prevail in the Alaskan part of the Arctic basin, it is conceivable that the slight excess of heat in the Alaskan basin may have been accumulated from the earth's crust. Owing to the physical isolation of these dense waters, they do not contribute significant amounts of bottom water to the world ocean.

This role is played rather conspicuously by the Antarctic circumpolar ocean. The influence of cold climates extends to relatively low latitudes in the Southern Hemisphere, because there are no warm currents with strengths corresponding to the Gulf Stream and Kuroshio of the Northern Hemisphere oceans. This fact tends to encourage the formation of Antarctic circumpolar water, which sinks abundantly during the winter months in the offing of the Weddell and Ross Seas. Winter cooling and ice formation increases the salinity of shelf water to 34.62 $^0/_{00}$, which at temperature —1.9°C produces water with density of near 1.02789 gm/cm^3. This dense water is produced by this and other mechanisms[18] in such great volumes that it extends throughout the bottom layers of the South Pacific, southern Indian Ocean, the South Atlantic, and portions of the North Atlantic Oceans (Fig. 7–7).

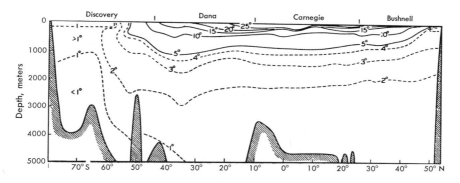

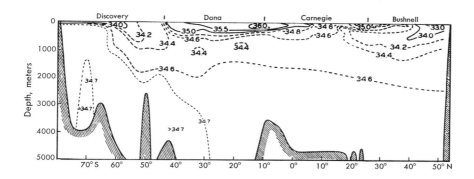

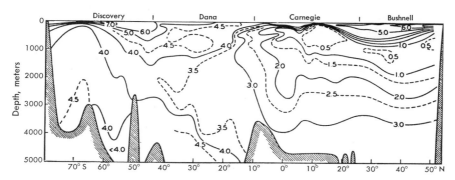

Fig. 7–7.   Vertical sections showing distribution of temperature, salinity, and oxygen in the Pacific Ocean, approximately along the 170° meridian. [After H. U. Sverdrup, M. W. Johnson, and R. H. Fleming, 1942, *The Oceans, Their Physics, Chemistry, and General Biology*, New York: Prentice-Hall.]

Less than correspondingly severe subpolar climates are found in the Irminger and Labrador Seas of the North Atlantic. Here, it is believed that Gulf Stream water of salinity above 35 $^0/_{00}$ mixes with sub-Arctic water of salinity near 34 $^0/_{00}$ carried southward by the East Greenland-Labrador Current system. The mixture of these two water types attains a salinity of about 34.9 $^0/_{00}$, a temperature near 3°C, and a density of 1.02782 gm/cm^3, which is sufficiently high to sink and flood the depths of the North Atlantic basin (Fig. 7–8). Sverdrup (1954)[19] estimates that the average flow toward the south during the year is about 4×10^6 m^3/sec. This volume is evidently sufficient not only to maintain bottom water having these characteristics in the northern North Atlantic but to override the more dense Antarctic Bottom water as the Deep water of both the southern North Atlantic and northern South Atlantic Oceans.

In addition to this strong influence on the deep structure of the Atlantic Oceans, there is a lesser outflow from the European Mediterranean of 2×10^6 m^3/sec of warm (13°C) and highly saline (38.6 $^0/_{00}$) water which flows into the Atlantic basin from the Straits of Gibraltar. Mediterranean water finds its equilibrium depth at about 1200 meters, and spreads out to form a thin tongue of anomalously warm and highly saline water which can be traced, some say, as far south as latitude 50° in the western offing of Africa.*

Deep water is not formed in the North Pacific except for a small contribution from the Sea of Okhotsk. The bulk of Pacific Deep water (of salinity 34.66 to 34.69 $^0/_{00}$, temperature 1.5° to 2.0°C, and density near 1.02775 gm/cm^3) is derived from a mixture of South Atlantic Deep water with Antarctic Circumpolar water, and makes its way into the Pacific by way of the southern Indian Ocean. The sub-Arctic water of the North Pacific formed in the Bering Sea is discharged by the Oyashio at a salinity under 34 $^0/_{00}$ and a temperature near 3.5°C along the northern margin of the Kuroshio extension. This flow tends to collect in the Gulf of Alaska to form an isolated water mass in slow cyclonic rotation under the eastern part of the Aleutian low. The lack of a characteristic bottom water mass in the Pacific has been ascribed to the failure of the Kuroshio to carry water of high salinity to northern latitudes, and to the great excess of precipitation over evaporation in the Bering Sea.

Above the deep and bottom water masses of the oceans lie the intermediate water masses, which are thought to be associated with the subpolar convergences in both hemispheres. The most conspicuous of these, the Antarctic convergence, has been traced all around the Antarctic continent,

---

* An important physical effect related to the discharge of basins like the Mediterranean and Red Seas has been discovered by A. J. Faller, 1960, *Tellus*, 12(2): 159–171.

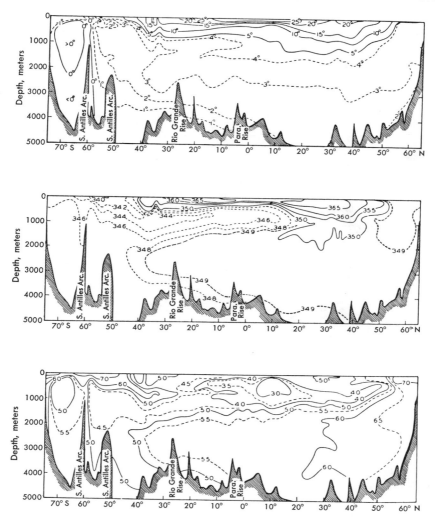

Fig. 7–8.   Vertical sections showing distribution of temperature, salinity, and oxygen in the Western Atlantic Ocean (after Wüst).

on an irregular course between latitudes 48°S and 60°S according to Deacon (1937)[#] and Mackintosh (1946).[20]   Circumpolar water (Fig. 7–9) sinking in the vicinity of the Antarctic convergence forms an Intermediate-water mass having a salinity of 33.80 °/₀₀, a temperature of 2.2°C, and a density of 1.02702 gm/cm³, which spreads northward at depths of from 700 to 800 meters to about 30°N in the Atlantic Oceans but only to 20°S in the Indian and Pacific Oceans.

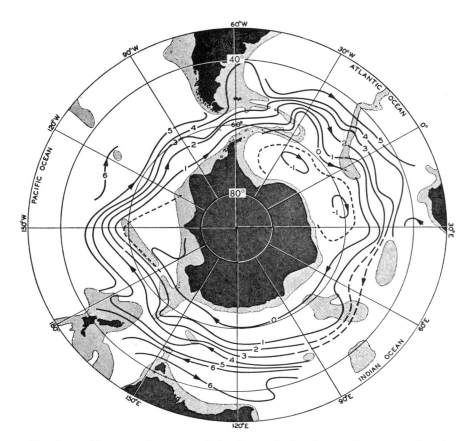

Fig. 7–9.   Transport lines around the Antarctic Continent.   Between each pair of lines the transport relative to the 3000-decibar surface is about 20 million m³/sec. [From H. U. Sverdrup, 1942, *Oceanography for Meteorologists*, New York: Prentice-Hall.]

Laboratory experiments performed to study the wind-driven circulation of the Southern Hemisphere oceans reveal a counterpart of the Antarctic convergence which is not a continuous ringlike feature, but rather consists of two spiral arms. One arm is rooted at the tip of the Palmer Peninsula and extends eastward; the other is rooted near the Wilkes coast of Antarctica and extends eastward nearly to the Drake Strait. The roots of the so-described Antarctic convergence fit the sites of the principal sources of Antarctic Bottom water, namely, off the Ross Sea in the Pacific area and off the Weddell Sea in the Atlantic sector. Were the Antarctic convergence a continuous feature around the Antarctic continent, it would seem less reasonable to suppose that Antarctic water tends to be dispersed from these particular locations.

Nested under the atmospheric subtropical high-pressure cell in each ocean compartment is the central water mass (Fig. 7–10). Owing to the desert conditions and clear air in these latitudes, the central water mass in each ocean tends to develop high salinity and temperature and therefore to be underlain by a strong vertical gradient of salinity and temperature, generally at depths of several hundred meters. The salt concentration in North Atlantic Central water is high, $35.0\ ^0/_{00}$ to $36.5\ ^0/_{00}$ at the surface, and in South Atlantic Central water only a little less so, $34.4\ ^0/_{00}$ to $35.6\ ^0/_{00}$. Indian Ocean Central water is generally fresher than that in the Atlantic, ranging from $34.5\ ^0/_{00}$ at depth to near $35.5\ ^0/_{00}$ at the surface.

The Pacific Ocean system is complicated by differences between the eastern and western halves, as well as between Northern and Southern Hemispheres. North Pacific Central waters are of higher salinity than South Pacific Central waters, and in each case the salinity and temperature decrease with depth at different rates. More than this, the Pacific is dominated by an Equatorial water mass which on the surface is like that of the equatorial Indian Ocean ($35\ ^0/_{00}$ and 15°C). The salinity is higher in the depths of the Indian Equatorial water mass because of the discharge of the Red Sea, which, as a consequence of the large excess of evaporation over precipitation in that region, has a salinity initially near $40\ ^0/_{00}$.

The main thermocline underlying the central water masses in the subtropical oceans is generally deepest under the zone of maximum heating and evaporation. In the North Atlantic the 10°C isotherm, which has been considered to be near the center of the main thermocline layer, is found at a depth near 800 to 900 meters just south of Bermuda, and near the 500-meter level in the central South Atlantic. In the Pacific Oceans, the same isotherm is nowhere deeper than 500 meters. Above the main thermocline in the North Atlantic there is an extensive body of water having a temperature of 18°C and a salinity $36.5\ ^0/_{00}$ which underlies the seasonal thermocline.

The tropical thermocline is everywhere shallowest near the position of the meteorological equator. For the Atlantic Ocean system, the *Meteor* observations (Figs. 7–11 and 7–12) reveal a second shoaling of the main thermocline under the geographic equator as well, perhaps as a result of dynamical causes since here the Coriolis parameter falls to zero. In subpolar latitudes the main thermocline becomes indistinct owing to the very small vertical gradient of temperature, and for practical purposes can be considered to have reached the surface along with the 10°C isotherm. Polar water remains liquid at temperatures well below 0°C as a result of the depression of the freezing point by dissolved salts.

Some 93.3% of the ocean volume is colder than 10°C, and 76.5% is colder than 4°C, the temperature at which fresh water reaches maximum

198

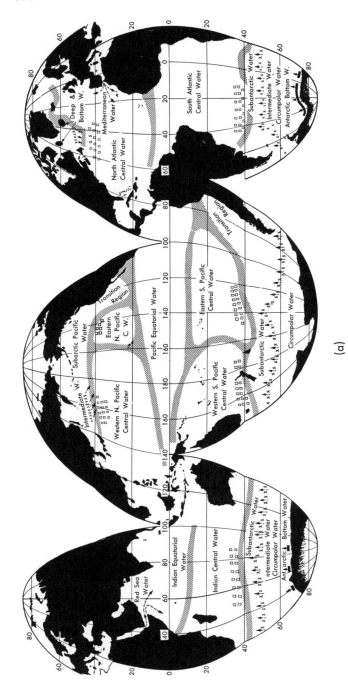

(a)

Fig. 7-10. (a) Approximate boundaries of the upper water masses of the ocean. Squares indicate the regions in which the central water masses are formed; crosses indicate the lines along which the Antarctic and Arctic intermediate waters sink. (b) Temperature-salinity relations of the principal water masses of the Indian, Pacific, and Atlantic oceans. [After H. U. Sverdrup, M. W. Johnson, and R. H. Fleming, 1942, *The Oceans, Their Physics, Chemistry, and General Biology*, New York: Prentice-Hall.]

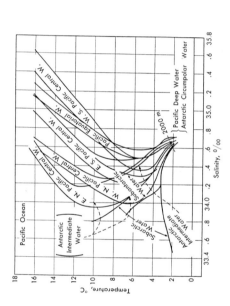

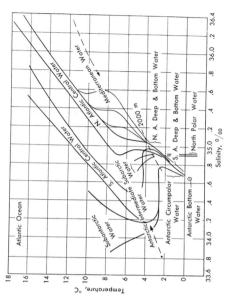

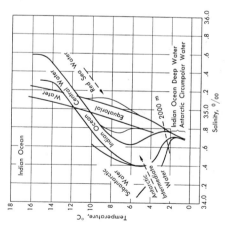

(b)

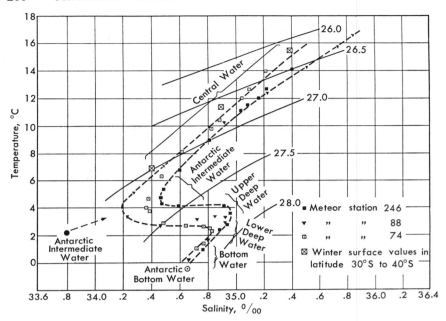

Fig. 7–11.   Temperature and salinity of the water masses of the Atlantic Ocean, as derived from *Meteor* stations 74, 88, and 246. [After H. U. Sverdrup, M. W. Johnson, and R. H. Fleming, 1942, *The Oceans, Their Physics, Chemistry, and General Biology*, New York: Prentice-Hall.]

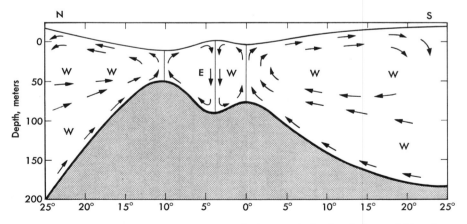

Fig. 7–12.   Schematic representation of the vertical circulation within the equatorial region of the Atlantic. The main direction of the currents is indicated by the letters W and E. The water below the discontinuity surface, which is considered to be at rest, is shaded. (After Defant), [adapted from H. U. Sverdrup, 1942, *Oceanography for Meteorologists*, New York: Prentice-Hall.]

density. The temperature of maximum density for sea water varies with its salt content. When its salinity is 24.70 $^0/_{00}$, the maximum density point falls at the freezing point, $-1.332°C$. In most sea water the temperature at which maximum density would be attained is below the freezing point. This fact considerably simplifies the analysis of hydrostatic layering.

Climatic overturning

The relationship between the layered structure of ocean water masses and climate implies that there is some form of continuous or intermittent meridional circulation. This convective circulation should be limited to single hemispheres in some oceans, but permit communication between hemispheres in others. For example, there seems to be a more voluminous exchange of water between the Northern and Southern Hemispheres in the Atlantic Ocean than in the Pacific or Indian Oceans. A physical explanation of this exchange in the Atlantic Ocean system may be found in the northward transport of surface water across the equator as a result of the trade winds of both hemispheres driving water against the inclined northern boundary of the South American continent. Water driven by the wind into the North Atlantic basin must return to the south by another route, presumably at depth. There is biochemical evidence to support this view.

The North Atlantic Ocean tends to be poorer in phosphate (Fig. 7–13) and richer in oxygen at intermediate depths than is the North Pacific. Redfield (1958)[21] suggests that this is explained by the fact that the biological populations in the northward-moving surface waters of the South Atlantic steadily deplete the supply of phosphate, and populations die before the water is carried into the North Atlantic. The rain of decaying organisms into the depths of the South Atlantic consumes oxygen, and the phosphate released at depth is carried southward in the subsurface return flow. The biological populations of the Atlantic Ocean system are considered to be encouraged by the high fertility of Antarctic surface waters but to fall off on the way northward as though the organisms were traveling on a badly perforated belt conveyor. The oxygen concentration in the southward-flowing water below mid-depths remains high and its phosphate concentration low in comparison with that of the North Pacific because of its origin at the surface in high latitudes. Conditions in the southern Pacific Ocean, Indian Ocean, and southern Atlantic Ocean are more nearly comparable with each other, biologically and chemically.

The physical motions associated with climatological overturning of the oceans are obscure at best, and may be either intermittent or so slow that they are more aptly classed as processes of eddy diffusion. It may also be that both the vertical and horizontal motions are developed as highly localized streams. The latter possibility is discussed in the next section.

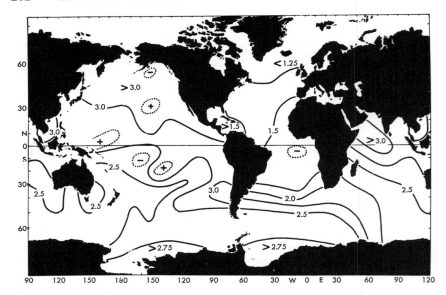

Fig. 7–13.   Distribution of phosphorus at 2000-meter depth in the oceans of the world.   Contour interval is 0.25 mg atoms/m³.   This diagram is based on some 1600 measurements between depths of 1900 and 2100 meters.   Data are available from about 75% of the 10° squares.   About half the observations fall within the designated contour interval, the remainder within adjacent intervals.   [From A. C. Redfield, 1958, *Amer. Scientist*, 43(3).]

There are at present no satisfactory methods for measuring vertical motions in the sea which would permit direct observations of convective overturning.   However, water moving horizontally along the very bottom of ocean basins might tend, very gradually, to accumulate heat from the earth's interior.

The flux of heat through the sea floor has been measured by Revelle and Maxwell (1952)[#] with the aid of a thermometric probe driven into sediments at various points in the Pacific basin.   The average flux of heat is found to be directed upward at the rate of 1.2×10^{-6} gm·cal/cm²/sec, which does not differ significantly from the average of nearly six-dozen measurements in bore holes and mines on the land which have a mean of 1.23×10^{-6} gm·cal/cm²/sec.   Bullard (1954)[22] found a mean value of 1.0×10^{-6} gm·cal/cm²/sec for the Atlantic.   If we assume the average is somewhere near 1.2×10^{-6} gm·cal/cm²/sec over the whole surface of the earth, the total amount of energy conducted outward from the interior of the earth is in the vicinity of 6×10^{12} gm·cal/sec.   This is about 1/25,000 of the average heating rate provided by the sun.

The production of heat in the sea floor and the attendant slow heating of the bottom water masses have been used to measure the rate of flow of deep water across the ocean floor.[23] Some of these estimates suggest that roughly one thousand years may be required to return bottom water to the surface layers, thus completing one convective cycle. This assumes, of course, that the motion of the deep water is uniform over the surface of the earth.[24]

In other respects, the horizontal motion of the deep layers must be partly tidal and partly barotropic responses to meteorological pressure changes which extend from the sea surface to the very bottom. The tidal velocities in the open sea are necessarily quite small, amounting to perhaps a few centimeters per second or less, but are probably large in comparison with the mean motion. Various estimates of the rate of mean motion in the abyssal circulation have been made on the basis of continuity. These estimates range from millimeters of displacement per day to perhaps an order of magnitude larger, based on direct measurements made during the *Meteor* expedition.

It was not until 1957, when J. C. Swallow used neutrally buoyant free-floats to trace the motions in the deep layers of the sea, that the magnitude of the motions could be more fully appreciated as being composed of both elliptical components presumably related to the tide and rather incoherent displacements that occurred at rates quite surprisingly in the order 1 to 10 cm/sec and more. *

Thermohaline motions

The relative importance of wind torque and differential heating in the maintenance of the total circulation in the oceans is conjectural. Thermal and evaporation-precipitation balances seem, in some ways, to serve as a driving force nearly as well as winds. Indeed, it was C.-G. Rossby's contention that the nature of the forces driving the oceans is of far less consequence than their distributions.

If the overturning circulations of the oceans are maintained by the thermal influence of the sun, the resulting motions would be upward in low latitudes and downward in high latitudes. If, on the other hand, the evaporation in low latitudes exceeds that in high latitudes, the reverse sense of motion would occur. Since there are insufficient data to determine whether it is the thermal or haline mode of overturning that predominates, the entire process is somewhat ambiguously called the *thermohaline circulation*.

---

* The first few observations were made in the offing of Gibraltar and Bay of Biscay from *Discovery II* of the National Institute of Oceanography. Later observations made in the vicinity of Bermuda show that the motion in the deep layers is often in the range 10 to 30 cm/sec in the western parts of the North Atlantic basin. See J. C. Swallow, 1957, *Deep-Sea Res.*, 4: 93–104. Recent observations in the same area have shown that the motion at depth tends to have a similar direction to that near the surface and occasionally reaches speeds as great as 30 to 40 cm/sec.

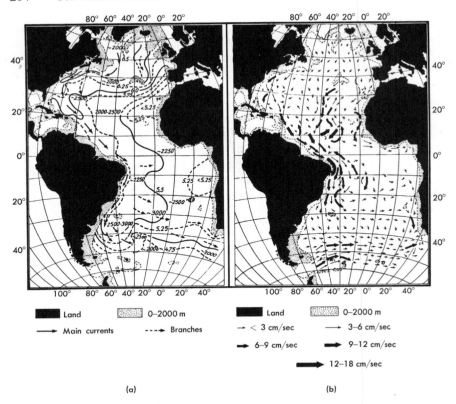

Fig. 7-14. (a) Flow of the middle North Atlantic deep water, as represented by the oxygen content in layers of maximum oxygen (after Wüst). (b) Current flow at the 2000-meter depth based on dynamical computations (after Defant). [From G. Wüst, 1951, *Naturwissenschaftliche Rundschau*, 3.]

During glacial times the concentration of salt in the oceans must have been nearly 5% higher than it is at present, owing to the fresh water that was extracted from the sea to form the continental ice sheet. Much of this water was probably distilled from the sea surface in low and middle latitudes so as to cause the haline circulation to be more strongly opposed to the thermal circulation than is the case at present. The formation of brines under sea ice in high latitudes probably would not counteract the effects of rapid distillation in lower latitudes. The total effect may have been to depress the main subtropical thermocline and to constrict the dimensions of the subtropical gyre of wind-driven circulation as the atmosphere polar front moved equatorward with the advancing ice front. The effects of recovery from a glacial regime are more difficult to imagine because the latitude zone of principal meltwater discharge is unknown, but a freshening of the surface layer may have been quite noticeable in coastal regions adjacent to the mouths of glacial rivers.

Until only very recently there has been no satisfying theory of the global ocean circulation. Suggestions had been made that in addition to a general rotation of the surface layers with the prevailing winds, the oceans overturned very slowly in a thermally driven meridional cell extending symmetrically from the equator toward each pole. Various evidences were cited to support this view, but none of the suggestions carried the weight of firm conviction. It had also been recognized that the circulations in the oceans of the Southern Hemisphere, particularly the South Atlantic and South Pacific, are far less intense on their western sides than their counterparts in the Northern Hemisphere, even though the zonal winds of the Southern Hemisphere are stronger than those in the Northern Hemisphere.

In a study of the *Meteor* observations, Georg Wüst (1955)[25] found, as shown in Fig. 7–14, that the principal meridional currents in the South Atlantic Ocean are to be found against the margin of South America, more or less as narrow, relatively swift water streams hugging the coast in a fashion similar to the westward intensified surface circulations so well known from studies of the Gulf Stream and Kuroshio. In these studies Wüst calculated that Atlantic Bottom water made its way northward as a narrow current in the depths of the western basin of the South Atlantic, with a volume transport of 2×10^6 m^3/sec. He also found that at a somewhat shallower depth, North Atlantic Deep water made its way southward with a volume transport of 27×10^6 m^3/sec, while above the Deep water sub-Antarctic Intermediate water flowed northward with a volume transport of 7×10^6 m^3/sec. This picture was rather different from the interpretation that had been held previously in which the intermediate, deep, and bottom water layers were supposed to move northward and southward as continuous sheets, with uniform velocity across the full width of each ocean basin.[26] The velocities calculated by Wüst were on the order of centimeters per second rather than the millimeters per day derived earlier. This result, together with evidence from other expeditions, led Stommel (1957)[27] to seek and find a physical explanation for intensified circulations adjacent to the western boundaries at depth (Figs. 7–15 and 7–16).

This explanation takes its clues from the shape of the oceanic thermocline. The main thermocline, it will be remembered, lies deepest in the vicinity of the mid-latitude high-pressure cells of the atmosphere, is shallow under the equatorial low-pressure cell, and rises toward the surface in the vicinity of the maritime low-pressure centers clustered around the polar caps. On the poleward side of these low-pressure centers, the main thermocline is essentially nonexistent, owing, it is thought, to the tendency for the severe climate in these regions to produce significant amounts of high-density water which sinks to the sea bottom.

From continuity considerations, it may be expected that, if sinking occurs in high latitudes, there may be slow rising motions everywhere else in the

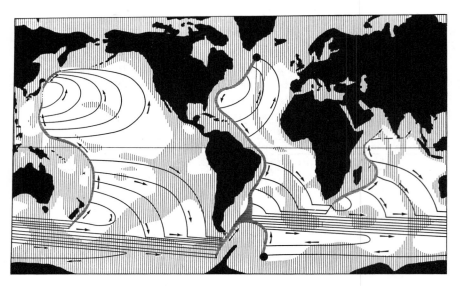

Fig. 7–15. The deep circulation of the world ocean according to Stommel. [From H. Stommel, 1957, *Deep-Sea Res.*, 4.]

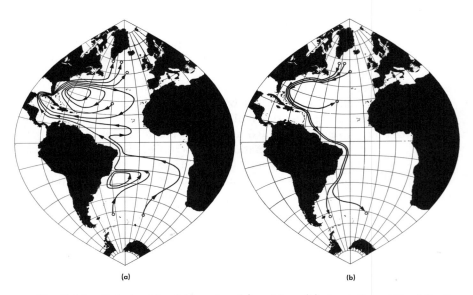

(a)   (b)

Fig. 7–16. Relationship of the surface (a) and deep (b) circulations of the Atlantic Ocean according to Stommel. [From H. Stommel, 1957, *Deep-Sea Res.*, 4.]

world ocean. This rising motion has been regarded by Robinson and Stommel (1959)[28] as sufficient to arrest the downward eddy diffusion of heat from the surface layers, and indeed to oppose the sinking of warm surface water produced in association with the Ekman convergences in middle latitudes. According to this view, the main thermocline provides a floor at well-known levels beneath the wind-driven circulation, and a ceiling above the circulation of water in the abyss.

In this theory it is assumed that the motion in all parts of the interior of the oceans, except in the Ekman layer, is geostrophic and that the relative vorticity is negligibly small, so that $f/D = $ a constant. It is also assumed that the western boundary current can develop at any depth and can be employed in the ocean circulation processes to maintain continuity.

From these assumptions it can be reasoned that columns of water fed into the abyss in high latitudes are stretched vertically by the slow diffusion required to hold the main thermocline in place in other parts of the ocean. With stretching there is a requirement, if potential vorticity is to be conserved, that the columns move into higher latitudes where the value of f is as much greater as is the increase of D. But there are barriers to flow in this direction in both hemispheres. Alternatively, then, the deep current turns equatorward, and moves as a narrow stream, frictionally coupled to the western boundary of the ocean basin. Flow of this sort can cross the equator if continuity so requires. With change of latitude the columns tend to lose mass in lateral (mainly zonal) geostrophic flows which supply the gentle upward flow required to support the main thermocline. These processes are not intuitively obvious, but the patterns of motion predicted from theory have been shown to be physically possible in a series of laboratory experiments recently conducted by A. J. Faller[29] (Figs. 7–17 and 7–18).

With this encouraging result to depend upon, Stommel has extended these ideas to suggest a physically consistent hypothetical circulation at abyssal levels throughout the world ocean in a manner that is also compatible with the wind-driven circulation above the main thermocline. In this remarkable study one of the two principal sources of abyssal water is situated in the region of climatic subsidence in the Labrador and Irminger Seas. This subsidence gives rise to a narrow deep current flowing along the east coast of North and South America to the latitude of the southern subtropical convergence. Since the poleward flow in the western South Atlantic is strong at depth, the western boundary (Brazil) current of the wind-driven surface circulation is not necessarily as strong as its counterpart (the Gulf Stream) in the North Atlantic, where the density-induced deep-current flow is opposed to the wind-driven circulation.* Upon reaching

---

* A strong western countercurrent at depth can also be accounted for in a wind-driven ocean. See E. M. Hassan, 1958, *Deep-Sea Res.*, 5: 36–43.

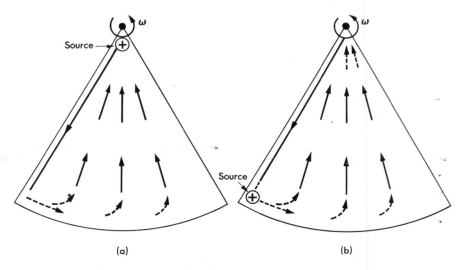

(a)   (b)

Fig. 7–17.   (a) Sketch of the flow pattern to be expected with a source of mass ⊕ at apex of sector and with the surface of the fluid rising uniformly.   Dashed arrows indicate those portions of the flow required by continuity.   (b) Sketch of the flow pattern to be expected with a source of mass ⊕ at the western corner of the basin and with the surface of the fluid rising uniformly. [From H. Stommel, A. B. Arons, and A. J. Faller, 1958, *Tellus* 10(2).]

Fig. 7–18.   Verification of the experiment suggested in Fig. 17–7 (a). The mass added at ⊕ is tinted with ink. [From H. Stommel, A. B. Arons, and A. J. Faller, 1958, *Tellus* 10(2).]

the subtropical convergence of the Southern Hemisphere, the deep current is presumed to turn eastward, cross the Indian Ocean, and contribute South Atlantic Deep water to the Pacific Deep water mass.

The sinking of waters in the offing of the Weddell Sea is regarded as a second source for a second western boundary current in the South Atlantic which flows northward to the latitude (40°S) of the subtropical convergence to meet the southbound flow. At this juncture, in the vicinity of the LaPlata estuary, Antarctic Intermediate water is mixed with South Atlantic Deep water. After joining, the two boundary currents make their way across the South Atlantic and Indian Oceans to supply the Pacific Ocean with its Deep water.

According to this model, the western boundary currents on the surface are weak or strong according to whether or not flow in the deep western boundary current aids or opposes their motions. In the North Atlantic, the deep current is imagined to flow in the direction opposite to the Gulf Stream, so that the surface transport toward the north must be strong. In the North Pacific, the deep current is imagined to move with the Kuroshio to about the latitude of Japan, but is thereafter opposed by some equatorward motion from the north. This would explain why the Kuroshio may be somewhat less vigorous than the Gulf Stream. Several analyses have shown this to be the case even though the North Pacific is more than twice as wide and the winds are nearly as strong as those crossing the North Atlantic. In the South Atlantic, the transport of the Brazil Current is thought to be aided by the transequatorial deep current from the North Atlantic, so that it may be expected to be indefinite and weak. This fits Wüst's analysis and is in accord with surface observations.

The thermohaline model of deep circulation places strong emphasis on the circumpolar circulation of the Antarctic Ocean. Clowes (1933)[30] and later Deacon (1937)# estimated the maximum zonal transport of the Antarctic circumpolar current to be as great as 120×10^6 m^3/sec, which is the most voluminous transport to be found in any current on earth. This is accommodated in Stommel's model as the principal avenue of exchange between the major ocean basins.

Seeking data for a further comparison of this model with nature, Swallow and Worthington (1957)[31] found evidence, using both dynamical and direct[32] methods, that the hypothetical deep flow southward under the Gulf Stream is probably present, and is significantly swift in the vicinity of the Blake Plateau. If this current extends, as Stommel suggests,[33] across the equator into the western reaches of the South Atlantic Ocean, we have for the first time a satisfying explanation of why the volume transport of the Brazil Current is so small in comparison with that of the Gulf Stream. That is, the poleward transport is shared between the bottom and surface currents in the South Atlantic, as originally suggested by Wüst

(1955),[34] while it occurs in the surface layers of the North Atlantic, as is now amply shown in the dynamical observations by Worthington.

As a consequence of these recent thoughts, experiments, and observations, we are left with the idea that the general circulation of the oceans consists of a wind-driven gyre under each pair of the major zonal wind belts, and a global pattern of thermally excited transport systems which work in the same sense as the wind-driven circulations in the Northern Hemisphere but more generally oppose them in the Southern Hemisphere. The details of this interaction are still obscure, but the broad pattern fits the observed motions of the sea fairly well, and is supported by a physically consistent dynamical theory.

STUDY QUESTIONS

1. What factors make the atmosphere more effective than the oceans in producing a poleward flux of heat?

2. The oceans evaporate about 1 meter of water per year. Calculate the average fraction of solar energy incident on the sea surface that is expended in the production of water vapor.

3. What processes cause the flux of sensible and latent heat between the oceans and the atmosphere to be mainly upward? Under what conditions can heat be transferred from the atmosphere to the oceans?

4. Calculate, to an order of magnitude, the total heat produced if the primary ocean circulation were suddenly brought to a standstill relative to the earth. Do the same for the atmosphere. Compare these results with the total heat each receives from the sun each day. (Take the mean speed of the ocean to be 10 cm/sec, that of the atmosphere to be 10 m/sec, and the mass of the atmosphere to be 10^{22} gm. Recall that $J = 4.2 \times 10^7$ ergs/cal.)

5. Using comparable units compare the ratio of the kinetic energies of the oceans and atmosphere, respectively, to the energy of latent heat resident in atmospheric water vapor. (Assume that the atmosphere contains 3 gm of water vapor above each square meter of the earth's surface.)

6. Show by means of a diagram of the earth how the distribution of salinity and temperature in the surface layer of the oceans is related to the climatological mean circulation and transparency of the atmosphere.

7. Draw a meridional section of the North and South Atlantic Oceans showing the distribution of water masses, and the position of the main thermocline. What dynamical arguments have recently provided some insight into the processes which maintain the thermocline in its present configuration?

8. Schmidt's method has been used as an indirect means for estimating evaporation from the sea. What are its basic assumptions and procedures?

9. It has often been suggested that radiogenic heat from the earth's crust might produce a measurable change in the temperature of bottom water which, in turn, could be interpreted as a measure of its rate of horizontal motion and of the rate of climatic overturning. How would the existence of western boundary currents in bottom water alter the prospects of success for this method, and how would they change the expected pattern of horizontal temperature gradients?

REFERENCES

1. J. S. Malkus, 1958, *Pap. phys. Oceanogr.*, 8(2): 47 pp.

2. R. B. Montgomery, 1938, *Pap. phys. Oceanogr.*, 6(2): 55 pp.

3. W. H. Hobbs, 1926, *Glacial Anticyclones*, New York: Macmillan.

4. H. Mohn, 1883, pp. 135–150, *Meteorology* (*The Norwegian North-Atlantic Expedition 1876–1878*, Vol. II), translated by John Hazeland, Christiania: Grøndahl & Son.

5. W. C. Jacobs, 1942, *J. Mar. Res.*, 5: 37–66.

6. F. Albrecht, 1949, *Ann. Met., Hamburg*, 2: 129–143.

7. P. A. Sheppard, H. Charnock, and J. R. D. Francis, 1952, *Quart. J. R. met. Soc.*, 78: 563–582; H. Charnock, J. R. D. Francis, and P. A. Sheppard, 1956, *Phil. Trans.* A, 249: 179–234.

8. W. Schmidt, 1915, *Ann. Hydrogr., Berlin*, 43: 111–124, 169–178.

9. H. Mosby, 1936, *Ann. Hydrogr., Berlin*, 64: 281–286.

10. G. Wüst, 1936, *Länderkundliche Forschung*, Festschrift Norbert Krebs: 347–359.

11. I. S. Bowen, 1926, *Phys. Rev.*, 2, 27: 779–787.

12. V. P. Starr and J. Peixoto, 1958, *Tellus*, 10: 188–194.

13. O. Krümmel, 1907, p. 334, *Handbuch der Ozeanographie*, Vol. I, Stuttgart: J. Engelhorn.

14. G. Wüst, 1936, *Länderkundliche Forschung*, Festschrift Norbert Krebs: 347–359; 1954, *Arch. Met.*, Wien A, 7: 305–328.

15. G. Wüst, W. Brogmus, and E. Noodt, 1954, *Kieler Meeresforsch.*, 10: 137–161.

16. L. V. Worthington, 1954, *Deep-Sea Res.*, 1: 244–251.

17. L. V. Worthington, 1953, *Trans. Amer. geophys. Un.*, 34: 543–551.

18. N. P. Fofonoff, 1956, *Deep-Sea Res.*, 4: 32–35.

19. H. U. Sverdrup, 1954, Ch. 5, pp. 215–257, *The Earth as a Planet*, G. P. Kuiper, editor, Chicago: University of Chicago Press.

20. N. A. Mackintosh, 1946, pp. 177–212, *Discovery Reports*, Vol. XXIII, Cambridge: University Press.

21. A. C. Redfield, 1958, *Amer. Scient.*, 46: 205–221.

22. E. Bullard, 1954, *Proc. Roy. Soc.*, A, 222: 408–429.

23. W. S. Wooster and G. H. Volkmann, 1960, *J. Geophys. Res.*, 65(4): 1239–1249.

24. K. F. Bowden, 1954, *Deep-Sea Res.*, 2: 33–47.

25. G. Wüst, 1955, *Deep-Sea Res.*, Suppl. 3: 373–397; 1957 *Stromgeschwindigkeiten und Strommengen in den Tiefen des Atlantischen Ozeans* (*Wissenschaftliche Ergebnisse der Deutschen Atlantischen Expedition auf dem Forschungs-und Vermessungsschiff "Meteor" 1925–1927*, Bd. 6(2, 6) Berlin: Walter de Gruyter & Co.

26. A. Merz, 1925, *S. B. preuss. Akad. Wiss.*, 31: 562–586; A. Defant, 1930, *Die Ozeanische Zirkulation*, Madrid: Gráficas Reunidas, S. A.

27. H. Stommel, 1957, *Deep-Sea Res.*, 4: 149–184.

28. A. Robinson and H. Stommel, 1959, *Tellus*, 11: 295–308.

29. H. Stommel, A. B. Arons, and A. J. Faller, 1958, *Tellus*, 10: 179–187.

30. A. J. Clowes, 1933, *Nature, London*, 131: 189–191.

31. J. C. Swallow and L. V. Worthington, 1957, *Nature, London*, 179: 1183–1184.

32. J. C. Swallow, 1955, *Deep-Sea Res.*, 3: 74–81.

33. H. Stommel, 1958, *Deep-Sea Res.*, 5: 80–82.

34. G. Wüst, 1955, *Deep-Sea Res.*, Suppl. 3: 373–397.

SUPPLEMENTARY READING

BROOKS, C. E. P., 1949, *Climate Through the Ages*, 2nd ed., New York: McGraw-Hill.

BYERS, H. R., 1944, *General Meteorology*, 2nd ed., New York: McGraw-Hill.

FALLER, A. J., 1960, "Further Examples of Stationary Planetary Flow Patterns in Bounded Basins," *Tellus*, 12(2): 159–171.

FUGLISTER, F. C., 1960, *Atlantic Ocean Atlas of Temperature and Salinity Profiles and Data from the International Geophysical Year of 1957–1958*, Woods Hole Oceanographic Institution Atlas Series, Volume 1, Woods Hole, Mass.

MALONE, T. F., editor, 1951, *Compendium of Meteorology*, Boston: American Meteorological Society.

MILANKOVITCH, M., 1936, "Mathematische Klimalehre und Astronomische Theorie des Klimaschwankungen," *Handbuch der Klimatologie*, W. Köppen and R. Geiger, editors, Bd. I, Teil A, Berlin: Gebrüder Borntraeger.

MONTGOMERY, R. B., 1958, "Water Characteristics of Atlantic Ocean and of World Ocean," *Deep-Sea Res.*, 5: 134–148.

ROSSBY, C.-G., 1941, "The Scientific Basis of Modern Meteorology," Part 4, pp. 599–655, *Climate and Man* (1941 Yearbook of Agriculture), Washington, D.C.: U.S. Department of Agriculture, U.S. Government Printing Office.

ROSSBY, C.-G., 1959, "Current Problems in Meteorology," pp. 9–50, *The Atmosphere and the Sea in Motion* (The Rossby Memorial Volume), Bert Bolin, editor, New York: The Rockefeller Institute Press.

SHAPLEY, H., editor, 1953, *Climatic Change*, Cambridge, Mass.: Harvard University Press.

STARR, V. P., and J. P. PEIXOTO, 1958, "On the Global Balance of Water Vapor and the Hydrology of Deserts," *Tellus*, 10: 188–194.

STARR, V. P., J. P. PEIXOTO, and G. C. LIVADAS, 1958, "On the Meridional Flux of Water Vapor in the Northern Hemisphere," *Geofis. Pur. Appl.*, 39: 174–185.

SVERDRUP, H. U., 1957, "Oceanography," pp. 608–670, *Handbuch der Physik*, S. Flügge, editor, Bd. XLVIII, Geophysik II, Berlin: Springer-Verlag.

SWALLOW, J. C. and L. V. WORTHINGTON, 1960, "An Observation of a Deep Countercurrent in the Western North Atlantic," *Deep-Sea Res.*, 8: 1–19.

WÜST, G., and A. DEFANT, 1936, *Atlas zur Schichtung und Zirkulation des Atlantischen Ozeans, Deutsche Atlantische Expedition auf dem Forchungs- und Vermessungsschiff "Meteor."* Band VI, Atlas.

CHAPTER 8

Current Measurements by Direct Methods

Up to this point we have been concerned with the theories and descriptive aspects of physical oceanography. The science also has its practical side, especially in the measurement of water velocities and of the volume transported by a flow per unit time.

Given knowledge of the field of motion that exists in the sea, many of the questions we now ask about the chemical, physical, and biological properties of the oceans could be answered. But the measurement of ocean current velocities is essentially an art. There are several approaches to the problem, but for one reason or another they all fail to give absolute results, that is, motion measured with respect to a coordinate system at rest in the earth.

The problem of current measurements arises only partly from the fact that the magnitude of water motion at sea is relatively small compared with environmental disturbances. The latter consist of the heaving motion of the observing ship in a seaway, ship drift, interference due to living organisms, and variabilities or incoherence of the water motions themselves which disturb efforts to inquire into the character of the so-called "mean motion." However, in seeking the mean motion it must be recognized that the sea moves in a variety of quasi-periodic ways, with frequencies ranging from those associated with molecular motion to those resulting from secular change in the climatic regime of the whole earth. Therefore it is not to be expected that a single physical method can deal with all of the frequencies and scales of motion that are present; and since there is energy in all parts of this very broad spectrum, the definition of "noise" varies with the frequency favored in each set of observations. A wide range of water speeds are encountered at sea.

The speeds of the surface-layer motions, those induced by wind stress and horizontal pressure differences, range upward to 3 m/sec but average less than 0.5 m/sec. The speeds of tidal currents, on the other hand, are not well known in deep water, but appear to increase toward land where, in certain shallow estuaries, they may approach critical values and even produce a kind of moving hydraulic jump or shock wave (the tidal bore). Although vertical velocities in the sea remain to be measured by direct methods, continuity considerations support the belief that these motions are from one to three orders of magnitude smaller than the horizontal motions with which they are associated.

The horizontal motions that occur in the sea vary more or less systematically with depth. Near the free surface of the ocean, velocities are generally one or two orders of magnitude greater than those in the abyss. Abyssal motions have recently been observed to have values mainly on the order 1 to 10 cm/sec in association with tidal influences. The motions associated with the abyssal parts of the climatological mean circulation may range from possibly an order of magnitude smaller than this to the next larger.

The procedures that have been tried or used to measure the motions of sea water are so numerous that they nearly exhaust the roster of physical possibilities. They can, however, be divided into direct and indirect categories, which are treated in this chapter and the next.

Direct methods. (1) *Eulerian* methods in which the velocity of flow past a fixed geographical point is measured as a function of depth and time, and (2) *Lagrangian* methods in which the trajectories of tagged particles or tracers at several depths in the water are plotted with respect to time.

Indirect methods. (1) The *geostrophic* method in which the density distribution observed in the sea is used to estimate the horizontal component of the field of pressure, and this, in turn, is related to idealized fluid motions on a rotating earth, and (2) the *electromagnetic* method in which the gradient of electric potential in the sea is associated with water motion through the earth's magnetic field.

Units of measurement

The direction of an ocean current is always recorded in the downstream sense—the compass heading toward which the water and anything adrift upon or within it is carried by the flow. This convention is opposite to that used in meteorology, but comes just as naturally from the concern of the navigator to know the direction in which his ship is being set by the current. Direction is ordinarily given in degrees, measured from 0° at the north point to 90° at the east point, 180° at the south point, 270° at the west point, to 360° or 0° again at the north point. To correct for local magnetic declination, *magnetic* headings are distinguished from the *true* bearings given by a gyrocompass by a suffix M or T, respectively, following the bearing in degrees. Bearings may also be expressed by compass points, such as NE or SW by W, or by an angle measured from the north or south point toward the east or west, as for example N 30°E or S 20°W. Any notation is

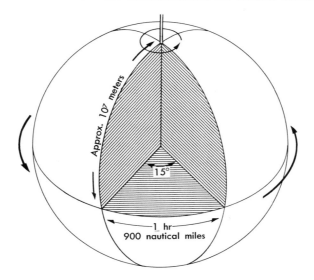

Figure 8–1

acceptable that specifies clearly a reference point or origin which can be related to true north.

Ordinarily the nautical and metric systems of units are used to express current speed in the open sea. The English system is more often reserved for discussion of the flow of rivers and estuaries, particularly in engineering contexts.

The nautical system of units is based on the fact that the earth rotates once in 24 mean solar hours, making its angular motion 15° per mean solar hour (Fig. 8–1). By convention each of these degrees is subdivided into 60-minute intervals, and each minute into 60 seconds of arc. Measured along the equator of the earth, two points separated by one angular minute of arc are, by definition, one nautical mile apart.

The nautical mile is a valuable unit length for practical purposes, because an angular measurement of the great-circle separation of any two points on the earth in minutes of arc is, to a spherical approximation, numerically equal to the linear separation of these points in nautical miles.* It is useful to remember that along the equator

$$1 \text{ hr} \quad = \quad 15° = \quad 900 \text{ nautical miles,}$$
$$4 \text{ min} = \quad 1° = \quad 60 \text{ nautical miles,}$$
$$4 \text{ sec} = \quad 1' = \quad 1 \text{ nautical mile.}$$

---

* The ellipticity of the earth causes the arc length intercepted by a unit angle to decrease with increasing latitude.

The metric unit of length is based on an arbitrary standard, but this is within 0.02% of 10^{-7} of an earth quadrant. Therefore the metric system can be conveniently related to the nautical system, within this degree of approximation.

The units in all three systems are tabulated below:

Nautical	Metric	English
naut. mi/hr (knots)	cm/sec	ft/sec
naut. mi/day	m/sec	sta. mi/hr

The nautical mile is the length of one minute of arc on the earth's surface at the equator, and is equivalent to 6080.2 ft, 1.1516 sta. mi, or 1.85325 km. Thus with mean solar time common to all, the transformation of measurements in nautical units to other units becomes:

$$
\begin{aligned}
1 \text{ knot} &= 51.48 \text{ cm/sec} \\
&= 1.689 \text{ ft/sec} \\
&= 1.1516 \text{ sta.mi/hr} \\
&= 1.853 \text{ km/hr} \\
&= 6080.2 \text{ ft/hr, and so on.}
\end{aligned}
$$

Volume transport or discharge units are much more varied but rarely expressed in the nautical system. The common units are shown in the tabulation.

Metric	English
cm³/sec	ft³/sec or ft³/min
m³/sec or ton/sec	gal/min

Transformation of these units can be effected through the following relationships:

$$
\begin{aligned}
1 \text{ gal/min} &= 0.002228 \text{ ft}^3/\text{sec} = 63.08 \text{ cm}^3/\text{sec}; \\
1 \text{ cm}^3/\text{sec} &= 0.0021186 \text{ ft}^3/\text{min}; \\
1 \text{ ft}^3/\text{min} &= 0.1247 \text{ gal/sec} = 472 \text{ cm}^3/\text{cm, and so on.}
\end{aligned}
$$

These expressions for velocity and transport ordinarily refer to the mean motion of the fluid. The scale of turbulence in the open sea may be so large, however, that such a distinction cannot always be made with adequate justification.

Eulerian methods

Eulerian methods consist of mechanical or dynamical measurements of the flow past a geographically fixed point. The "fixed point" is a most difficult element to provide at sea. In many cases the ship is anchored or the current meter is moored on the sea bed, but in shallow water an oil-drilling platform, a Texas tower, or possibly a bridge can be used as a fixed point of reference. The flow of water past an instrument mounted above the bottom or suspended on a cable in the sea can be measured by: (1) counting against time the *rotations* of a suitable free-turning propeller, Pelton wheel, Rauschelbach or Savonius rotor (Fig. 8–2), (2) measuring the *torque* of an arrested propeller or rotor, (3) measuring the *ram pressure* on a plate, membrane, sphere, or Pitot orifice, (4) measuring the *slope of wire* supporting a known drag, (5) measuring the change of the *velocity of sound* between two points a known distance apart, and (6) measuring the *motional emf* of the flow through a known natural or artificial magnetic field. The direction of flow is usually determined through direct or remote reference to a magnetic or gyrocompass mounted in the instrument or aboard the attending ship.

The various mechanisms designed for Eulerian measurements of current velocity are generally called *current meters*, and are usually thought of primarily in connection with the general problem of velocity determinations at sea. There are perhaps five or six dozen recognized forms of current-metering mechanisms, each being designated by the name of its inventor and perhaps some descriptive adjective.

The most famous of all current meters is that invented by Ekman[1] (Fig. 8–3). This now old-fashioned but highly ingenious instrument has been modified slightly and improved through the years, but is still manufactured by Bergen Nautik in Norway and used throughout the world. Its

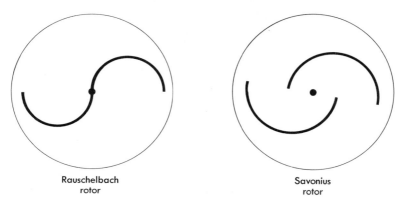

Rauschelbach
rotor

Savonius
rotor

Figure 8–2

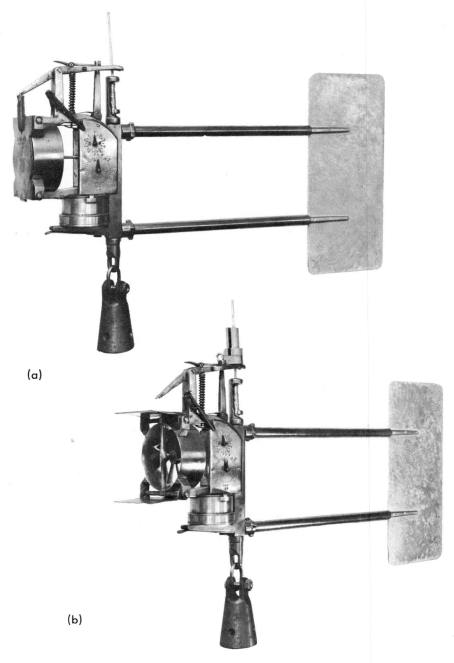

Fig. 8-3.   (a) Ekman meter cocked.   (b) Ekman meter open.

mechanism consists of a simple propeller mounted in agate bearings and connected by a worm to a succession of gears and dials similar to those found in the household electric or gas consumption meter. The Ekman current meter can be attached to a standard 5/32 in. hydrographic wire and lowered to any depth, since it contains no pressure-sealed compartments. When it reaches the desired level, a messenger* is used to open the shutters and release the current-metering mechanism for a period of time, after which the propeller is locked by a second messenger. Once the instrument has been calibrated, the difference between the initial and final readings on the counter dials indicates the length of the water column that has passed through the propeller housing. Since the times of freeing and arresting the propeller rotations are known, the average velocity of flow can be computed.

The most delightfully ingenious feature of this instrument is the means by which the direction of flow is recorded. With each revolution of one of the metering shafts, a small passage is opened and closed to permit one small bronze pellet to drop from a magazine into a cup mounted on the pivot of a compass needle. One side of the cup communicates with a groove on the compass needle which guides the pellet into one or another of 36 chambers arranged in a circle (Fig. 8–4). Each chamber represents one 10° increment of the angle between magnetic north and the orientation of the current meter. During the course of a measurement a dozen or so pellets will have been dropped into the compartmented box, so that when

Fig. 8–4. Ekman compass and crate (left).

---

* A messenger is a small metal weight designed to slide down the wire on which various instruments are suspended in the sea. The speed of messenger descent varies with its shape from 150 to 300 m/min.

the instrument is retrieved, the pellets will be found distributed within a greater or smaller number of compartments indicating both the mean and the extremes of the instrument orientation while it was registering the flow at a given depth.

Simple as this mechanism may be and limited though it is to a single observation beneath an attending ship, it has probably produced more direct measurements of ocean-current velocities than any other single design. Other instruments, both similar to the Ekman meter and very different from it, provide a means for making repeated measurements at the same depth, quasi-continuous measurements as a function of depth, and long series of observations through automatic mechanisms contained within or by means of radio links with shore stations. These are but a few of a very large variety of alternative technical possibilities. Accounts of many of these instrument designs have been summarized by G. Böhnecke (1955)[2] whose paper brings up to date an earlier publication by Thorade (1934)[3] which presents a nearly complete survey of the current meters in use up to that time. In view of Böhnecke's and Thorade's summaries as well as those in Sverdrup, et al., 1942,# The Oceans, and the further descriptions of current metering instruments to be found in the recent summary by Johnson and Wiegel (1959)[4], no roster of individual instruments will be given here.

Current meters can be described according to the various physical means by which flow is detected, or instruments can be classified roughly in three categories: (1) those registering or recording internally, (2) those which indicate the speed and direction of flow to a remote observer or recording apparatus, and (3) those designed for unattended measurements taken over long periods of time. The instruments also differ in that some will withstand pressure to a limited depth, while others are essentially free flooding. Older instruments are designed to be suspended on ordinary hydrographic wire and operated by messengers. Ingenious devices have been employed to permit messenger-triggered mechanical meters to be operated repeatedly, and also in vertical series on a continuous length of wire. More recent designs favor electric cable both as the means for suspending the instrument and for transmitting the information to the ship's deck[5] (Fig. 8–5), to floating radio transmitters,[6] or to shore stations.

However these things are accomplished, we are well counseled by Ekman, who suggests that "so long as the permanent great currents in the sea . . . are concealed by irregular movements with greater velocities, we have not attained all that may be desirable when the velocity and set of the average current has been determined. A knowledge of the magnitude of the more rapidly changing movements would also be of very great importance in several respects. A proper current meter ought not therefore to give means, but . . . make distinct, repeated determinations of velocity; and the principles

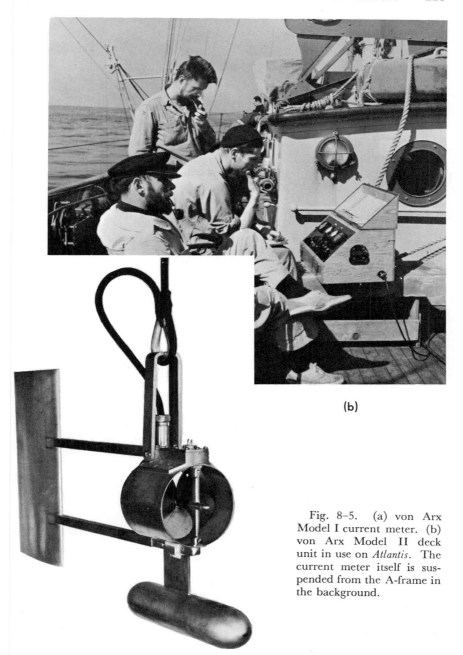

(b)

Fig. 8–5.   (a) von Arx Model I current meter. (b) von Arx Model II deck unit in use on *Atlantis*. The current meter itself is suspended from the A-frame in the background.

(a)

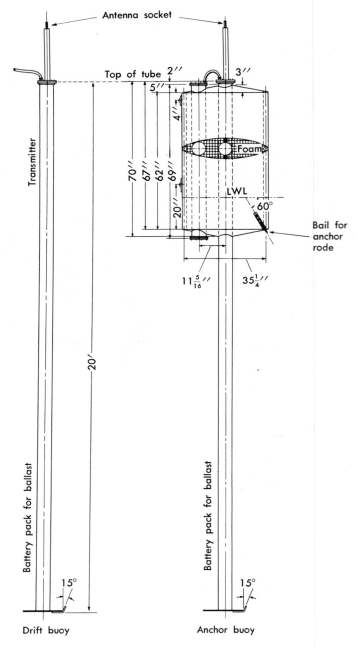

Fig. 8–6. General form of some contemporary transponding radio buoys.

upon which this instrument is based should be the guiding principles for future work in this all-important problem of thoroughly investigating the movements of the water in the ocean."[7]

Lagrangian methods

In this class of current-measuring methods the trajectories of water parcels are tracked and plotted in space and time with the aid of tracers, which may consist of the following devices and practices:

1. *Drift bottles*, for release at sea, from ships at known positions and at recorded times, are ballasted and stoppered glass bottles that contain a numbered card offering a small reward to the finder who fills in the date and position of recovery and returns the card by mail. *Drift cards*[8] in double plastic envelopes have also been dropped in rapidly developed patterns at sea from aircraft.

2. *Radio buoys*,[9] often of a transponding variety (Fig. 8–6), are released at sea from ships, and tracked by direction-finding radio receivers on ships, aircraft, and shore stations.

3. Contaminants or dye stuffs, such as *fluorescein, radioactive materials*, and *chemical wastes*[10-14] having sufficiently powerful responses to physical or chemical tests at extreme dilution are sometimes released at sea, and their progress followed by a suitable program of sampling.

4. *Current poles*[15] are vertical shafts of wood or metal tubing suitably ballasted to drift with only a small portion of their lengths exposed to the wind.

5. *Deep drogues*[16] (Fig. 8–7) may consist of a metal or wooden cross, fish net, aviator's parachute, or other device having a large drag at the level of measurement, connected by a very fine piano wire or nylon filament to a small buoy floating on the surface which in turn may support a flag, a radar reflector and a light for nighttime observation.

6. *Neutrally buoyant floats*[17] have been constructed of aluminum alloy tubing to form a chamber which is less compressible than sea water and which will therefore hover at a level in the sea determined by the amount of ballast added. These floats have been equipped to emit sonic signals at regular intervals, so they can be tracked by acoustic ranging techniques as they drift with the water motion at depth (Fig. 8–8).

7. *Ship drift*[18] employs the ship itself as a tracer whose motions are measured with the aid of the navigating equipment on board.

Lagrangian methods require that the successive positions of the tracer be known relative to the point of release. With an anchored buoy to mark this spot, it is possible, on an attending ship, to determine the time rate of change of range and bearing to the tracer by dead-reckoning navigation runs between the buoy and the tracer. The time elapsed and courses

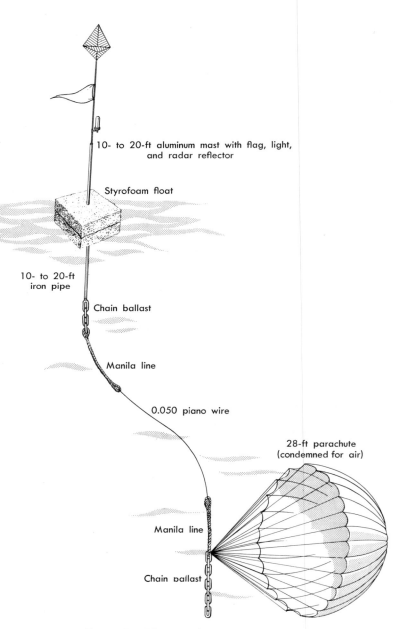

Fig. 8–7.   The parachute drogue.

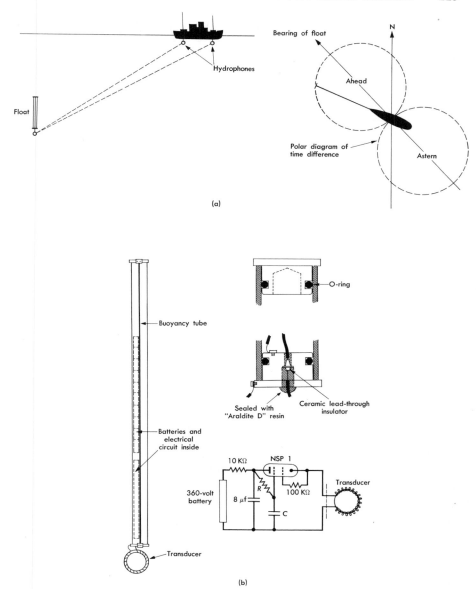

Fig. 8–8.   (a) System used for locating Swallow floats. (b) Sketch of a Swallow float, end plugs, and circuit diagram of acoustic transmitter.   In the transmitter, R and C are chosen to give the required pulse repetition rate. [From J. C. Swallow, 1955, *Deep-Sea Res.*, 3.]

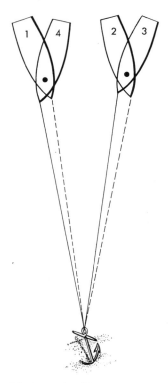

Fig. 8–9.   Motions of a ship at a windy anchorage.

steered are noted in both directions, and are averaged so as to cancel the effects of surface current on the dead reckoning.   The variable influence of wind and waves on the ship's progress can introduce serious error. Moreover, the motions of a reference buoy anchored in several hundred or even thousands of meters of water may be large, as much as 30% of the depth, unless special precautions are taken[18, 19] (Fig. 8–9).   Ordinarily the displacement of the tracer is allowed to accumulate over a period of hours so that individual measurements may be smoothed in extracting an estimate of the water motion as a function of time.

The tracer may emit visual, radio, or sonic signals for tracking by cross bearings obtained from direction-finding techniques, or the tracer may be tracked by radar while the ship is held at the release point.   These techniques permit the paths of several tracers to be measured at the same time.   However, the difficulty with anchored buoys applies equally well to anchored ships.   Maneuvering over a fixed point for direction finding or over the release point for radar ranging requires ready reference to an anchored buoy, an external navigation system, radar ranges to shore points, or possibly acoustic ranges to conspicuous natural or artificial bottom features to fix the origin of measurements.

Alternatively the ship itself can be used as a surface-current tracer, and the electronic navigation facilities on board can be used to find the drift. Windage under these conditions can be measured by a current meter suspended over the lee rail and below keel depth. The drift of deep-drogue tracers may be tracked by maneuvering the ship alongside the surface-indicator float so as to measure the drift by means of the ship's radio navigation facilities.

In good weather the drift of a ship making its way under power across an ocean current can be estimated from a comparison of the course made good with the course steered. The accumulated discrepancy between these two navigational reckonings is an indication of the average speed and direction of the surface current. * If a succession of such integrated displacements is plotted with respect to time, the local velocity can be estimated to within the limits of error of the ship's navigation system. With Loran, Decca, and astronomical methods, a ship's position is known with a precision rarely better than ± 1 km. With Loran-C, Shoran, Lorac, and other high-precision radio techniques, this method becomes much more sensitive but no more reliable because of wave action and the contribution of windage. When the sea is overtaking a vessel, it becomes particularly difficult to keep the ship's head on a steady course. While the instantaneous direction of progress is known by compass to about $1°$, the yawing motion of the vessel may contribute an uncertainty of as much as $\pm 5°$ in the average heading.

———————————

The Loran-A system of radio navigation provides, through time-difference measurements, a pattern of hyperbolic lines of position at sea. Pairs or trios of intersecting lines of position (Fig. 8–10) define the position of the ship with an accuracy somewhat better than the order of 1% of the range from the transmitters to distances as great as 600 nautical miles. It is the advantage of range that makes Loran-A widely used on the coasts of North America in preference to Decca, which is commonly employed along the western European coasts. Decca is similar to Loran-A in geometric principles, but differs greatly in operational detail, being a continuous wave system rather than a pulsed system and having the capability of locking in, presenting thereby continuous as well as periodic indications of position. The errors of hyperbolic radio navigation arise in the timing of signals at the transmitters, in reading the time delay at the receivers, and in the absolute rate of propagation of the signals from the transmitters to the receivers through the earth's lower atmosphere. These problems are discussed by J. A. Pierce, A. A. McKenzie, and R. W. Woodward, 1948, *Loran*, M.I.T. Radiation Laboratory Series Vol. 4, New York: McGraw-Hill.

———————

* The displacement of a moving ship from the course steered can also be measured by an electromagnetic method, an indirect technique discussed in Chapter 9.

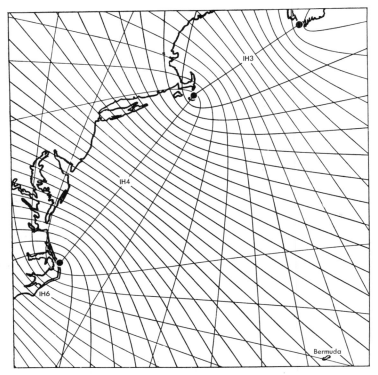

Fig. 8–10. Schematic diagram of a Loran service area. [Generalized from information on Chart 1000-L of the U.S. Coast and Geodetic Survey.]

Experience at sea indicates that the standard error in a Loran-A fix made in the range 300 to 500 miles from transmitters is approximately 0.7 nautical mile or 1 km while that for Loran-C is one-tenth as great. The hyperbolic lines of position frequently cross at acute angles which make the uncertainty in the direction of the smaller angle between the hyperbolas larger than in the direction included by the larger of the two angles. Strictly speaking, then, the figure of error is a parallelogram.

When the velocities of ocean currents are computed through a comparison of Loran data with dead-reckoning information, conditions frequently provide for a greater error in the Loran data if the current flows in the direction of the major diagonal of the parallelogram of error than if it flows along a minor diagonal. Thus it is possible to have a greater probable error in data on current magnitude than in the distance run across current between fixes, or vice versa. Only rarely are the two the same. For simplicity, however, we will regard the parallelogram of error (Fig. 8–11) as approximated by a circle of the same perimeter.

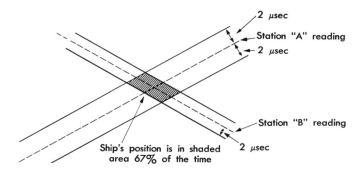

Fig. 8–11. The parallelogram of uncertainty in fixing a ship's position from two sets of time-delay hyperbolas.

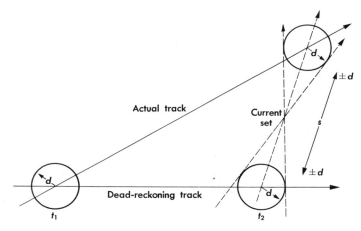

Figure 8–12

The estimate of current speed C from displacements measured by Loran is $C \pm \Delta C = S \ (\pm 2d/\Delta t)$, where S is the actual displacement, d is the radius in the equivalent circle of standard error around each Loran fix, C is the current speed in knots, and t is the time in hours between fixes. The value of d is nearly constant over a small region; hence the error in estimated current speed will diminish only as t is increased. It is assumed that Δt is known without appreciable error.   If, for example, Loran-A fixes are compared at hourly intervals for current measurements some 300 to 500 miles from shore, the probable error ΔC in the current magnitude becomes approximately $\pm 2d/\Delta t = \pm 0.9/1$ or ± 0.9 knot. If half-hourly readings are compared, the average error in current magnitude rises to $\pm 2d/\Delta t = \pm 0.9/0.5 = \pm 1.8$ knots.   The average error in direction can be appreciated from the geometry of Fig. 8–12.

The errors of Loran should generally tend to favor neither positive nor negative deviations from the mean. Therefore either the continuous data of Loran-C or a great number of Loran-A measurements should have a statistically normal distribution around the mean. Since Loran-A fixes for current measurements are made in statistically small numbers, it is entirely reasonable to expect that chance may provide erroneous trends in the mean of a succession of measurements. Current directions are less affected because the direction is computed in terms of distance run and distance set in the same interval of time. The length run is known with better precision than the drift, because it is usually several times greater in magnitude than the average error $\pm 2d$. Still, a calculated current direction is meaningless if the drift is less than $2d$, and also if the distance run in the interval between fixes is less than $2d$.

If a starting point at rest or in a known state of motion is given, currents can also be measured by the time integration of the rudder angles used to steer a given compass heading through a current. When a ship passes from still water into an ocean current, the bow will at first be set downstream, turning the hull bodily unless the rudder angle is changed appropriately. With a gyrocompass or other directional reference, it is possible to maintain a given heading across a shearing current even though the hull is being set. In the absence of wind and wave action, the rudder angles required to maintain a given heading, together with knowledge of the speed of advance of the hull through the water, can be interpreted as a measure of the horizontal shear of the current. This method is rarely feasible on the surface, but can be applied in a submarine when it is running steadily and in perfect trim and when the rotation rate of the hull is known as a function of rudder angle and forward speed.[20]

Measurement of currents in the deep ocean can also be obtained from drifting ships by measuring the difference in velocity between the surface and some other level. If the surface motion of the vessel is known with respect to geographic coordinates and if a shallow current meter is used to measure the ship's wind-drift, the velocity of the surface current is determined by subtracting one vector measurement from the other. Alternatively if a current meter is lowered to a very great depth where it can be assumed that the water motion is small or nil, the difference in velocity between the deep and shallow instruments can be interpreted as a measure of the velocity of the shallow layer without reference to navigational aids. Since, at present, astronomical navigation can be effected with greatest precision only at certain times of day[21] and since there are still large areas of the earth beyond the service of shore-based radio navigation transmitters, the differential measurement is used quite frequently. While it has many virtues, the technique is very time consuming, and for this reason it is difficult to make observations with sufficient rapidity in a given area to arrive at a synoptic pattern of motions. It is possible, however, with certain types of current meters to obtain vertical profiles of velocity at many depths between the surface and bottom of the deep ocean. These measurements remain relative because of the assumption that the bottom current meter is being towed through motionless water.

The Lagrangian methods of current measurement are time consuming, because displacements at sea must be allowed to accumulate to be measured with reasonable accuracy. This tends to blunt their capabilities for studying changes of velocity with time. In general they provide only a limited view of the pattern of motion within a given area. Except under circumstances where the synoptic pattern of motion is already known and the trajectory of an individual particle is in question, one would generally prefer that synoptic data be accumulated in an area. Some of the work on the equatorial undercurrent of the Pacific and on the trajectory of particles in the Gulf Stream and of the deep counter-current beneath the Gulf Stream are cases where Lagrangian techniques have been used to advantage. But the generality of such observations is always open to question unless many tracers are followed at the same time in a significantly large volume of water.

Position

Any observation at sea is incomplete unless its position is specified. Despite long-continued effort, the problems of navigating vessels and of mapping the earth have not been completely solved.

A ship at sea lies in the center of a moving disc of heaving water on which there are few clues to indicate absolute motion or orientation. The changing color of the water from a plankton-bearing green near the coast to the luminous ultramarine of the open sea, the clouds, the air temperature, and the freshness of the wind all suggest a change of season more than of travel. Beyond the sight of land there is no real sense of directed progress until the ship's position is plotted in the chartroom and found to be in rough agreement with expectations for the course being steered and the time spent making a known speed of advance. Travel at sea becomes a microcosm of measurements and computations. An observer can piece together his direct impressions of the natural world only by assembling them on a chart. Even from the air, indications of progress are few; haze is a serious obstacle and the sea surface is completely changed in appearance, seeming remote and almost unreal. The grand view exists only in the mind and this view is vague without the aid of maps.

An exact map of the earth is a manifest impossibility because a spherical shell is nondevelopable; that is, it cannot be reduced to a flat plane without tearing, folding, bending, or stretching the surface in some arbitrary manner.

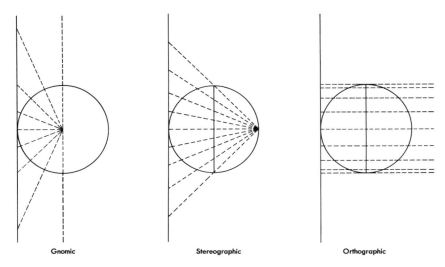

Gnomic   Stereographic   Orthographic

Fig. 8–13.   The perspective equatorial projections.

The distortions needed to reduce a spherical surface to a plane may involve a change of areas, shapes, scales, and the bearings of points relative to one another* (Fig. 8–13).

Dead-reckoning navigation is the simplest of the arts by which man can find his way over a chart that, hopefully, has close correspondence with the earth. It is the method by which the position of a vessel is found by extrapolation of the speed maintained through the water and the direction in which it has been steered. This is different from *piloting* in which the motion of a vessel is determined with reference to landmarks and soundings. Today, electronic aids to navigation are used extensively, and these are actually an extension of the piloting process. Electronic position-finding methods have become so important that they virtually constitute a separate method of navigation. *Celestial navigation* is the art of finding an observer's position with reference to visual sights on such celestial objects as the sun, moon, planets, and stars for which astronomical positions and relative motions are known as functions of time. Recently it has become possible to take angular observations of the sun and "radio stars" in any kind of weather through their radio-frequency emission.

Navigational reckonings are most often plotted on charts constructed in the Mercator projection, but occasionally a simple cartesian grid is

* See W. Chamberlin, 1947, *The Round Earth on Flat Paper*, National Geographic Society, Washington, D.C., and special publications of the U.S. Coast and Geodetic Survey, particularly No. 205, 1943, *Cartography, a Review and Guide*, 2nd ed., by C. H. Deetz and No. 68, 1944, *Elements of Map Projection*, 5th ed., by C. H. Deetz and O. S. Adams.

Fig. 8–14. Gerardus Mercator. [From "The Macpherson Collection," London: Halton and Co., Ltd.]

used. The latter serves where there is concern for so small an area of the earth that both the earth's curvature and the convergence of meridians can be neglected.

Synoptic charts

The Mercator projection is usually adopted in cases where synoptic observations are made over a large area of the sea. The Mercator chart is customarily drawn as an equatorial projection which, in Mercator's (Fig. 8–14) own words, is simply "a new proportion and a new arrangement of the meridians with reference to the parallels," that is, a conventional grid. The system of lines is designed to permit a loxodromic curve, or rhumb line, to become a straight line. The loxodrome is defined as a line which crosses successive meridians at the same angle. If such a line were to be continued indefinitely across the earth it would form a spiral closing at a pole.

The Mercator chart (Fig. 8–15) is in no sense "projected" but is constructed from a set of rules. These rules require that at each point the meridional scale be equal to the zonal scale of the map. Thus if the meridians are spaced at equal intervals around the equator, an equatorial scale is established on which 60 nautical miles correspond to each degree. Near the equator the meridional spacing of latitude circles will be to this same scale. Elsewhere the meridional scale is varied approximately as the secant of the latitude with corrections for the ellipticity of the earth's figure.*

Distances can be measured on the Mercator projection by transferring the separation of any two points to the latitude scale given at the edge of the chart. The measurement of angles must be limited to small areas of the Mercator projection, so that the great-circle propagation of light and radio waves is not significantly different from the loxodromic curve, and the change of scale with latitude in a measured distance does not contribute significant error.†

While the Mercator is the most widely used projection for navigating over very nearly the entire surface of the earth, the distortions increase with latitude, so that it is not well suited as a base chart for representations of large-scale natural phenomena. Similar objections can be raised against the cylindrical projections, such as Gall's, and the cylindrical gnomonic projection. In equatorial cases these all considerably distort the polar extremities of the earth, and lead to false impressions of the relative size of ocean and land masses. Other systems serve to cover the entire surface of the earth in more useful ways but involve distortions of another nature.

If we consider the near hemisphere of the earth to be represented by a circle, and the central meridian and equator to be diameters of this circle at right angles to each other, then the back hemisphere of the earth can be represented by an extension of the chart into an elliptical figure with its major axis on the equator. The major axis of the external ellipse is twice the length of the minor axis, and latitude circles are represented by straight lines parallel to the major axis and spaced so as to produce equal areas between uniformly spaced meridian circles. The result is a pleasant,

---

* Tables exist for the meridional spacing of latitude lines to form the Mercator projection. These are given by Deetz and Adams, *Elements of Map Projection*, 1944, 5th ed., U.S. Coast and Geodetic Survey Publication No. 68, reprinted from *Traite d'Hydrographie* by A. Germain, Paris, 1882, for all latitudes between the equator and 80°.

† Tables for conversion of true bearings to Mercator bearings are available in the *United States Coast Pilot* series issued by the U.S. Coast and Geodetic Survey, and in *Radio Navigational Aids*, 1955, U.S. Navy Hydrographic Office Publication 205.

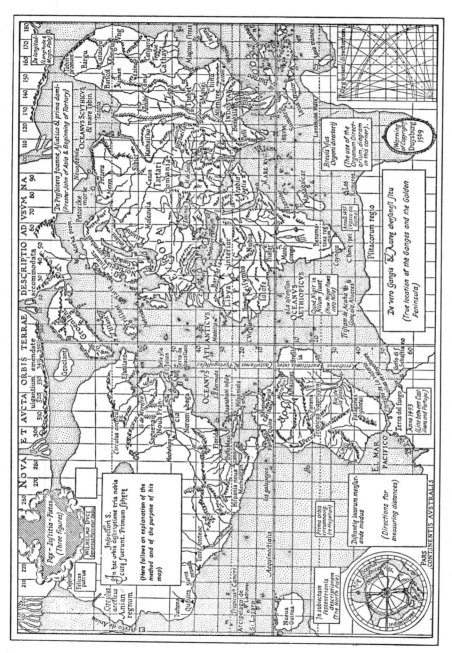

Fig. 8-15. Mercator's chart of the world (1569). [From *Encyclopaedia Britannica*, Vol. 14.]

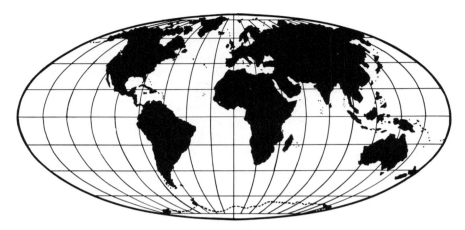

Fig. 8–16.   The Mollweide projection. [After J. Mainwaring, 1942, *An Introduction to the Study of Map Projection*, London: Macmillan.]

quite understandable representation of the entire surface of the earth*
(Fig. 8–16). This is a conventional projection which can be drawn with a
meridian coincident with the major axis of the format so as to include both
poles at the ends of a diameter of the interior circle, or as an oblique projection which results in the meridians and equator becoming sine waves within
the elliptical outline.

It is possible, in addition to enclosing the entire figure of the Mollweide
projection within an ellipse, to cut the equatorial projection from the poles
to the equator in such a way that the oceans are uninterrupted, as shown
in Fig. 8–17. This has some advantages in showing the oceans as a whole,
except that the Southern Ocean is broken into segments. A useful version
of the elliptical format is the Adams' projection or Hammer-Aitoff projection in the oblique cases.† These are capable, as shown in Figs. 8–18 and
8–19, of representing the whole sphere and of being decentered or interrupted for the purpose of showing the oceans to good advantage. The
Adams' projection with the principal parting surface taken through the

---

* A description of the theory of this homolographic projection by Mollweide is
given by Dr. N. Herz, 1885, pp. 161–165, *Lehrbuch der Landkartenprojectionen*, and by
T. Craig, 1882, *Treatise on Projections*, U.S. Coast and Geodetic Survey, pp. 227–228.

† Adams' projection is described by O. S. Adams, 1925, in *Elliptic Functions
Applied to Conformal World Maps*, U.S. Coast and Geodetic Survey Special Publication 112. The Hammer-Aitoff projection is described in detail by E. Hammer,
1892, in *Petermanns geogr. Mitt.*, 39: 85–87, and certain oceanographic modifications
are discussed by A. F. Spilhaus, 1942, "Maps of the World Ocean," *Geogr. Rev.*, 32:
431–435.

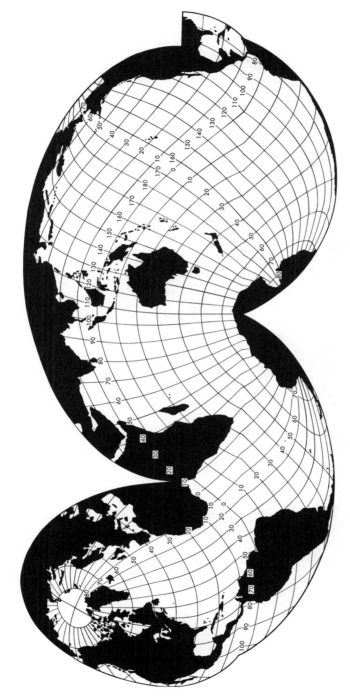

Fig. 8–17. Mollweide projection as modified by F. C. Fuglister to show the world ocean to best advantage.

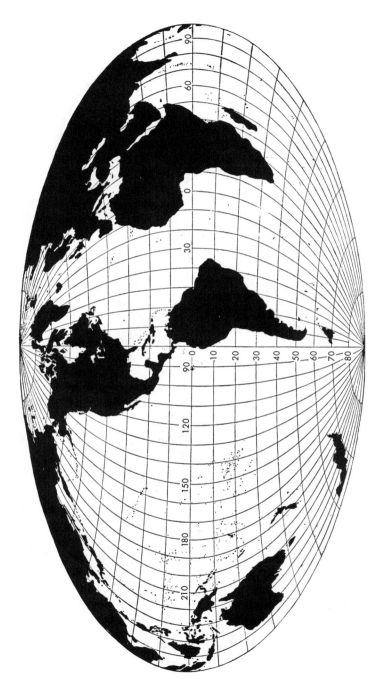

Fig. 8–18.   The Hammer-Aitoff equal-area projection of the sphere with the Americas in the center. [This is an authorized reproduction from U.S. Coast and Geodetic Survey Special Publication 68, 5th edition.]

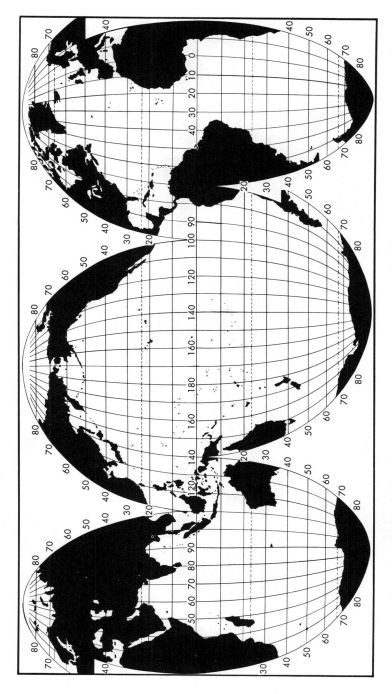

Fig. 8-19. Interrupted homolographic projection for ocean areas. [This is an authorized reproduction from U.S. Coast and Geodetic Survey Special Publication 68, 5th edition.]

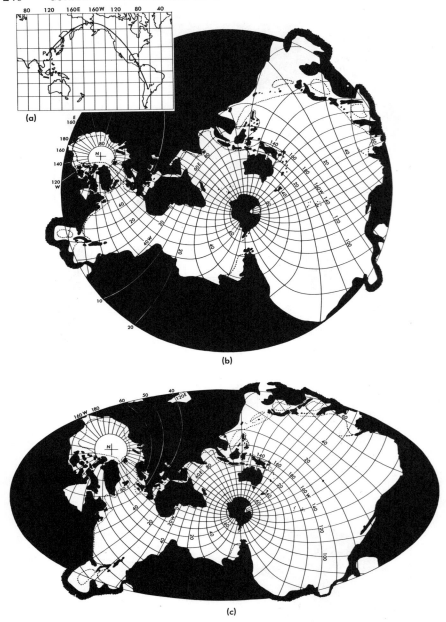

Fig. 8–20.   (a) Bonnefille's cut made preparatory to construction of his version of the Adams projection of the world ocean.   (b) Bonnefille's circular projection of the world ocean.   (c) Bonnefille's elliptical projection of the world ocean.   [From R. Bonnefille, 1957, *Avant Projet de Modele Hydraulique Fixe Representant les Marees Oceaniques Semi-Diurnes*, Chatou, France: Laboratoire National d'Hydraulique.]

Bering Straits and extending from Central Argentina to Southeastern China is shown in Fig. 8–20. This modification of the Adams' projection and of circular projection was made by R. Bonnefille of Laboratoire National d'Hydraulique in Chatou, France, in connection with a study of global tides.

Maps of the whole globe have also been constructed in the equidistant projection; that is, a radially symmetrical projection which represents the bearing and geodesic range from the center of projection to all other points on the earth. Such charts have been used in studies of great-circle processes, such as the propagation of radio waves from or to a station at the center of projection.

STUDY QUESTIONS

1. Distinguish between piloting, dead reckoning, and celestial navigation. What are the basic limitations of each method?

2. Describe the essential principles of a Loran-A navigation system. Why are the time-delay hyperbolas in a Loran-A service area ordinarily published on Mercator charts?

3. Distinguish between the Mercator chart and the perspective cylindrical projections which it resembles most closely. Can the Mercator chart be modified to an oblique form and still be useful for navigation?

4. If you were given the task of observing the Ekman spiral in mid-ocean, what methods and instruments would you use?

5. How could you measure the transverse velocity profile at a given level in the Gulf Stream from a submarine headed across the current at constant speed and depth, but equipped with only a gyrocompass and a rudder-angle indicator?

6. While steering a heading 090°T, a ship's track established by Loran showed a steadily increasing departure northward from the dead-reckoning plot, which reached a maximum of 12 nautical miles in 5 hours, remained constant for 2.5 hours, and then as steadily returned to coincide with the dead reckoning in the next 11 hours. What were the width, speed, and direction of the currents crossed if the ship's forward speed was 12 knots? What courses should be steered and how long would it take to return along a straight line to the starting point with the engines making turns for 8 knots?

7. Suppose that a long, thin, rigid rod connects two sizable spheres of equal drag but of unequal weight, and that suitable devices in one sphere record the amount and direction of inclination of the rod as a function of pressure. The system is designed to sink slowly to the bottom, rest there a moment, and then return to the surface. How could the vertical profile of velocities be computed from the resulting data? What assumptions would you have to make?

REFERENCES

1. V. W. Ekman, 1926, *Publ. Circ. Cons. Explor. Mer*, 91: 27 pp.

2. G. Böhnecke, 1955, *Publ. sci. Ass. Océanogr. phys.*, 14: 28 pp.

3. H. Thorade, 1934, pp. 2865–3095, *Handbuch der biologischen Arbeitsmethoden*, E. Abderhalden, editor, Abt. 2, Teil 3, Berlin: Urban & Schwarzenberg.

4. J. W. Johnson and R. L. Wiegel, 1959, *Investigation of Current Measurement in Estuarine and Coastal Waters*, Publication No. 19, State Water Pollution Control Board, Sacramento, California.

5. W. S. von Arx, 1950, *J. Mar. Res.*, 9: 93–99.

6. K. T. Adams, 1942, pp. 416–417, *Hydrographic Manual*, 2nd ed., Special Publication No. 143, U.S. Government Printing Office, Washington, D. C.

7. V. W. Ekman, 1901, *Nyt Mag. Naturv.*, 39: 163–187.

8. P. Hughes, 1956, *Quart. J. R. met. Soc.*, 82: 494–502. F. C. W. Olson, 1951, *J. Mar. Res.*, 10: 190–193.

9. D. H. Frantz, Jr., 1956, pp. 137–142, *Aspects of Deep-Sea Research*, W. S. von Arx, editor, National Academy of Sciences–National Research Council Publication 473, Washington, D.C.; R. G. Walden, D. D. Ketchum, and D. H. Frantz, Jr., 1957, *Electronics*, 30: 164–167.

10. T. R. Folsom and A. C. Vine, 1957, Ch. 12, pp. 121–132, *The Effects of Atomic Radiation on Oceanography and Fisheries*, National Academy of Sciences–National Research Council Publication 551, Washington, D.C.

11. B. H. Ketchum and W. L. Ford, 1952, *Trans. Amer. geophys. Un.*, 33: 680–684.

12. F. W. Moon, Jr., C. L. Bretschneider, and D. W. Hood, 1957, *Publ. Inst. Mar. Sci.*, Univ. Tex., 4: 14–21.

13. A. C. Redfield and L. A. Walford, 1951, *A Study of the Disposal of Chemical Waste at Sea*, National Academy of Sciences–National Research Council Publication 201, Washington, D.C.

14. W. S. von Arx, 1954, pp. 265–273, *Bikini and Nearby Atolls*, Part 2, Prof. Pap. U.S. Geol. Surv., 260-B-I, U.S. Government Printing Office, Washington, D.C.

15. F. J. Haight, 1938, *Currents in Narragansett Bay, Buzzards Bay, and Nantucket and Vineyard Sounds*, U.S. Coast and Geodetic Survey Special Publication No. 208.

16. T. Cromwell, R. B. Montgomery, and E. D. Stroup, 1954, *Science*, 119: 648–649. D. W. Pritchard and W. V. Burt, 1951, *J. Mar. Res.*, 10: 180–189. G. Volkmann, J. Knauss, and A. Vine, 1956, *Trans. Amer. geophys. Un.*, 37: 573–577.

17. J. C. Swallow, 1955, *Deep-Sea Res.*, 3: 74–81; 1957, *Deep-Sea Res.*, 4: 93–104.

18. W. S. von Arx, 1952, pp. 13–35, *Oceanographic Instrumentation*, J. D. Isaacs and C. O'D. Iselin, editors, National Academy of Sciences–National Research Council Publication 309, Washington, D.C.

19. W. N. Bascom, 1953, *A Deep-Sea Instrument Station*, Scripps Institution of Oceanography Ref. 53-38, unpublished manuscript.

20. W. S. von Arx, 1963, "Measurement of Subsurface Currents by Submarine," *Deep-Sea Res.*, 10: 189–194.

21. W. S. von Arx, 1963, "Applications of the Gyropendulum," Vol. 2, pp. 325–345, *The Sea*, New York: Interscience.

SUPPLEMENTARY READING

BÖHNECKE, G., 1955, "The Principles of Measuri ng Currents," *Un. géod. géophys int., Ass. Oceanogr. phys. Publ. sci.*, 14: 28 pp.

BOWDITCH, N., *American Practical Navigator*, 1958, U.S. Navy Hydrographic Office Publication No. 9.

CHAMBERLIN, W., 1947, *The Round Earth on Flat Paper*, National Geographic Society, Washington, D.C.

FISHER, I., and O. M. MILLER, 1944, *World Maps and Globes*, New York: Essential Books.

JOHNSON, J. W., and R. L. WIEGEL, 1959, *Investigation of Current Measurement in Estuarine and Coastal Waters*, Publication No. 19, State Water Pollution Control Board, Sacramento, California.

McBRYDE, F. W., and P. D. THOMAS, 1949, *Equal-Area Projections for World Statistical Maps*, U.S. Coast & Geodetic Survey Special Publication No. 245, U.S. Government Printing Office, Washington, D.C.

RAISZ, E., 1960, *General Cartography*, McGraw-Hill series in geography, third edition in preparation, New York: McGraw-Hill.

THORADE, H., 1934, "Methoden zum Studium der Meeresströmungen," pp. 2865–3095, *Handbuch der biologischen Arbeitsmethoden*, E. Abderhalden, editor, Abt. II. Teil 3, Berlin: Urban & Schwarzenberg.

CHAPTER 9

Current Measurements by Indirect Methods

Application of the geostrophic approach to oceanography was made during the early decades of the present century. The general circulation of the oceans and the approximate transports of the surface currents had by 1940 been explored, measured, and described in what are essentially present-day terms. It was inevitable that the climatological mean circulation would receive almost exclusive attention because of the steady-state assumption of the geostrophic equation.

Emphasis on geostrophic oceanography was encouraged by the need to protect ships in the Atlantic sea lanes from the menace of drifting ice and by the fisheries problem which was of such great concern to the peoples of western Europe. To these ends the International Council for the Exploration of the Sea was formed in 1902[#], with headquarters in Copenhagen, Denmark. The influence of the International Council has been far-reaching. It first coordinated investigations into the biology of the North and Norwegian Seas, primarily for the conservation of those fisheries.[*] In 1914[#], as a consequence of the *Titanic* disaster, the Council inaugurated the International Ice Patrol.

---

[*] The fisheries problem, a source of scientific stimulation and capital investment, has affected oceanography in much the same way that the weather forecast problem has both advanced and distracted the scientific development of meteorology. Considerably larger numbers of people have been concerned with a study of the weather forecast problem and the fisheries problem than have devoted effort to fundamental and objective inquiry into the phenomena and properties of the oceans and atmosphere. Since the end of the Second World War, however, more emphasis has been placed on objective studies of both fields, and with the renaissance of interest in classical physics this trend may continue to the point where the attention given to the economic aspects of these activities will become a secondary concern. This development may prove to be of benefit to both points of view, since any increase in physical understanding almost inevitably leads to more intelligent economic exploitation and more effective prediction of change.

The geostrophic method

The geostrophic method provides a means for computing the field of relative (geostrophic) motion in a fluid from a knowledge of the internal distribution of pressure. Historically it is derived from the theorems on circulation due to Stokes (1845)[#] and Kelvin (1867)[#] which were broadened by V. Bjerknes (1898)[#] to treat a class of fluid motions having geophysical relevance. For present purposes, however, the essentials of the geostrophic method can be drawn directly from the hydrostatic and geostrophic equations discussed in Chapter 4.

When the density of ocean water is observed as a function of depth, it is possible to compute the pressure as a function of depth from the hydrostatic equation

$$\delta p = \rho g \, \delta z. \tag{9-1}$$

From the levels at which certain standard pressures are attained, it is possible at any locality to assign a thickness to the layers of water bounded by these arbitrarily chosen standard pressure surfaces. If at another locality in the ocean the thickness of these water layers is found to be different, it is evident that the isobaric surfaces defining their vertical limits are inclined to one another. If one of the isobaric layers can be assumed to be exactly horizontal, it is possible to calculate the slopes of all the other isobaric surfaces relative to it. Horizontal planes passing through the inclined isobaric surfaces at a succession of levels reveal by the spacing of intersections the horizontal gradient of pressure at each level in the ocean. From this measure of the horizontal pressure gradient and from a knowledge of the latitude of observations, it is possible to calculate the horizontal component of geostrophic motion at depth z from the finite-difference form of the geostrophic equation

$$c_g = \frac{1}{\rho f}\left(\frac{\Delta p}{\Delta n}\right)_z. \tag{9-2}$$

If we combine the geostrophic equation with the hydrostatic equation $\Delta p = \rho g \, \Delta Z$, we obtain an expression for the geostrophic velocity on a given pressure surface p:

$$c_g = \frac{g}{f}\left(\frac{\Delta Z}{\Delta n}\right)_p. \tag{9-3}$$

As it is not known in advance which horizontal direction, n, lies normal to the geostrophic flow and since soundings of the water structure are taken at points a distance L apart, the expression is written more realistically in terms of the depths at which a given isobaric surface p is observed to lie at two places, say A and B:

$$c_g = \frac{g}{f}\left(\frac{z_A - z_B}{L}\right)_p,$$ (9–4)

where c_g refers to the horizontal component of geostrophic motion on the isobaric surface p and at right angles to the line joining the two points of observation.

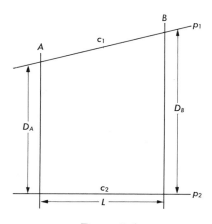

Figure 9–1

Since g varies in a systematic way over the surface of the earth, the depth at which a given hydrostatic pressure is attained is somewhat greater in low latitudes than near the poles. It is more convenient, therefore, to measure the vertical separation of isobaric surfaces in terms of the work done per unit mass against gravity, $g(z_A - z_B)$, than to retain the linear measurement. To preserve numerical correspondence between the depth of an isobaric surface in linear meters and the work done, the unit of specific work has been chosen as 10^5 dyne-cm/gm. Since the expenditure of this amount of work requires that a unit mass be lifted very nearly one linear meter against gravity, the unit $D = 10^5$ dyne-cm/gm is called the *dynamic meter*. In terms of dynamic meters, the vertical separation of any two isobaric surfaces in a perfectly motionless hydrostatic ocean would be uniform at all latitudes. Geostrophic motion can therefore be related to the horizontal variation of dynamic thickness of the water layer contained within any arbitrary pressure interval. The difference in geostrophic velocity between the upper and lower isobaric surfaces bounding the layer (Fig. 9–1) is given by

$$(c_1 - c_2)_g = \frac{\Delta D_A - \Delta D_B}{Lf},$$ (9–5)

where ΔD_A and ΔD_B are the dynamic thicknesses of the same pressure interval measured at two stations separated by a horizontal distance L. This is the formula derived by Sandström and Helland-Hansen in 1903[#] from the Bjerknes (1898)[#] circulation theorem. The same formula was given by H. Mohn in 1885,[#] some thirteen years before Bjerknes' paper and eighteen years before Sandström and Helland-Hansen's work, but Mohn's efforts somehow escaped recognition.

The geostrophic motions computed from the Mohn-Sandström-Helland-Hansen formula remain relative unless some valid observation is made to find a level where the motion can be assumed to be zero or is otherwise known. In such a case the field of relative geostrophic motion is calibrated and becomes an approximation of the actual velocity field.

The *method of dynamic sections*, as the foregoing procedure is sometimes called, incorporates the assumptions of geostrophic motion, (1) that the flow is unaccelerated and (2) that the flow is frictionless, and further assumes (3) that the pressure field is essentially in hydrostatic equilibrium and (4) that observations at successive stations are either simultaneous or that water properties do not change in the time interval between successive sampling operations. When a line of stations has been occupied, an extended vertical section is developed in the ocean which reveals the structure in terms of the distributions of temperature and salinity and, indirectly, density, from which one can find the horizontal component of relative pressure and finally of geostrophic motion passing at right angles through the section. If the geostrophic velocity is computed for each successive layer, the volume transport normal to the line joining each pair of stations can be computed by simply summing up the products of velocity and cross-sectional area for the successive pressure intervals. When the method is applied to a grid of stations, the water properties and relative horizontal geostrophic motions can be developed for the volume delimited by the grid.

The computational practices for applying the geostrophic method at sea are now well worked out and are concisely summarized by Eugene C. LaFond (1951).[11] Measurements are made of the vertical gradient of salinity and temperature beneath the ship at each successive station (Fig. 9–2), because pressure itself as a function of depth cannot be measured precisely enough by direct methods. Instead, pressure is computed from the integrated weight of the sea water as a function of temperature, salinity, and hydrostatic pressure.* It is also possible to calculate the geopotential thickness of a given pressure interval. This is most conveniently done from consideration of the integrated specific volume of the intervening water column.

---

* In fresh water the problem is simplified to a consideration of only temperature and pressure effects on water density. See J. C. Ayers, 1956, *Limnol. Oceanogr.*, 1: 150–161.

Fig. 9–2. L. V. Worthington making a cast from R. V. *Albatross III*. [Photograph by J. Hahn.]

The change in geopotential dD with a change in pressure dp in a perfectly hydrostatic ocean is given in dynamic meters by

$$dD = \alpha_{s,t,p}\, dp, \qquad \alpha = \frac{1}{\rho} \qquad (9\text{--}6)$$

where the factor of proportionality α is the specific volume, which is a function of temperature t, salinity s, and pressure p. When the equation is integrated between two pressure surfaces p_1 and p_2 we have

$$D_1 - D_2 = \int_{p_1}^{p_2} \alpha_{s,t,p}\, dp, \qquad (9\text{--}7)$$

to show that the difference in geopotential of any two pressure surfaces in the ocean can be determined from suitable measurements of the specific volume of the water as a function of depth between them.

To avoid computations involving small differences between large quantities the integral of Eq. (9–7) is usually broken into two parts, one containing the contributions of a *standard ocean* having a uniform salinity of 35 $^0/_{00}$ and temperature of 0°C and another containing the departures of the real ocean from the standard:

$$(D_1 - D_2)_s + \Delta D = \int_{p_1}^{p_2} \alpha_{35,0,p} \, dp + \int_{p_1}^{p_2} \delta \, dp. \qquad (9\text{–}8)$$

$$\Delta D = \int_P^{P_2} \delta \, dp$$

The first integral in Eq. (9–8) gives the standard geopotential interval between isobaric surfaces p_1 and p_2, and the second integral contains the departural change of geopotential between these same pressure limits. Bjerknes (1910)[2] computed the standard geopotential change between isobaric surfaces at various intervals, taking into account the effects of compressibility. The distribution of properties within the standard ocean can be computed once and for all.

The differences between the properties of the real ocean and those of the standard ocean are expressed as *anomalies* (Fig. 9–3). The standard ocean is cold and salt enough for the differences between it and the real ocean to result in small positive anomalies. The standard ocean is considered

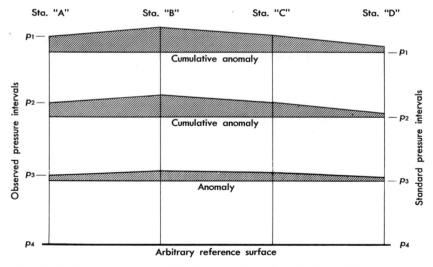

Fig. 9–3.   Pressure anomaly (shaded areas) in schematic form. Pressure changes more slowly with depth where water is warmer or fresher than that of surrounding columns.

to be motionless, so that geostrophic effects in the real ocean can be directly related to the anomalous distribution of density, or more usually the specific volume anomaly of the water.

The specific volume of sea water (in cm^3/gm), a quantity influenced by salinity, temperature, and pressure, can be evaluated from the value of σ_t contained in tables edited by Knudsen (1901, 1931),[#] or computed with the aid of the formula developed by V. W. Ekman.[3]

Separating the specific volume $\alpha_{s,t,p}$ of water *in situ* into that of the standard ocean $(\alpha_{35,0,0} + \delta p)$ and the anomaly of specific volume δ needed to describe the observed ocean we have

$$\alpha_{s,t,p} = (\alpha_{35,0,0} + \delta_p) + \delta, \qquad (9\text{--}9)$$

where δ itself is

$$\delta = (\delta_s + \delta_t + \delta_{s,t}) + (\delta_{s,p} + \delta_{t,p} + \delta_{s,t,p}). \qquad (9\text{--}10)$$

Because of its uniformity, $\alpha_{35,0,0}$ is a constant ($= 0.97264$ cm^3/gm) and δ_p takes account of the effects of pressure in the standard ocean. The sum $\alpha_{35,0,0} + \delta_p$ is the specific volume of the standard ocean given in Table 4 of LaFond.[1] In the expansion of the specific volume anomaly δ, the first group of three terms expresses the dependence of the specific volume anomaly on salinity and temperature and on their cross products. These are usually combined to form the single term

$$\Delta_{s,t} = \delta_s + \delta_t + \delta_{s,t}. \qquad (9\text{--}11)$$

Values of these combined terms for each degree centigrade and every 0.1 $^0/_{00}$ of salinity, through the ranges $-1°C$ to $29°C$ and from 21.00 to 37.00 $^0/_{00}$, are available together with values of differences in temperature and salinity for linear interpolation in LaFond's Table 5a. Other tables are also available for the evaluation of the products of $\delta_{s,p}$ and $\delta_{t,p}$, so that these terms, except for $\delta_{s,t,p}$ which is usually negligibly small, can be taken into account for the purpose of computing the specific volume for each standard pressure interval. From this the anomaly of dynamic spacing of isobaric surfaces beneath the sea surface can be found with the aid of the relation

$$\Delta D = \int_{p_1}^{p_2} \delta \, dp. \qquad (9\text{--}12)$$

From a comparison of dynamic depth anomalies at a pair of oceanographic stations separated by some known geographic distance, it is possible to compute the relative horizontal pressure gradient. With the Mohn-Sandström-Helland-Hansen formula the magnitude of the average relative geostrophic flow to be expected at right angles to the plane joining two water columns can be computed.

If a number of oceanographic stations have been occupied in an extended pattern or in serial sections through the volume of the sea, it is possible to develop for a succession of levels contour charts of the dynamic anomaly or, as it is also known, the *dynamic topography* of an area. Each of these can be looked upon as a topographic map of the relative pressure field at each level (Figs. 9–4 and 9–5). If the sea surface is considered to coincide with a geopotential surface, then the topography will become more and more rugged with depth. But if a reference level is chosen at some depth in the area where it can be assumed that the horizontal pressure gradients are either zero or small enough to be neglected, the geopotential anomalies of pressure surfaces can be recomputed upward and downward from this level so as to produce a topography that resembles, to a degree, that which might be expected to exist in the ocean. The choice of a *level of no motion* is often based on a subjective inspection of the temperature-depth and salinity-depth diagrams for each of the stations in the field, and made at the level where the water has the least change of temperature and salinity with depth (Defant's method). It is also possible to utilize direct current measurements for this purpose, and indeed this is very much to be preferred.

Once adjustment has been made of the relative geopotential of the succession of isobaric surfaces, the packing and curvature of the isobaric contours can be looked upon as resembling those to be seen on topographic maps of the land. If the curvatures are gentle, it is possible to relate the dynamic topography to geostrophic flow parallel with the contours. At any given latitude the speed of geostrophic flow is proportional to the intensity of the horizontal pressure gradient, measured from the packing of the contours. The flow is directed to the right of the direction of the measured slope in the Northern Hemisphere and to the left of this slope in the Southern Hemisphere. Since the latitude of observation is known, it is possible to superimpose on these topographic charts isopleths of current speed to produce what may be called an isokinetic diagram or current chart.

Current charts based on dynamic studies of the field of pressure have been used successfully for many years by the International Ice Patrol as a means for anticipating the mean drift of icebergs reported at different places in the Labrador Sea by search aircraft or passing ships. The method has also been used to estimate the drift of fish larvae and other planktonic forms of life, but experience has shown that these predictions are not wholly correct since productivity and mortality can shift the centers of these populations independently of the local current. There is also the further possibility of an error in the choice of level of no motion, so that a comparison of the relative field of motion with the actual path of any tracer may show considerable discrepancies.

Careful comparison of direct measurements of flow observed by Pillsbury (1890)[#] in the Straits of Florida with dynamic sections computed by Wüst

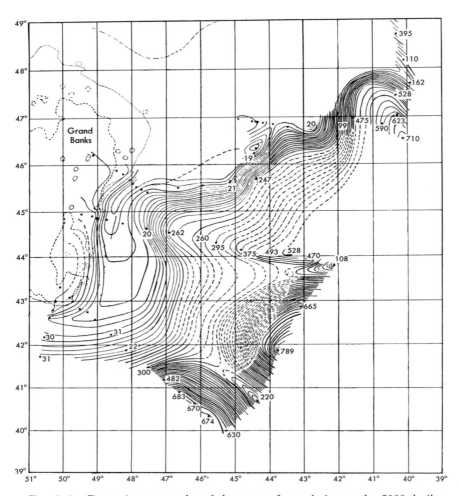

Fig. 9–4. Dynamic topography of the sea surface relative to the 2000-decibar reference surface in the Grand Banks area surveyed annually by the International Ice Patrol. [From F. M. Soule, 1939, *J. Mar. Res.*, 2(3).]

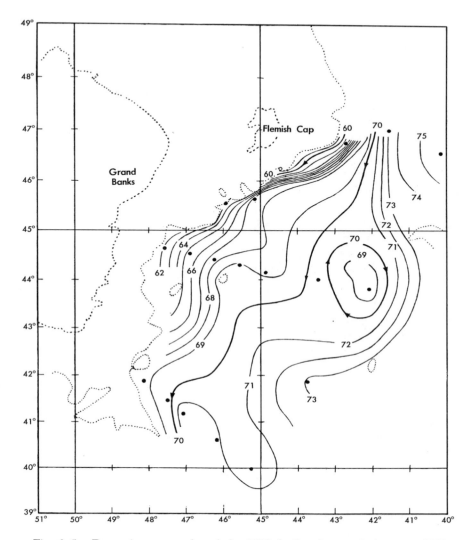

Fig. 9–5. Dynamic topography of the 3500-decibar layer relative to a 2000-decibar reference surface in the area of Fig. 9–4. [From F. M. Soule, 1939, *J. Mar. Res.*, 2(3).]

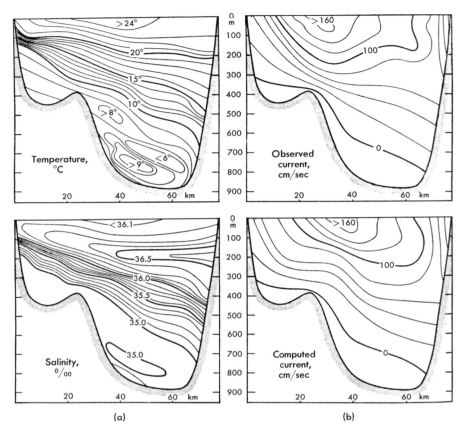

Fig. 9–6. (a) Observed temperatures and salinities in the Straits of Florida. (b) Velocities of the current through the straits according to direct measurements and computations on the distributions of temperature and salinity. [From H. U. Sverdrup, M. W. Johnson, and R. H. Fleming, 1942, *The Oceans, Their Physics, Chemistry, and General Biology*, New York: Prentice-Hall.]

(1924)[#] for the same area show that the relative field of motion obtained from dynamic sections, using Pillsbury's observed surface of no motion, lead to substantially the same results. This comparison is shown in the classic illustration reproduced in Fig. 9–6. Further comparisons of the field of motion inferred from the distribution of density by means of the geostrophic approximation have been made by the International Ice Patrol and by the *Meteor* expedition. In these cases it was found that the current observed by direct methods may depart from the geostrophic field of motion by amounts which range from 5% to 25%, with the mean departure in the neighborhood of 15%. Studies of the effects of curvature on the transport of the Gulf Stream were also made by the author, using data from the 1950 multiple-ship survey of that current off the coast of New England. These studies show that the difference between the computation based on geostrophic flow and that for meander flow can also amount to 15%. Parr (1938),[4] in a comparison of isopycnic and geostrophic analyses of the flows of the Labrador Current and the Gulf Stream system south of the Grand Banks of Newfoundland, showed that agreement "would be *utterly unobtainable*"[4] in some portions of the area. Defant (1950)[5] has shown that the errors produced by long-period internal waves are not negligible, and has suggested observational precautions which tend to suppress their influence.

But for all these quite real discrepancies, it should be remembered that the method of dynamic sections, as it is presently implemented, provides information on the *relative motion* to be expected in association with the *baroclinic* component of the field of pressure interpreted according to the assumptions of *geostrophic* flow. It pretends to achieve nothing more.

Observations. To obtain raw data for dynamic computations, research ships are equipped with water-sampling equipment and navigational aids to find the position of each sampling site. The sampling equipment consists of a dozen or more reversing water bottles fitted with reversing thermometers and, for deep-sea work, a winch holding as much as 10,000 meters of 5/32 in. 7 × 9 or 3 × 19 aircraft control cable made of the best quality plough steel.

The winch (Fig. 9–7) is driven by steam, hydraulic, or electric (d-c) motors to provide smooth control at 10 to 15 brake horsepower and to permit wire to be handled at rates anywhere between zero and about 3 m/sec. It is ordinarily fitted with a diamond thread level-winding traveler, and some sort of shock accumulator is either built in or appended to the machine to prevent the cable from going slack or sustaining heavy stresses under the influence of the ship's rolling motion. The wire cable is led over the ship's side at a point in full view of the bridge, so that the ship can be maneuvered to stay clear of the wire. The last sheave on the davit or A-frame from which the cable is suspended is usually a metering wheel to indicate the length of cable paid out. The end of the cable is carried down by a cylindrical sinker having a weight near 50 kgm. Since a deep cast may require some 5000 meters of cable and the attachment of a dozen water bottles, the dead weight stress may approach one-half ton. When a heavy sea is running, the stress on the cable may surge to a ton or more owing to the inertial reaction and frictional drag of the water on the hanging cable and suspended parts.

Samples of the sea water are obtained with reversing or remote-closing water bottles. These are usually made of bronze and lined with plastic to prevent contamination of the water sample. The reversing water bottle in

Fig. 9–7. Lidgerwood electric winch for making deep oceanographic soundings from *Atlantis*. [Photograph by J. Hahn.]

general use today is that designed by Fridtjof Nansen* (Fig. 9–8). It consists of a brass or bronze cylinder with a capacity of about 1.3 liters and fitted with a carefully machined and lapped wide-mouthed bronze stopcock at each end. The bottle is attached to the cable by a thumbscrew clamp at the lower end and by a triggerlike clip at its upper end. When a messenger sent down the cable strikes the trigger, the upper end of the bottle is freed from the cable, whereupon the bottle falls away to hang by the lower clamp

---

* For an early description of these reliable samplers see B. Helland-Hansen and F. Nansen, 1925, *Geofys. Publ.*, 4(2): 6–7.

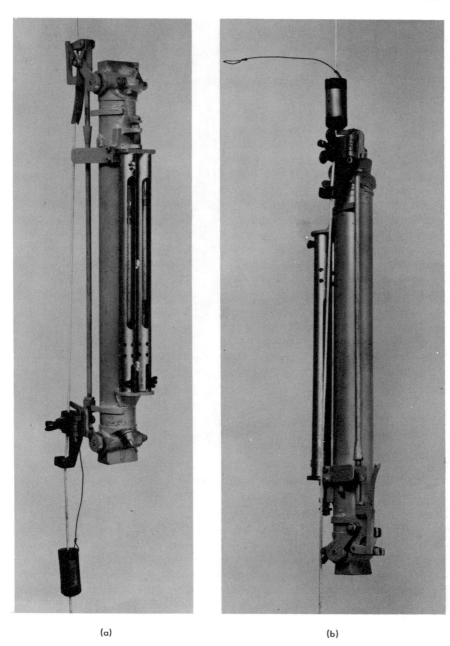

(a)

(b)

Fig. 9–8. Nansen bottle (a) cocked, (b) reversed and closed. [Photograph by C. Spooner.]

Fig. 9–9. D. F. Bumpus drawing oxygen and salinity samples from Nansen bottles aboard *Atlantis*. [Photograph by D. M. Owen.]

alone. This action involves rotation around the lower clamp which turns both stopcocks to seal off the sea-water sample. When the bottle is being sealed, a messenger hanging from the lower clamp is released to slide farther down the cable and trigger the next lower water bottle. The sliding messengers make a characteristic noise as they fall at speeds near 200 m/min and click as they trigger the water bottles, so that the unseen operation of sampling can be appreciated by holding the ear against the motionless cable. It takes several hours to make a deep cast and bring the samples to the deck laboratory, during which time the ship is hove to. To make two or three oceanographic stations and cover the miles between them is a day's work for a ship.

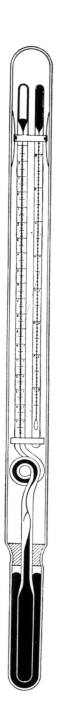

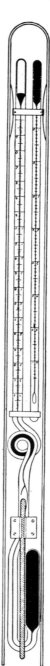

Fig. 9–10. Protected and unprotected reversing thermometers in set position, that is, before reversal. At the right is shown the constricted part of the capillary in set and reversed positions. [From H. U. Sverdrup, 1942, *Oceanography for Meteorologists*, New York: Prentice-Hall.]

When the water bottles have been brought aboard, they are set in racks so that the water temperatures may be read and water samples drawn off for general chemical analysis, titration of chlorides, or conductivity measurements (Fig. 9–9). This operation provides some of the data needed for finding the distribution of density and, from this, pressure in the sea-water column; but the actual depth of sampling is known only approximately from the bottle spacing and the length of wire paid out. Since the slope of the cable may vary with depth as a result of any number of unknown curvatures along its length, the cable angle at the surface is only a rough guide to the vertical distance between sampling bottles. The actual depths of sampling are obtained with a pair of reversing thermometers attached to each water bottle. The mercury threads in these instruments are designed to break when they are inverted by the rotation of the reversing water bottle.

One member of a pair of reversing thermometers is enclosed in a glass envelope to protect it against the ambient pressure of the sea (Fig. 9–10). This *protected thermometer* gives the temperature of the water at the level from which the accompanying sea-water sample is taken. The other thermometer is not protected against sea-water pressure. The *unprotected thermometer* therefore reads higher than the ambient temperature by some function of the hydrostatic pressure, which squeezes the mercury thread up the thermometer stem. The elevation of apparent temperature as a function of pressure is known for the unprotected thermometer, so that the difference between the readings of the protected and unprotected thermometers will indicate the pressure at the level of reversal. This simple technique has proved itself to be reliable to within ±5 meters for depths in the ocean less than about 1000 meters and to about 0.5% of greater depths. The accuracy of thermometric measurements of depth has been examined in detail by Wüst (1933).[6] The method has survived many efforts to replace it by more sophisticated techniques.

The electromagnetic method

The electromagnetic method has provided one means for observing from moving ships effects related to the velocities of surface currents in the open ocean. The basis for the method lies in the fact that sea water, containing an abundance of highly dissociated salts, is an electrolyte of reasonably high conductivity. The motion of an electrolyte through a magnetic field produces electromagnetic effects similar to those associated with the separation of charge in a solid conductor moving through a magnetic field. Motion of any part of a closed conducting circuit through a primary magnetic field will produce a gradient of electric potential and a negligibly small secondary magnetic field associated with the induced electric current. In the case of moving sea water, the amount of electric current flowing depends on the conductivity of the air above the sea and the sediments and rocks in the sea bed, and on the motions and conductivity of the sea nearby. The environmental effects of the air are very small, and the conductivity of the sea bed has much less effect on the motional electric potential than the variation of conductivity in the sea itself and of the distribution of motion from top to bottom. By reason of the latter effect electric potential measurements require careful interpretation before they can be called upon to give definitive indications of sea-water motion. Even so, the method has usefulness when properly applied and, being independent of the assumptions of the method of dynamic sections, serves as a basis for comparison.

The possibility of detecting sea-water motion by the effects of electromagnetic induction (Fig. 9–11) was foreseen by Michael Faraday in his Bakerian Lecture of 1832 before the Royal Society, in which he said:

> Theoretically, it seems a necessary consequence that where water is flowing, there electric currents should be formed: thus, if a line be imagined passing from Dover to Calais through the sea, and returning through the land beneath the water to Dover, it traces out a circuit of conducting matter, one part of which, when the water moves up or down the channel, is cutting the magnetic curves of the earth, whilst the other is relatively at rest.[7]

Faraday attempted to detect electric currents flowing in the earth in response to the water motions of the Thames, but the chemical potentials developed at the copper electrodes he used masked the effect. After the middle of the nineteenth century the effect was noted repeatedly between the local grounds and broken ends of submarine telegraph cables. These observations were summarized by M. S. Longuet-Higgins (1949).[8] Successful detection of the effects of electromagnetic induction in sea water over short distances was first made in 1918 in Dartmouth Harbor, England, by F. B. Young, H. Gerrard, and W. Jevons (1920),[9] who used both moored and drifting nonpolarizing electrodes to reveal the electric currents associated with tidal motions; but these results were not utilized in oceanographic research until late 1946, owing primarily to difficulty with instruments.[10] The method became practicable when the electronic continuous-balance potentiometer developed for industry was found to be well suited to work at sea. This instrument removed the necessity for a sensitive galvanometer in a Wheatstone bridge circuit, replacing it (as a null indicator) with a chopper amplifier to drive a small motor which maintains the system in continuous balance. This substitution permits almost no electric current to flow in the external circuit, and the ruggedness of the system permits some measurements formerly confined to physical laboratories to be made from mobile laboratories in trucks, ships, and aircraft.

Physical principles. The movement of sea water through the motionless magnetic field of the earth gives rise to electric potential gradients and

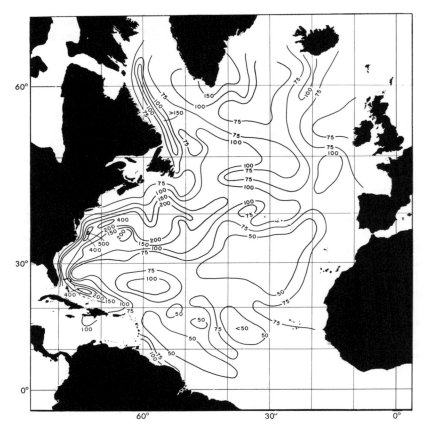

Fig. 9–11. Magnitude of the electric potential gradient on the surface of the North Atlantic, in millivolts per 10 kilometers.

electric currents. The character of the electric potential field can be deduced from Faraday's law of electromagnetic induction applied to a closed curve fixed in the fluid. If it is assumed that the earth's magnetic field is essentially uniform over the immediate area and that the resistivity of sea water varies only slightly with position and depth, then the gradient of potential $\nabla\phi$ is given by

$$\nabla\phi = (\mathbf{c} \times \mathbf{H}) - \rho\mathbf{i}, \qquad (9\text{–}13)$$

where $\mathbf{c}$ is the water velocity vector, $\mathbf{H}$ the magnetic field vector, ρ the scalar resistivity, and $\mathbf{i}$ the electric current density. In practice it is most convenient to measure the horizontal component of the motional electric potential gradient at the sea surface.   Expanding Eq. (9–13) we have

$$\nabla\phi = [\mathbf{I}(c_yH_z - c_zH_y) + \mathbf{J}(c_zH_x - c_xH_z) + \mathbf{K}(c_xH_y - c_yH_x)] - \rho\mathbf{i},$$
(9–14)

where $\mathbf{I}, \mathbf{J}, \mathbf{K}$ are unit vectors in the coordinate directions x, y, and z. Since the potential gradient is to be sampled in the horizontal xy-plane, H_x and H_y make no contribution; hence

$$\mathrm{grad}_h\ \phi = [\mathbf{I}(c_yH_z) + \mathbf{J}(c_xH_z)] - \rho\mathbf{i}_h,$$
(9–15)

or

$$\frac{\partial\phi}{\partial x} = c_yH_z - \rho i_x,$$
(9–16)

$$\frac{\partial\phi}{\partial y} = c_xH_z - \rho i_y,$$
(9–17)

where $\mathrm{grad}_h\ \phi$, the horizontal component of the electric potential gradient, is dependent only upon the strength of the vertical component of the magnetic field of the earth and the horizontal component of water motion. Thus if the measuring circuit is horizontal and at rest with respect to the earth's magnetic field, horizontal water motion past it yields a signal given by

$$\mathrm{grad}_h\ \phi = (\mathbf{c}_h \times \mathbf{K}H_z)_h - \rho\mathbf{i}_h,$$
(9–18)

where $\mathbf{c}_h$ is the horizontal water velocity vector, $\mathbf{K}$ the vertical unit vector, H_z the vertical component of the local magnetic field vector, ρ the scalar resistivity of the water and sea floor in the vicinity of the ocean current, and $\mathbf{i}_h$ the horizontal component of the electric current density.

When the measuring apparatus, having sampling points separated by a distance S, is drifting with the water through the magnetic field of the earth, the signal generated in the moving apparatus must also be taken into account; thus for the surface case

$$S\cdot\mathrm{grad}_h\ \phi = S\cdot\{[(\mathbf{c}_h \times \mathbf{K}H_z) - (\rho\mathbf{i}_h)]_o - [(\mathbf{c}_h \times \mathbf{K}H_z) - (\rho\mathbf{i}_h)]_m\},$$
(9–19)

where the subscripts o and m refer to the ocean circuit and measuring circuit respectively. This expression can be greatly simplified if a continuous-balance potentiometer is used to measure the potential gradient. In such a case $(\mathbf{i}_h)_m = 0$ and $(\mathbf{c}_h \times \mathbf{K}H_z)_o = (\mathbf{c}_h \times \mathbf{K}H_z)_m$, since the measuring circuit drifts with the ocean current at the same horizontal velocity $\mathbf{c}_h$ and the local vertical component of the earth's magnetic field is identical in both cross products. Thus the signal received per unit length is simply $(-\rho\mathbf{i}_h)_o$. If the measuring apparatus spans an ocean segment S units long,

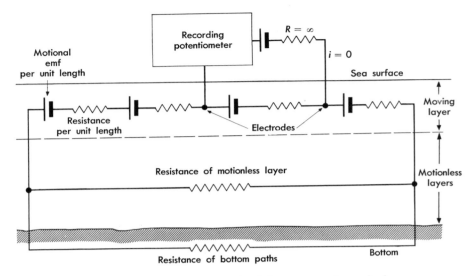

Fig. 9–12.   Equivalent circuit of electromagnetic method.

the horizontal potential difference in practical volts, V, measured by the potentiometer (Fig. 9–12) is

$$V = S \cdot (-\rho \mathbf{i}_h) \times 10^{-8} = k(SH_zC_h \times 10^{-8}), \qquad (9\text{–}20)$$

where 10^{-8} is the proportionality factor between electromagnetic and practical units, H_z the intensity of the vertical component of the earth's magnetic field in oersteds, C_h the speed of water motion at right angles to the sampling line segment S, and k an empirical correction factor (discussed in the next paragraph) which depends on oceanographic conditions. If C_h is positive toward the east, H_z is directed downward, and S is parallel with a meridian, V will have positive polarity (+) at the south and negative (−) at the north end of S. Where the direction of flow is unknown, it can be found by vector composition of two components of the horizontal electric potential difference measured at the sea surface between two points S centimeters apart. Since $\mathrm{grad}_h\,\phi = \mathbf{c}_h \times \mathbf{K}H_z$ and the vertical component of the earth's magnetic field is directed downward in the Northern Hemisphere, the direction of water motion lies 90° to the right of the positive sense of the horizontal component of the electric vector. In the southern magnetic hemisphere, the vertical component of magnetic flux is directed upward, so that the water motion vector lies 90° to the left of the horizontal electric vector. With the direction of surface-water motion determined, it remains to evaluate k for the site and compute the speed of flow.

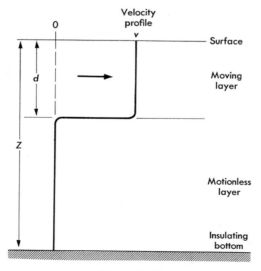

Figure 9–13

k-Factor. The scalar proportionality factor k is defined as follows:

$$k = \left| \frac{(\mathbf{c}_h \times \mathbf{K}H_z)}{(\rho \mathbf{i}_h)_o} \right| = \left| \frac{\text{true current speed}}{\text{observed current speed}} \right| . \qquad (9\text{–}21)$$

Absolute magnitudes are used because in actual practice the observed direction of water motion is very nearly correct. The observed current speed is usually deficient by an amount which is mainly dependent on the ratio of the *electrical thickness* of the ocean current d to the total depth of water $\mathcal{Z}$ (Fig. 9–13), that is,

$$d = \left(1 - \frac{1}{k} \right) \mathcal{Z}. \qquad (9\text{–}22)$$

Radio navigation can be used to provide an estimate of true current speed. This compared with the strength indicated by the electromagnetic method provides a value for k, and then $\mathcal{Z}$ can be obtained from the echo sounder to find d. The electrical thickness of ocean currents (Fig. 9–14) is not a measure of the depth of the level of no motion but rather of some isokinetic surface which usually has a value near 0.1 of the surface-water speed. From measurements of d, shallow wind-drift currents can be distinguished from more permanent baroclinic motions, and with passing time, both of these can be distinguished from tidal motions.

The value of k is typical of oceanographic provinces and is known fairly well in some areas. Off New England for example, $k = 1.5$ in the tidal

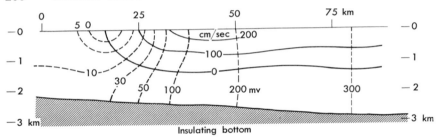

Fig. 9–14.   Potential field associated with a velocity field resembling that of the Gulf Stream.

waters on the continental shelf, 1.10 in slope water, 1.25 on either side of the current maximum of the Gulf Stream, 1.40 in the current maximum, and then drops abruptly to 1.05 in the Sargasso Sea toward Bermuda.   In the Straits of Florida values of k average near 1.4, since the current occupies an appreciable fraction of the total water depth, but in the equatorial current north of Brazil and in the equatorial Pacific k is very close to 1.0 owing to the shallowness of the current over the main thermocline and the great depths of water beneath.

When through a lack of navigational data there is no objective way to find k, it seems not far wrong to assume 1.10, the weighted average of several thousand open-ocean determinations.   On this assumption, the average error in current speed, compared with radio navigation data, turns out to be ±10%.   Making this assumption, and steering by compass with corrections for set due to local observed currents, ships have been held on straight courses through currents for several hundreds of miles.   The results of such experiments compare well with the landfalls that would have been made in still water, allowing for errors in steering and compass adjustment.[11]

Sailings.   The water motion at right angles to the ship's direction of progress is ordinarily measured by the electromagnetic method, because the fore and aft components include the ship's motion through the water. For this reason it has become a regular practice to tow the cable and electrodes astern and turn the ship and the towed cable through one right angle (Fig. 9–15) at intervals of about an hour to measure the component of sea-water motion along the ship's track, and from this to find the direction of flow relative to geographic coordinates.   If this evolution contains a 180° turn, it also provides the data needed to measure the electrochemical potential between the electrodes making contact with the sea.

Electrodes.   Contact between the measuring circuit and the ocean involves a transition from metal wire to sea-water electrolyte, which is the fundamental source of galvanic electricity in every battery.   If the two electrodes have exactly the same contact potential from the metal wire to the sea-water electrolyte, their opposing battery effects in the circuit will cancel out and

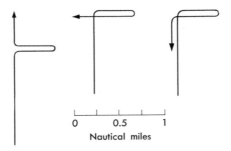

0   0.5   1
Nautical miles

Fig. 9–15.   Course changes used to obtain the zero point and a current fix.

no chemical voltage will be observed. However, two electrodes are seldom so much alike. Even if they are alike at one temperature, salinity, and partial pressure of oxygen, their contact potentials probably will not be alike at another.

The best of present electrodes will drift independently by as much as 0.001 volt under changing environmental conditions. This is the order of magnitude of the motional electromotive force produced by the motion of sea water through the earth's magnetic field. Therefore it is necessary to calibrate the *zero point* of electrical measurements periodically. This is done by sailing the ship on reverse courses through the same water so that on one course the motional emf is, for example, added to the electrochemical signal, and on the reverse course the difference is measured. The average of the two signals is taken as the electrical origin of motional emf measurements (Fig. 9–16), assuming that the water motion has not changed during the calibration. It has become routine practice to make the course reversal for zero point at right angles to base course, so that the component of water motion along the course is measured at the same time. The time spent off base course is limited to that required to allow the cable to tow dead astern for about two minutes following the necessary turns. In most ships the complete evolution requires about ten minutes.

Zero point. When the zero point of a pair of electrodes towed in tandem from a ship underway is known and the indicated rate of transverse motion is reduced to zero, the ship can be said to be steaming on a course which is parallel to the local surface-current direction in either the upstream or downstream sense. If, while tracking the current, the position of the ship is determined at regular intervals by means of a radio navigational aid, a line is generated on a chart which reflects certain properties of the flow. This line, in an accelerated system, does not fall within any of the existing hydrodynamic classifications: streamline, streakline, or trajectory. But it may be considered that the *line of zero set* has a family resemblance to these more conventional concepts, and can be a useful index in natural situations where the field of surface motion is otherwise undetermined.[12]

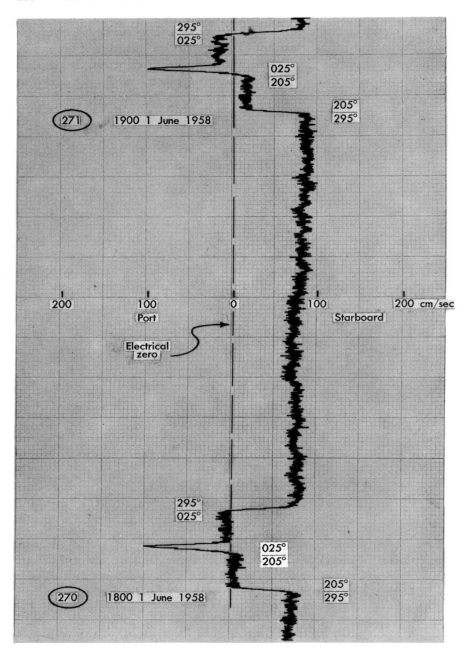

Fig. 9–16.   GEK trace, showing the results of two current-fix evolutions and the position of electrical zero.

Transport. When a ship is making serial electromagnetic measurements of current on a steady course, it is possible to estimate water transport in Loran areas. From observations of the Loran position of the ship and a knowledge of her forward speed, the water velocity can be estimated by the method described in Chapter 8. Since this navigational measurement of the water velocity can be equated to the signal that would be obtained with the apparatus at rest $[(\mathbf{i}_h)_o = 0)]$, it may be said to be equivalent simply to $(\mathbf{c}_h \times \mathbf{K}H_z)_o$. The volume transport of ocean currents can be obtained from a knowledge of the depth of water and the difference resulting from the suitably integrated value of $(\mathbf{c}_h \times \mathbf{K}H_z)_o$ less the integrated value of $(-\rho\mathbf{i}_h)_o$ determined by drifting apparatus between the same limits.

Malkus and Stern (1952)[13] have shown that where dl is an increment of the path traversed by a ship as it crosses an ocean current, a navigational determination of its actual path (by Loran, for example) compared with its dead-reckoning track can provide the integral

$$\int_{-\infty}^{+\infty} \mathbf{c}_h \times \mathbf{K}H_z \cdot dl, \tag{9-23}$$

while the measured potential indicated on the potentiometer recorder and adjusted to the dead-reckoning track provides

$$\int_{-\infty}^{+\infty} \rho\mathbf{i}_h \cdot dl. \tag{9-24}$$

Between finite limits of integration, which for practical purposes may be taken as the shore lines or the points on either side of a current where the current is nil or both small and parallel with the line of cross section, it can be said that the *mean velocity of flow* beneath the line element dl is given by the difference of integral (9-23) less integral (9-24). If dl is made finite and the depth of water z is known or measured, the volume transport through the area $z \cdot \Delta l$ is known. When the sea bottom is irregular or slopes across the direction of flow, the volume transport of current should be obtained by the summation of small increments. The integration can be carried out in one operation if the sea floor is level.

This method has been applied to the Gulf Stream, but the errors may be large since the observations are based on the measurement of small differences between large quantities, and windage causes serious trouble. The procedure is more reliable when observations are made from a submarine.

Motional electric potentials can also be observed with the measuring apparatus and line segments at rest on either side of flowing channels, for example the Straits of Florida and the English Channel.[14] From the observed motional potential difference, it is possible to infer the time rate of change of the volume transport through such channels, provided the

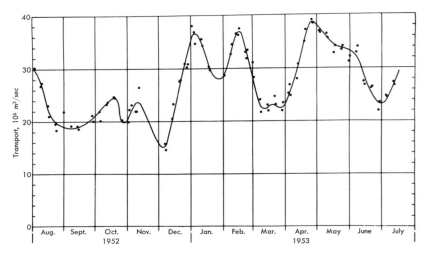

Fig. 9–17.   Mass transport of the Florida Current determined from the electric potential differences measured between Key West, Florida, and Havana, Cuba. [From G. K. Wertheim, 1954, *Trans. AGU*, 35(6).]

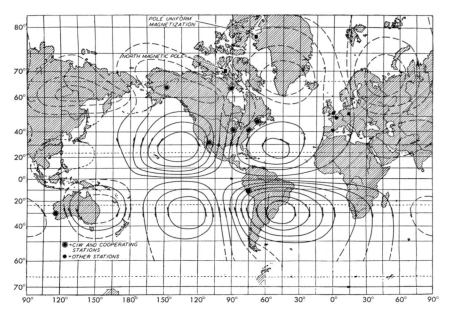

Fig. 9–18.   Pattern of earth currents associated with the momentary position of the subsolar point; equatorial view of the earth-current system at 18 hr GMT. [From *Terrestrial Magnetism and Electricity*, 1939, edited by J. A. Fleming, Dover Publications Inc., 180 Varick Street, New York 14, N.Y.]

resistivity of the bedrock can be estimated and the depth of water is fairly uniform.

If it is assumed that the bottom is flat and nonconducting and that the horizontal variation of the earth's magnetic field can be neglected, the transport is given by $T = Vd/H_z$, where T is the transport in m³/sec, V is the electric potential difference, d is the depth of water in meters, and H_z is the vertical component of the geomagnetic field in webers/m².[15]

The Gulf Stream is narrower than the width of the channel in the Florida Strait and, moreover, the bottom slopes between Key West, Florida, and Havana, Cuba, where measurements have been made (Fig. 9–17). Nevertheless Wertheim (1954)[15] has shown that the fluctuations of transport correspond closely to the tidal signatures of stations in the Gulf of Mexico. From this it is probably valid to reason that the indicated nontidal variations of transport are mainly real. Stommel (1954)[16] also made a series of observations between submarine telegraph relay stations at Bermuda, Horta in the Azores, and continental stations. Much of the theoretical interpretation of these observations from both fixed and moving stations has been provided by H. Stommel (1948),[17] M. S. Longuet-Higgins (1949),[18] W. V. R. Malkus and M. E. Stern (1952),[19] and by Longuet-Higgins, Stern, and Stommel (1954).[20]

Earth currents. From observations on hand, it has been inferred that there is a pattern of earth currents circulating in the crust and oceans, more or less in step with the diurnal motion of the subsolar point (Fig. 9–18). Earth currents due to magnetic storms and perhaps due to ionospheric induction can be recorded in the hydrosphere.*

Longuet-Higgins (1949)[18] indicated that earth currents tend to crowd into estuaries and avoid promontories, taking the path of least resistance between land and sea. It might be expected that the systematic patterns of circulating electric current, such as those described by Gish (1936),[21] would be more clearly developed in the more uniformly conducting medium of the ocean than in the fractured, inhomogeneous rocky masses forming the earth's crust. Whether it is also possible that these circulating currents penetrate so deeply into the outer layers of the earth that they find still more homogeneous conditions in the basement rock of the crust or outer layers of the mantle is difficult to say. There is some evidence, however, that the motional electric potentials developed by the ocean circulation (Fig. 9–19) are only very slightly influenced by the small conductivity of the sea bed or subjacent rocky structures despite their enormous cross sections.

---

* For their general nature see L. Cagniard, 1956, pp. 407–469, *Handbuch der Physik*, S. Flügge, editor, Bd. XLVII, Geophysik I, Berlin: Springer-Verlag; J. S. Fleming, editor, 1939, *Terrestrial Magnetism and Electricity* (Physics of the Earth VIII) New York: McGraw-Hill; and for disturbances observed in high latitudes see A. A. Lebedev, 1957, *Trudy Gos. Okeangr. Inst.*, 40: 50–56.

Fig. 9–19.   GEK recorder, showing controls and the trace obtained from a current fix.

Instrumentation. The instrumentation for electromagnetic observations has been developed almost exclusively around self-balancing potentiometers for both the fixed and drifting station cases.   Laboratory or industrial servo-potentiometers have been used for land-based measurements and modified industrial servo-potentiometers at sea.   The recorder system shown in Fig. 9–19 is known as the geomagnetic electrokinetograph, an unwieldy name usually abbreviated to GEK.   The circuit is shown schematically in Fig. 9–20.

The sensitivity of the system is arranged to keep the nominal signal $(V = H_z S c_h \times 10^{-8})$ for significant water motions (1 cm/sec) above the threshold of recorder response and large compared with the electrochemical drift of potential of the electrodes.

In round numbers the intensity of the earth's magnetic field can be considered to be 0.5 oersted, and a representative sea-water velocity can be taken as 50 cm/sec or about 1 knot, so that the corresponding magnitude

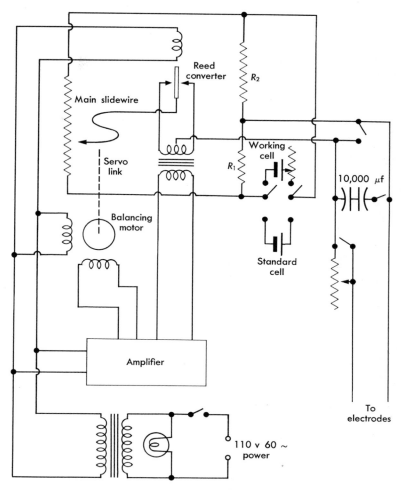

Fig. 9–20.   Schematic circuit diagram of GEK recorder.

of the maximum gradient is 25 emu/cm/knot or 25×10^{-8} volts/cm/knot. If the sampling interval S is as much as 100 meters, the signal intensity is 2.5 mv/100 meters/knot, which is well above the 0.005 mv threshold of modern electronic measuring techniques. It remains then to choose a practical sampling interval which will at the same time bring the signal strength due to water motions as small as 1 cm/sec within the range of sensitivity of a standard self-balancing potentiometer.

As shown in Fig. 9–21, the intensity of the vertical component of the earth's magnetic field ranges from zero at the magnetic equator to a maximum of about 0.600 at the north and 0.720 oersted at the south magnetic

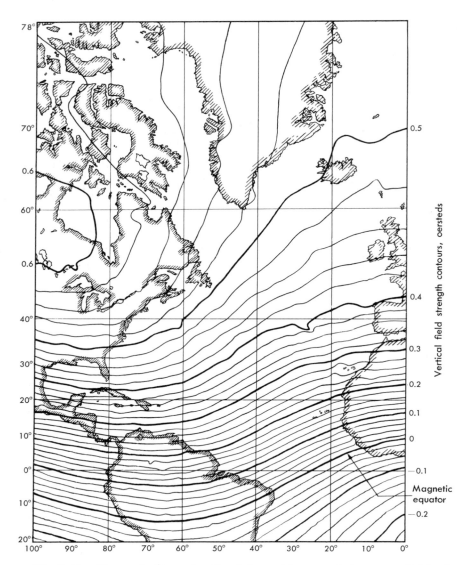

Fig. 9–21.   The vertical intensity, H_z, of the earth's magnetic force over the North Atlantic Ocean.   [Adapted from U.S. Navy Hydrographic Office Chart 1702.]

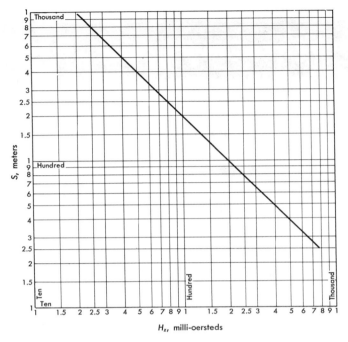

Figure 9–22

poles respectively. To maintain uniform instrumental sensitivity at all magnetic latitudes, it is required that the sampling points in the sea be separated by a distance S such that $H_zS =$ constant. This constant is conveniently chosen at perhaps 1 mv/knot in the nautical system (Fig. 9–22), or what is nearly the same as 0.05 mv/cm/sec in the metric system. This presents two additional choices: (1) to change the distance S between electrodes as observations are carried to different magnetic latitudes or (2) to change the recorder sensitivity. Near the 0.100-oersted isodynamic line the interelectrode length for a recorder having a fixed sensitivity of 1 mv/knot would be about 194 meters, while at 0.700 oersted the appropriate length is 27.6 meters, both of which are entirely practical lengths of cable to tow astern. Similarly, by changing recorder sensitivity at 0.100-oersted intervals and using 100 meters as a standard interelectrode length, the recorder scales of Table 9–1 are required to make observations anywhere on earth except at the magnetic equator.

Electrodes for GEK apparatus (Fig. 9–23) are presently patterned after A. S. Keston's (1935)* nonpolarizing silver-silver bromide electrode. This

---

* A. S. Keston, 1935, *J. Amer. Chem. Soc.*, 57: 1671–1673. This design was suggested to the writer by Professor H. S. Harned of Yale University.

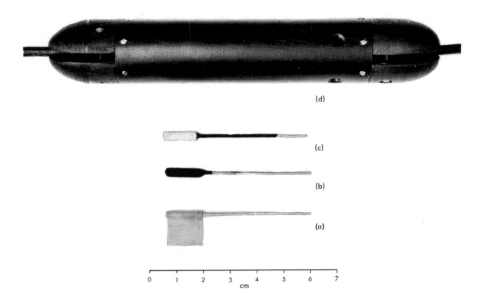

Fig. 9–23.   GEK electrode. (a) The uncoated silver gauze. (b) The gauze rolled and coated (wet). (c) The electrode after firing at 450°C. (d) Housing for electrode and splice to cable.

TABLE 9–1

Vertical magnetic intensity, oersteds	Recorder sensitivity, mv/50 cm/sec	Recorder span, millivolts	Recorder range, millivolts
0.100	0.500	5.00	±2.50
0.200	1.000	10.00	±5.00
0.300	1.500	15.00	±7.50
0.400	2.000	20.00	±10.00
0.500	2.500	25.00	±12.50
0.600	3.000	30.00	±15.00
0.700	3.500	35.00	±17.50

has been found to be adapted to marine purposes if made of silver-silver chloride. In the changing environment of temperature and salt concentration found in the wake of ships, certain pairs of these electrodes have been approximately one order of magnitude more stable than the best commercial alternatives built for laboratory use. Nevertheless, even the most stable pairs of electrodes constructed thus far lack, by perhaps an order of magnitude, sufficient stability to make the electromagnetic method suitable for routine navigational purposes or for the interelectrode length to be made sufficiently short to allow both longitudinal and transverse components of potential to be measured simultaneously. Greater electrode stability or rotation of more closely spaced pairs on a rigid arm might permit the design of equipment capable of measuring continuously both the transverse and longitudinal components of potential without altering course. Such information would yield, to a fair degree of approximation, knowledge of the motion of a vessel with respect to the sea bottom.

Keston-Harned silver-silver chloride electrodes utilize as raw ingredients chemically pure silver chloride powder and silver oxide powder. These can be mixed together in distilled water to form a thick paste which is applied to a silver or platinum helix, or a cylinder of silver or platinum wire gauze, approximately 2.5 cm long by 0.5 to 0.6 cm in diameter. After each of two applications, the paste is fired in a laboratory furnace for one-quarter hour at a temperature of 450°C. The firing decomposes the silver oxide, so that a spongy mass of silver metal and silver chloride is bonded with and encases the silver or platinum wire. It is found that approximately one part silver chloride to nine parts silver oxide is a suitable ratio. It is also found that porosity of 60% or more in this type of electrode increases the stability of a pair. Electrodes made of chemically pure lead also serve in waters where horizontal gradients of surface salinity, temperature, and oxygen are weak.

In the present mode of use GEK electrodes are towed astern to remove them from the disturbance of the earth's magnetic field by the ship's iron, and to avoid the galvanic potential field around the hull. Ordinarily at about one ship length astern the disturbances due to both these causes are quite small.

Long cables that are more dense than sea water introduce a vertical component of length z which can cut the horizontal component of the earth's magnetic field H_x, producing a signal proportional to $H_x(z \cdot v \sin \theta)$, where v is the speed of the ship and θ is the magnetic azimuth of the course steered. Since the horizontal component of magnetic flux has a fixed azimuth, the result confuses observations by adding a spurious signal interpretable as a component of water motion toward magnetic south in the northern magnetic hemisphere, and toward the north in the southern magnetic hemisphere. The effect is disturbing within approximately 30° of the magnetic equator.

Various attempts have been made to level the towed cable by means of plastic floats. The most promising development, however, appears to be the use of neutrally buoyant cable, together with either very small electrode housings or neutrally buoyant housings, which permit the cable to tow on the sea surface. Knauss and Reid (1957)[22] suggest that this arrangement

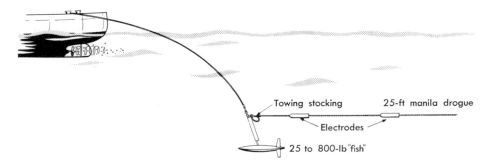

Fig. 9–24.   Method for towing a neutrally buoyant GEK cable horizontally below the surface.

Fig. 9–25.   The 400-pound "fish" in Fig. 9–24 being lowered from *Atlantis* for towing at a depth of 100 meters. The data in Fig. 11–8 were obtained with this apparatus. [Photograph by J. Hahn.]

subjects the cable to windage causing it to tow at an angle to the ship's way, but that this effect can be reduced by weighting the cable at a point near the ship (Fig. 9–24). At speeds above 5 knots a neutrally buoyant cable is observed to tow in a level attitude behind the weight. Figure 9–25 shows a streamlined 400-lb weight used by the author to tow electrodes to depths approaching 500 meters. Depressing the towed cable also reduces the signal due to surface waves.

In a seaway, waves move the cable and electrodes through the magnetic field of the earth and generate an alternating signal having a wavelike frequency and an amplitude which can be large compared with the steady signal due to ocean currents. Since this signal is a complicated function of the orientation of the ship and wave crests with respect to magnetic north, it has not been a useful measure of wave motions, and is generally filtered out of the record. A suitable low-pass filter is a simple *R-C* network having a variable time constant up to 20 sec. The potentiometer recorder input impedance ordinarily cannot exceed 2500 ohms, so that with this amount of variable resistance in series with the electrodes it is necessary to use 10,000 microfarads of capacitance across the electrodes to damp out the wave signals due to a long-period swell. A time constant of 6 to 12 sec is usually required, so that the full series resistance is not often used.

Combined uses of electromagnetic and geostrophic methods

The International Ice Patrol has made routine use of the electromagnetic method, in addition to geostrophic measurements of ocean currents in the North Atlantic Ocean and Labrador Sea where icebergs are often sighted (Figs. 9–26 and 9–27). Since the electromagnetic method is more reliable as a current-direction indicator than as an index of current speed, their practice has been to align the contours of dynamic height in maps of sea-surface topography with the direction of flow obtained from electromagnetic measurements along the courses between hydrographic stations. This procedure has simplified the contouring process and reduced the uncertainty of interpolation between widely spaced stations and between lines of stations, but makes the assumption that tidal and wind-driven motions are absent.

The electromagnetic method has also been used to simplify the problems of contouring synoptic bathythermograph surveys of complicated water temperature structures in and near the Gulf Stream. It is often assumed that in this part of the North Atlantic the isotherms parallel the direction of water motion, so that a chart of the horizontal distribution of temperature is representative of the current pattern. On several occasions the results of an electromagnetic survey have been added to a thermal survey of the Gulf Stream to aid in drawing the crossing angles of the isotherms and the

Fig. 9–26.   *Albatross III* "GEKing" in the North Atlantic.   [Photograph by J. Hahn.]

ship's tracks, with some improvement in general confidence concerning the interpretation of both kinds of data.   It is found that the corrected current speeds also agree with geostrophic expectations from the temperature field, being strongest in the regions of closely spaced isotherms and weak or variable where the thermal structure is open or indefinite.

Prospects

Because there is no wholly satisfactory method for making current measurements at sea, fresh approaches are always being sought.   Among these

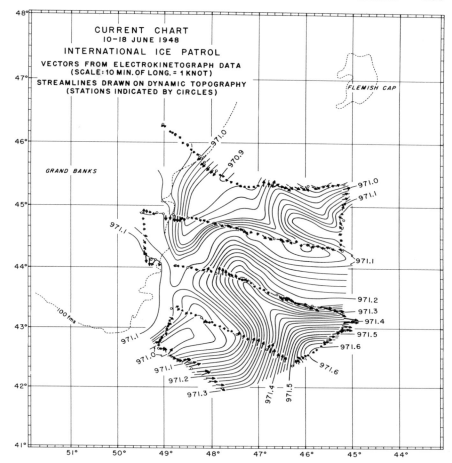

Fig. 9–27.   Coordination of GEK vectors with dynamic contours.

in view at present are the modern developments in inertial navigation systems[23] which seem to offer promise as a means for sensing both horizontal and vertical accelerations upon entering ocean currents in submerged vehicles. With the aid of the precise vertical reference these systems provide, it may also be possible to measure the topography of the sea-surface slopes accompanying both barotropic and baroclinic flows. If this step can ever be taken (for a discussion of the problem, see Reference 20 of Chapter 8), it may constitute a major advance in the technical resources of physical oceanography.

STUDY QUESTIONS

1. Calculate the work done per unit mass in lifting a water parcel 9.80 linear meters at latitudes 0°, 30°, and 75°. Express these results in dynamic meters.

2. Give definitions for the following surfaces: geopotential, isobaric, isosteric, isopycnal.

3. What is the standard ocean? What service does the concept of the standard ocean perform?

4. Two forms of the geostrophic formula are

$$C_g = \frac{g}{f} \left(\frac{\delta z}{\delta n} \right)_p \quad \text{and} \quad C_g = \frac{10}{f} \frac{(\Delta D)_A - (\Delta D)_B}{L} .$$

(a) Given the dynamic depth anomaly at two stations, 50 km apart, as 0.837 and 0.941 dynamic meters, calculate C_g relative to the 500-decibar surface at latitude 30°. Assume that there is no motion at the level of the reference surface.

(b) At each station, what is the depth of the 500-decibar surface in linear meters?

(c) Determine the slope of the sea surface between the two stations (cm/km).

5. Given the temperature 4.16°C and salinity 34.96 $^0/_{00}$ of a water parcel at a pressure of 1000 decibars, find its specific volume anomaly and density *in situ*. (See Reference 1.)

6. Given a sounding as shown in the tabulation, (a) prepare a *T-S* diagram and make a defensible guess as to the site of the sounding; (b) calculate σ_t and the cumulative dynamic depth anomaly for each level of sampling.

Depth, meters	Temperature, degrees centigrade	Salinity, $^0/_{00}$
0	18.17	33.46
10	18.17	33.46
20	18.17	33.46
30	17.92	33.45
50	17.06	33.42
75	15.85	33.42
100	14.17	33.42
150	10.62	33.49
200	9.10	33.78
250	8.10	33.95
300	7.39	34.03
400	6.40	34.10
500	5.80	34.20
600	5.27	34.26
700	4.90	34.34
800	4.55	34.40
1000	3.95	34.43

7. Show that, under certain assumptions, a GEK measures the iR drop in the ocean water between the towed electrodes.

8. Why does the GEK give reliable indications of surface-current direction in the open sea but generally underestimate the speed of motion? Under what circumstances will the indicated speed of flow be too high?

9. What signals would you expect to receive on a GEK recorder mounted in a submarine headed at right angles to the water motion (a) in the body of the Gulf Stream, (b) beneath the Gulf Stream, and (c) when the boat has been brought to rest on the bottom beneath the Gulf Stream?

10. What signals would be measured if a pair of electrodes and connecting cable extend from surface to bottom and if (a) they were moored vertically in the ocean, (b) they were free to drift with the mean motion?

11. Suppose a pair of electrodes were mounted a certain distance apart at right angles to the direction of motion of a ship. What signals would be received? Determine whether the system could be used to measure the ship's speed of advance over the bottom, its speed relative to the water, or some other quantity.

12. Calculate the surface-current velocities from the GEK record reproduced in Figure 9–16. (Assume $k = 1.10$.)

REFERENCES

1. E. C. LaFond, 1951, *Processing Oceanographic Data*, U.S. Navy Hydrographic Office Publication No. 614.

2. V. Bjerknes, 1910, "Hydrographic Tables," *Dynamic Meteorology and Hydrography*, Carnegie Institution of Washington, Publication No. 88.

3. V. Bjerknes and J. W. Sandström, 1910, p. 31, "Part I: Statics," *Dynamic Meteorology and Hydrology*, Carnegie Institution of Washington, Publication No. 88, 146 pp.

4. A. E. Parr, 1938, *J. Mar. Res.*, 1: 119–132, 133–154.

5. A. Defant, 1950, *J. Mar. Res.*, 9: 120–138.

6. G. Wüst, 1933, *Hydrogr. Rev.*, 19(2): 28–49.

7. M. Faraday, 1832, *Phil. Trans.* for 1832, Part 1, p. 176; 1839, p. 56, *Experimental Researches in Electricity*, Vol. 1, London: Richard and John Edward Taylor.

8. M. S. Longuet-Higgins, 1949, *Mon. Not. R. astr. Soc. geophys.*, Suppl. 5: 285–307.

9. F. B. Young, H. Gerrard, and W. Jevons, 1920, *Phil. Mag.*, 6, 40: 149–159.

10. W. S. von Arx, 1950, *Pap. phys. Oceanogr.*, 11(3): 62 pp.

11. W. S. von Arx, 1951, *J. Inst. Navig.*, 4: 117–125.

12. W. S. von Arx, 1960, *Deep-Sea Res.*, 7: 219–220.

13. W. V. R. Malkus and M. E. Stern, 1952, *J. Mar. Res.*, 11: 97–105.

14. M. S. Longuet-Higgins and N. Barber, 1948, *Nature, London*, 161: 192–193.

15. G. K. Wertheim, 1954, *Trans. Amer. geophys. Un.*, 35: 872–882.

16. H. Stommel, 1954, *Arch. Met.*, Wien, A, 7: 292–304.

17. H. Stommel, 1948, *J. Mar. Res.*, 7: 386–392.

18. M. S. Longuet-Higgins, 1949, *Mon. Not. R. astr. Soc. geophys.*, Suppl. 5: 285–307.

19. M. V. R. Malkus and M. E. Stern, 1952, *J. Mar. Res.*, 11: 97–105.

20. M. S. Longuet-Higgins, M. E. Stern, and H. Stommel, 1954, *Pap. phys. Oceanogr.*, 13(1): 37 pp.

21. O. H. Gish, 1936, *Sci. Mon., N.Y.*, 43: 47–57; 1936, *J. Wash. Acad. Sci.*, 26: 267–289.

22. J. A. Knauss and J. L. Reid, 1957, *Trans. Amer. geophys. Un.*, 38: 320–325.

23. P. J. Klass, 1956, *Aviat. Week*, Special Report: 19 pp.; M. Schuler, 1923, *Phys. Z*, 24: 344–350.

SUPPLEMENTARY READING

BÖHNECKE, G., 1932, pp. 208–244, *Ozeanographische Methoden und Instrumente* (*Wissenschaftliche Ergebnisse der Deutschen Atlantischen Expedition auf dem Forschungs-und Vermessungsschiff "Meteor" 1925–1927*, Bd 4, Teil 1), by Georg Wüst, Gunther Böhnecke, Hans H. F. Meyer, Berlin und Leipzig: Walter de Gruyter & Co.

CAGNIARD, L., 1956, "Electricité tellurique," pp. 407–469, *Handbuch der Physik*, S. Flügge, editor, Bd. XLVII, Geophysik I, Berlin: Springer-Verlag.

DEFANT, A., 1929, "Stabile Lagerung ozeanischer Wasserkörper und dazugehörige Stromsysteme," *Veröff. Inst. Meeresk.*, Univ. Berl., A. 19: 33 pp.

FLEMING, J. A., editor, 1939, *Terrestrial Magnetism and Electricity* (Physics of the Earth VIII), New York: McGraw-Hill.

Instruction Manual for Oceanographic Observations, 1955, 2nd ed., U.S. Navy Hydrographic Office Publication No. 607.

LaFOND, E. C., 1951, *Processing Oceanographic Data*, U.S. Navy Hydrographic Office Publication No. 614.

PROUDMAN, J., 1953, *Dynamical Oceanography*, London: Methuen.

SVERDRUP, H. U., M. W. JOHNSON, and R. H. FLEMING, 1942, *The Oceans*, New York: Prentice-Hall.

Tables for Sea Water Density, 1952, U.S. Navy Hydrographic Office Publication No. 615.

CHAPTER 10

Laboratory Models

The use of laboratory models to aid in the study of ocean currents is a new aspect of physical oceanography that has developed with the growth of interest in estuarine circulations and in synoptic changes in the ocean. Two quite different experimental approaches are employed: one favors physical study of the validity of analytical models and the other (with which this chapter is concerned) attempts to produce simple analogues of nature. The study of analogues has an element of subjectivity, for with changes of scale from the dimensions of estuaries through those of seas to oceans, different sets of forces become important.

Three classes of circulations can be distinguished by the predominant manner in which the pressure gradient forces are opposed, namely, (1) by friction, (2) by a combination of friction and Coriolis force, and (3) by the Coriolis force alone. In a broad way the first applies to small-scale flow in rivers, narrow estuaries, and some lakes; the second to the circulations in marginal and mediterranean seas and some very large lakes; and the third to ocean-current systems. The three cases can be characterized more or less as shown in Table 10–1.

Similarity and characteristic numbers

Similarity in models can be considered as being graded from *geometrical* through *kinematic* to *dynamic*. A geometrically similar model has similarity of form, and the ratios of all homologous dimensions are equal. In a model having kinematic similarity the paths and patterns of motion are geometrically similar to those of homologous occurrences in the prototype, and the ratios of homologous velocities are equal at all times. Dynamic similarity is achieved when the ratios of homologous masses and forces affecting motion are equal at all times. In fluid models kinematic similarity, of necessity, involves elements of dynamic similarity. Complete dynamical similarity is difficult to achieve, but workable approximations can often be managed.

TABLE 10-1

Local circulations	Regional circulations	Planetary circulations
Bays, rivers, narrow estuaries, shallow lakes under 100 km in size.	Deep bays, large lakes, marginal and mediterranean seas, from 100 to 1000 km in extent.	Oceans and ocean systems over 1000 km in extent.
Pressure gradient force opposed by friction; effects of earth rotation negligible.	Pressure gradient force opposed by friction and deflecting force of earth rotation.	Pressure gradient force is opposed by Coriolis force alone.
Pressure gradients due to: river flow, runoff, evaporation, seepage, local wind, tides, salt-water intrusion.	Pressure gradients due to: synoptic winds, synoptic pressure gradients (air), balance of runoff, evaporation, precipitation, tides.	Pressure gradients due to: planetary winds, climatic distribution of seasonal winds, insolation, evaporation, precipitation.
Motion directly across isobars predominates. Seiching is rectilinear.	Motion has a cross-isobar component. Seiching tends to be amphidromic.	Motion predominantly geostrophic. Inertia circles negligibly small.
System approximated on a flat, stationary reference frame using ordinary Froude (and Reynolds) modeling criteria.	System approximated on a flat, rotating reference frame using rotating Froude (and Reynolds) modeling criteria.	System approximated on a curved, rotating reference frame using vorticity equation as a modeling criterion.
$L_r = \dfrac{1}{10^3}$ to $\dfrac{1}{1.5 \times 10^4}$	$L_r = \dfrac{1}{10^4}$ to $\dfrac{1}{5 \times 10^5}$	$L_r = \dfrac{1}{10^7}$
$T_r = \dfrac{1}{10^3}$	$T_r = \dfrac{1}{10^3}$	$T_r = \dfrac{1}{10^3}$ to $\dfrac{1}{10^4}$
$d_r = 10$ to 10^2	$d_r = 10^2$	$d_r = 10^5$
$\Omega_m = 0$	$\Omega_m = \dfrac{\Omega_p (\sin\phi)_p}{T_r}$	$\Omega_m = \sqrt{\dfrac{2gD_{90}(e-1)}{R^2}}$

The ideal in modeling is to construct a system which is like its proto-type in all physically significant respects. These factors are represented in several terms of the equation of motion (Chapter 4).

$$\frac{d\mathbf{c}}{dt} = -2\mathbf{\Omega} \times \mathbf{c} - \frac{1}{\rho}\nabla p + \mathbf{g} + \mathbf{F}, \qquad (10\text{--}1)$$

or in words,

Inertial term $=$ $-$ Coriolis term

$-$ pressure gradient term

$+$ gravity term

$+$ friction term. $\qquad (10\text{--}2)$

Since the totality of forces cannot be easily reconciled in laboratory models, a few important terms are usually singled out.

Ratios of terms are used to define dimensionless numbers, for example:

Euler number, $\qquad$ Eu $\equiv \dfrac{\text{inertial term}}{\text{pressure gradient term}} = \dfrac{c}{(2\,\Delta p/\rho)^{1/2}}$,

Froude number, $\qquad$ Fr $\equiv \dfrac{\text{inertial term}}{\text{gravitational term}} = \dfrac{c}{(L\,\Delta\gamma/\rho)^{1/2}}$,

Rossby number, $\qquad$ Ro $\equiv \dfrac{\text{inertial term}}{\text{Coriolis term}} = \dfrac{c}{\Omega R}$,

Reynolds number, $\qquad$ Re $\equiv \dfrac{\text{inertial term}}{\text{frictional term}} = \dfrac{cL}{\nu}$,

Ekman number, $\qquad$ Ek $\equiv \dfrac{\text{Coriolis term}}{\text{frictional term}} = \dfrac{\Omega R^2}{\nu}$,

where c is a characteristic velocity, L a characteristic length, R the length of an equatorial radius, ϕ the geographic latitude, ρ the fluid density, ν the kinematic viscosity μ/ρ, Ω the angular velocity of rotation, p the pressure, and γ specific weight. These ratios are characteristic of flow conditions in which the designated forces predominate. For similarity to prevail, the numerical value of the nondimensional number must be the same for the model as for the prototype. A model based primarily on one or another of these dimensionless numbers is usually referred to as a Froude model, Reynolds model, etc., according to the case. (A picture of William Froude is reproduced in Fig. 10–1 and of Osborne Reynolds in Fig. 10–5.)

As a preliminary step in scaling a natural phenomenon down to labora-tory size, it is useful to construct a ratio of units which by convention is

$$\mathcal{J}_r = \frac{\mathcal{J}_m}{\mathcal{J}_p}, \qquad (10\text{--}3)$$

Fig. 10–1. William Froude. [From H. Rouse and S. Ince, 1957, *History of Hydraulics*, Iowa City: Iowa Institute of Hydraulic Research, State University of Iowa.]

where J_m is a number representing the dimension of a certain property of the model and J_p the dimension of the homologous property in the prototype. When the characteristic length ratio of a model is fixed by the available laboratory space or equipment, the ratios of the other units included in a dimensionless group can be altered to hold the numerical value of each dimensionless group constant. In this way time may have to be counted in larger or smaller units, or the unit dimensions of other related properties may have to be altered to permit the fluid motion in the model to be regarded as behaving in the same physical manner as that in the prototype.*

It is often difficult to maintain similarity where the ratios of two or more dimensionless numbers must be held constant. There is sometimes no other alternative than to make compensatory adjustments of the properties or units in two dimensionless numbers to make the system behave in some desirable manner. When this approximation is made, similarity *per se* is lost. The system is then analogous to nature only in the respect specifically desired.

In the case of a Froude model, which reproduces with kinematic similarity the pattern and velocities of currents associated with shallow-water gravity waves propagated at a speed $c = \sqrt{gd}$, we are concerned with

---

* The problems of dimensional analysis and modeling have been studied since the rise of classical physics. Recent discussions of the subject are to be found in H. L. Langhaar, 1951, *Dimensional Analysis and Theory of Models*, New York: Wiley, and in Chapter 1 of *Advanced Mechanics of Fluids*, 1959, H. Rouse, editor, New York: Wiley.

the ratio of inertial terms to gravity; that is, with the dimensionless quantities $(c^2/Lg)_m$ and $(c^2/Lg)_p$. The velocity ratio can be written

$$c_r^2 = \frac{c_m^2}{c_p^2} = \left(\frac{L_m}{L_p}\right)\left(\frac{g_m}{g_p}\right). \tag{10-4}$$

Since the model will be operated in air under the same gravitational field as its prototype, $g_m/g_p = 1$. This leaves L_r, the length ratio, and the velocity ratio of the gravity wave c_r as adjustable parameters. If now we distinguish between d as the vertical length and L as the horizontal length, then c_r, the wave velocity ratio, which is itself L_r/T_r, where T_r is the time scale, becomes

$$c_r = \frac{L_r}{T_r} = \sqrt{d_r}. \tag{10-5}$$

If the Froude model is geometrically similar to its prototype, then $d_r = L_r$. But it may be necessary to depart from similarity by distorting vertical dimensions such that $d_r > L_r$. Table 10–2 gives a partial list of scale ratios in terms of distorted and undistorted length. Many others can be derived.

TABLE 10–2*

Property	Undistorted Froude model	Distorted Froude model
Horizontal length	L_r	L_r
Vertical length	L_r	L_r
Time	$(L_r)^{1/2}$	$L_r(d_r)^{-1/2}$
Velocity	$(L_r)^{1/2}$	$(d_r)^{1/2}$
Total discharge	$(L_r)^{5/2}$	$L_r(d_r)^{3/2}$

* Assuming that the model and prototype are operating under the same gravitational acceleration, and that the natural and experimental fluids are identical.

Models of small inshore areas

The small-scale oceanographic model, such as that in Fig. 10–2, may be defined as one in which gravity and friction are dominant forces and the direction of flow is primarily down the gradient of pressure. This is the case in most rivers, bays, and narrow estuaries. The treatment of these small-scale oceanographic problems follows engineering precedent closely except for the introduction of forces due to local wind stress and external flow fields. Such forces have been applied most successfully to Froude models which involve a single layer of liquid.

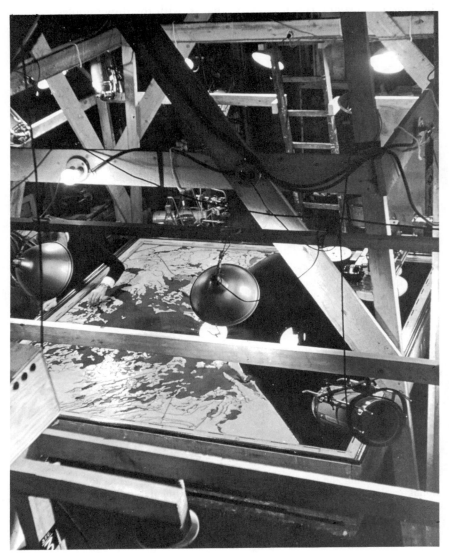

Fig. 10–2.   Barataria Bay model as seen from the western side.   [From Chapter 1, *Physics and Chemistry of the Earth*, Vol. 2, 1957, L. H. Ahrens, *et al.*, editors, London: Pergamon.]

The choice of a suitable horizontal scale for a Froude model depends primarily on fitting the geographic area into the dimensions of the laboratory. In the absence of density contrasts in oceanographic models, L_r may range from 1/5000 to as little as 1/200,000. In characteristic examples, the salinity distribution and current system of Puget Sound have been studied in a model with a horizontal scale of 1/40,000,[1] while Bikini, the Rongelap Atolls, and surrounding parts of the Pacific Ocean were represented for purely kinematic analysis on horizontal scales of 1/100,000 and 1/150,000 respectively.[2]

The vertical scale must generally be chosen so that the least depth of water in which the flow is to be studied is approximately 1 cm to prevent capillary effects from being important. Because of the complexity of some shore lines, it is usually imprudent to attempt to build a model of an estuary on a horizontal scale smaller than $L_r = 1/20,000$. In general the vertical exaggeration may be larger than that usually regarded as acceptable for engineering purposes. Values of $d_r/L_r = 100$ may be considered.

In simplest terms the Froude law concerns the rate of motion of a long gravity wave in shoal water. Since the acceleration due to gravity is the same in both the model and the prototype, the velocity ratio c_r is entirely controlled by the square root of the vertical scale, $\sqrt{d_r}$. Thus

$$c_r = \sqrt{d_r} = \frac{L_r}{T_r}. \tag{10-6}$$

It can be seen from this equation that the time scale shortens with increasing depth and lengthens with an increase in the horizontal scale of the model. One is led to a choice of scales by a path which touches first the minimum depth requirement and then an adjustment of the length L_r and time T_r scales such that either

$$\frac{L_r}{T_r} = \text{constant} = \sqrt{d_r}, \tag{10-7}$$

or

$$\frac{L_r}{\sqrt{d_r}} = \text{constant} = T_r, \tag{10-8}$$

where L_r is adjusted to the dimensions of the laboratory. The time ratio is usually the most flexible parameter available and determines the frequencies of the tides.

The circulation in an estuary is often dominated by the tides. The reproduction of tides in oceanographic models of estuaries may be simplified if taken from broader considerations than those relating to the tidal waves in the estuary itself. When the tide generator is located offshore (Fig. 10–3) and arranged to act along a cotidal line in a modeled portion

Fig. 10–3.   Barataria Bay model seen from above, showing tidal streams. [From Chapter 1, *Physics and Chemistry of the Earth*, Vol. 2, 1957, L. H. Ahrens, *et al.*, editors, London: Pergamon.]

of the ocean, the tidal wave will advance into shoal water with the proper phase differences and heights determined by the sequence of depths along each route.   However, the tidal coefficients are less well known offshore than at coastal stations.   As a first approximation, the offshore tidal components can be estimated from a weighted average of the tidal coefficients at the most seaward of a number of shore stations some distance away in both directions along the coast.   Further empirical adjustments are then made as comparison of the tide records from the model and those from nature may suggest.

The complexity of a tide generator depends on the fidelity with which the tidal trace at any point is to be reproduced.   The important tidal species having speed numbers sufficiently close together can be collected to simplify the mechanism.   Separate tidal components must be set apart if their amplitudes and speed numbers are such that they largely determine the characteristic sequence of tides.

If neither the diurnal inequality nor the semimonthly range sequence is of importance to the experiment, the tide can be approximated by a single element driven at the weighted mean rate for all the significant semidiurnal species. If the diurnal inequality is too great to be neglected but the diurnal terms are small, two semidiurnal elements will suffice. If the diurnal species and semimonthly species are both significant, three elements will be required, as in Fig. 10–4(a). Only rarely will four or more elements be necessary.

The volume of water to be displaced in the model, to effect the rise and fall of tides, is the volume ratio $(L_r^2 d_r)$ times the volume change (tidal prism) in the corresponding areas of the prototype. To calculate the volume contribution of each of the tide-generating elements in the model, it is convenient to scale the mean volume of the tidal prism and then to prorate this among the important harmonic constituents in proportion to the ratio of their amplitudes to the sum of all terms. By this means the contributions of the neglected constituents are distributed impartially among other terms to effect the necessary total rise and fall.

The mechanical means by which tides have been introduced vary in their complexity. Less complicated equipment includes adjustable weirs in river models or simple cylindrical drums driven by Scotch yokes at appropriate speed numbers such as shown in Fig. 10–4(b) for a system of bays. More complicated equipment includes electromechanical servo-mechanisms controlling either a single plunger or a pump, in turn controlled by a cam representing a day's or fortnight's sequence of tides actuating a cam follower. The effects of evaporation in the model can be compensated automatically if the cam follower is coupled to a point or capacitance gauge or other device (sometimes optical) for detecting the rise and fall of the free surface with respect to an arbitrary datum. Such servo apparatus has become common in hydraulic laboratories during the past decade.

The tidal circulation in estuaries often involves the interaction of fresh or brackish water with sea water of higher density. When water is used in the Froude model, these density differences should be maintained one to one, that is, $\rho_r = 1$ and $\Delta\rho_r = 1$. This leads to satisfactory results when the problem being studied involves only the kinematics of flow. For example, in the intrusion of a large mass of salt water flowing in from the sea under fresh water in an estuary, the velocities, volumes, and times of arrival of the main water masses may be correctly represented by following the Froude law, but the mixing and mutual dilution of the two water masses will usually be quite dissimilar and unrealistic, particularly if the model is distorted vertically. In such situations one must either satisfy both Reynolds (Fig. 10–5) and Froude criteria or depart from strict scaling procedures.

(a)

(b)

Fig. 10–4. (a) A simple form of tide generator having three displacement cylinders. (b) Scotch yokes and drive of the tide generator in (a). [From Chapter 1, *Physics and Chemistry of the Earth*, Vol. 2, 1957, L. H. Ahrens, *et al.*, editors, London: Pergamon.]

Fig. 10–5.   Osborne Reynolds. [From H. Rouse and S. Ince, 1957, *History of Hydraulics*, Iowa City: Iowa Institute of Hydraulic Research, State University of Iowa.]

It is possible through a choice of fluids other than water and through suitable linear dimensions to satisfy both the Froude and Reynolds criteria simultaneously.   As fortunate as this possibility seems, it is rarely utilized because few fluids have the necessary properties when the model is small compared with its prototype.   Specifically, equality of the Froude number in model and prototype requires that the time ratio $T_r = L_r^{1/2}(g)_r^{-1/2}$ while the Reynolds law requires that $T_r = (L_r^2\rho/\mu)_r$ where L is the length scale, ρ the fluid density, and μ the dynamic viscosity.   Equating these two requirements leads to the relationship

$$T_r = L_r^{1/2}(g)_r^{-1/2} = \left(\frac{L^2\rho}{\mu}\right)_r. \qquad (10\text{–}9)$$

If the model and prototype operate under the same gravitational acceleration, the ratio g_r equals 1, and a length scale satisfying both criteria at the same time scale will be

$$L_r = \left(\frac{\mu}{\rho}\right)_r^{2/3}. \qquad (10\text{–}10)$$

Unless L_r is nearly 1 or chosen such that a real fluid other than water satisfies the above criterion of similarity, it is difficult to meet its requirements.

As a less satisfactory alternative one may reduce the density contrast between the fresh and salt layers, and thus reduce the buoyancy forces tending to cause the two layers to stratify.   This permits the vertical component of the existing level of turbulence to cause the layers to mix more realistically. Equalizing the force per unit area acting in the horizontal and vertical directions on the vertically distorted unit volume

$$\Delta\rho_m = \Delta\rho_p \left(\frac{L_r^3}{L_r^2 d_r} \right) = \Delta\rho_p \left(\frac{L_r}{d_r} \right) \qquad (10\text{--}11)$$

amounts to a reduction of density contrast by the reciprocal of the vertical exaggeration. For practical reasons this cannot be carried to extremes. In oceanographic models a vertical exaggeration of 100 is often necessary; hence the normal difference in density between ocean and river water, 0.025 gm/cm^3, would thereby be reduced to 0.00025 gm/cm^3, which is beyond experimental control. Difference in density amounting to 0.001 gm/cm^3 can be maintained experimentally, and if it is sufficient to represent approximately the mixing of a salt wedge intrusion and its rate of motion, it is possible to choose a horizontal length scale and/or a suitably adjusted vertical exaggeration d_r/L_r such that the corresponding difference in density in the fluids is greater than 0.001 gm/cm^3.

The effects of wind stress on the circulation in estuaries and bays are often important and sometimes predominate over the tide. To scale the wind-driven circulation in a model requires more knowledge of the dynamics of momentum transfer across the air-sea interface than we now possess. Thus empirical procedures must be made to suffice.

These procedures involve the calibration of a set of blowers arranged to move air across the model at rates which will produce circulations in the surface layer of the model which are "in scale" and in accordance with field observations under given wind conditions. It is desirable that the field data represent conditions where wind is the only important influence. Alternatively one may assume that where the fetch is greater than an equivalent of 5 to 10 thousand meters and a steady state exists, a layer of surface water will be moved downwind at a speed between 2 and 3% of the steady wind speed measured at some level between 3 and 10 meters above the natural water surface. In practice the actual speed of the wind on the model is disregarded and only the velocity of the wind-driven motion of the surface layer is scaled. The calibration of blowers supplying the wind-drift effects are referred to a clean water surface having some standard surface tension.

Wind waves on an oceanographic model are generally slight, and those present are usually dominated by capillary rather than gravitational forces.*

---

* However, internal waves can dominate the circulation. See C. H. Mortimer, 1951, *Un. géod. géophys. int., Ass. int. Hydrol.*, 3: 335–349 ;1951, *Verh. int. Ver. Limnol.*, 11: 254–260.

Capillary waves, in contrast to gravity waves, move faster as their lengths diminish. To be in scale a modeled wave would have a height d_r and a length L_r in a distorted open channel model. Such dimensions soon become physically impossible as the vertical exaggeration is increased. In cases where the roughness of the sea surface has been accounted for through direct scaling of the wind-drift current, it is unnecessary to be concerned with waves. But if waves contribute to mixing, a small model with vertical distortion cannot be used.

Because the natural situation is never a pure case to which any special modeling law applies exactly and because of scale effects (the consequences of providing unavoidably different combinations of forces than those in nature), a new model cannot initially be relied on to represent nature quite faithfully. The model must be compared with field data from its prototype.

To begin, field observations may be made repeatedly at a number of points believed to be representative of the general character of the fluid motions. The model is verified against these as a first approximation. Following this, the truly critical points usually reveal themselves in the model, and with further field observations at these new points verification can proceed. When this cycle of comparisons has been carried forward to a practical conclusion, it is assumed that the behavior of the model at all intermediate points is also essentially correct.

Verification in its simplest form consists of roughening and smoothing the courses of flow. In a distorted open channel model, it is usually difficult to make the bottom rough enough, and in the undistorted model it is equally difficult to make the bottom smooth enough. The optimum distortion for matching the roughness of the chosen bottom materials can be determined from the Manning formula, which is discussed in some detail by Rouse (1938, 1946) [3] and applied to specific problems by Stevens (1942). [4]

In oceanographic models the minimum depth restriction (1 cm) may preclude application of the Manning formula. Vertical exaggeration must then be increased, in which case the bottom roughness usually becomes too slight. Roughness elements may be placed in a suitable pattern on the bottom, or if the horizontal scale is very small and connecting channels are narrow, it may be more convenient to interpose wire screens of suitable meshes in the channels to retard the flow.

Models of intermediate size areas

The circulation in marginal seas and the smaller mediterranean seas of the earth are influenced by the same forces considered for estuaries, and yet may be large enough for the effects of the earth's rotation to appear. This would be manifested in shallow seas by a tendency for a tidal flow or impulsively generated wind currents to follow inertial paths closing a circuit once each half-pendulum day, and for other kinds of motion to be developed at an angle *cum sole* from the direction of the pressure gradient. This angle should increase as the frictional influence of the bottom grows smaller in comparison with that of the Coriolis force. Ideally, the angle increases to 90° when frictional influences are negligible and the state of motion is steady enough to permit the flow to become quasi-geostrophic.

If the area to be modeled has dimensions on the order of 1000 km on a side, and if in nature the frictional influence of the bottom on the flow is small compared with the Coriolis force, then it becomes desirable to rotate the Froude model in order to provide conditions favoring the development of Kelvin waves and amphidromic tidal motions. These conditions are related to inertial motion in which the deflecting force of earth rotation is balanced by the centrifugal force accompanying the apparent curvature of the particle's trajectory, that is,

$$\frac{c^2}{r} = (2\Omega \sin \phi)c, \tag{10-12}$$

where c is the velocity of the parcel with respect to the earth and r is the radius of curvature of the path.

We will assume that the north-south dimensions of the model in degrees of latitude are small compared with a right angle, so that the variation of Coriolis parameter with latitude can be neglected. The model can then be centered on a turntable. We will also assume that the model is being operated under the same gravitational acceleration as that acting on the prototype, and that water will be used to fill it. To provide the effects of earth rotation in proper scale, it is required that scaling ratios be formed from

$$\mathcal{J}_r = \frac{[(c^2/r) = (2\Omega \sin \phi)c]_m}{[(c^2/r) = (2\Omega \sin \phi)c]_p}, \tag{10-13}$$

where the subscripts m and p again refer to the model and the prototype respectively. If the Coriolis force ratio be defined as $f_r c_r = f_m c_m / f_p c_p$, we can write the balance of specific inertial forces as follows:

$$\frac{c_r^2}{L_r^2} = \frac{(c^2/r)_m}{(c^2/r)_p} = \frac{(2\Omega_m \sin \phi_m)c_m}{(2\Omega_p \sin \phi_p)c_p}. \tag{10-14}$$

When the axis of rotation of the model is vertical, $\sin \phi_m = 1$. In terms of the dimensional ratios of Table 10–2 it can be said that

$$\frac{c_r}{L_r} = \frac{\Omega_m}{\Omega_p \sin \phi_p}, \tag{10-15}$$

and we know from Table 10–2 that $c_r = (d_r)^{1/2}$ and $T_r = L_r(d_r)^{-1/2}$;

therefore

$$\frac{1}{T_r} = \frac{(d_r)^{1/2}}{L_r} = \frac{\Omega_m}{(\Omega \sin \phi)_p}, \qquad (10\text{--}16)$$

and finally

$$\Omega_m = \frac{(\Omega \sin \phi)_p}{T_r}. \qquad (10\text{--}17)$$

Such rotation will result in the mean sea level of the model being para-boloidal, with the vertex centered over the axis of rotation. It is necessary to refer all vertical dimensions in the model to this surface.

If toward the margins of the model the slope of the paraboloid reference surface is so great that the sine of this slope cannot be considered essentially equal to the slope itself, neither the earlier assumption that $\sin \phi_m = 1$ nor the assumption that the gravitational acceleration in the prototype is sub-stantially the same as that in the model is valid. In such a case it would be desirable to reduce the vertical scale and hence lengthen the time scale to diminish the required rate of rotation.

The foregoing can only be applied to models filled with a single homo-geneous fluid. Shallow-water gravity waves move on a rotating model with a celerity

$$c = \sqrt{g'D + \frac{\Omega^2\lambda^2}{4\pi^2}}, \qquad (10\text{--}18)$$

where Ω is the angular velocity of the tank, λ is the wavelength, D is the depth, and $g' = (\Delta\rho/\rho)g_0$ the effective gravitational acceleration. In the case of surface waves, $\Delta\rho/\rho = 1$ essentially and the rotational term can be neglected, but for internal waves g' may be quite small in comparison with g_0 and the effects of the rotational term become dominant. The hydrodynamical phenomena associated with two-layer systems in rotation are not well understood, but are being given experimental and theoretical study in several laboratories. It will take time for these abstract studies to be developed and translated into forms which permit them to be used in analogue models.

Single-layer rotating Froude models provide a severe technical chal-lenge.[5] An electrical analogue model which includes the effects of earth rotation has been proposed by Ishiguro.[6] Both analogue and computa-tional models have been widely used in connection with flood prediction and storm surge problems, but these generally do not take the Coriolis terms into account.

Models of oceans

On a planetary scale where internal friction is of small importance it is possible to study fluid motion with the aid of rotating tanks in which effects analogous to both the *rotation* and the *curvature* of the earth are present. Models of planetary fluid motions were first conceived in connection with

Fig. 10–6. Optical system and controls mounted above the parabolic basin of 2-meter diameter and 0.5-meter focal length shown on page 301. This apparatus has been used for vorticity models of each hemisphere of the world ocean.

atmospheric circulations. These attempts antedate oceanic studies by many years. Vettin (1884)[#] was probably the first to study the motions in a rotating fluid heated zonally from below. Since then Exner (1923,[#] 1925[7]) and Rossby (1926),[8] and more recently Fultz (1951)[9] and Long (1951)[10] and their associates, have made intensive studies of laboratory models in which the regime of motion is analogous to that in the upper levels of the atmosphere. Faller (1956)[11] has succeeded in simulating low-level frontal structures of the atmosphere using similar methods. Studies by Hide (1953)[12] made with reference to the fluid motions of the earth's mantle have been of interest to geomagneticians as well as to meteorologists. These methods are essentially independent realizations of a suggestion by James Thomson (1892)[#] who, with remarkable insight, foresaw both the experimental scheme and the results that might be obtained.

    The first attempt to simulate the ocean circulation may have been that by Lasareff (1929)[#] in which the trade winds were simulated over a plaster

model at rest. Spilhaus (1937)# studied the motion of a jet entering a fluid in solid rotation. Since 1950 several attempts have been made by the author (1952, 1957)[13] to represent the oceanic circulation in the laboratory, taking account of the rotation of the earth, certain effects due to the curvature of the earth, aspects of the geometry of the ocean basins, and the stress of the mean zonal wind field (Fig. 10–6). Thus far all experimental efforts have been concerned with barotropic flows in homogeneous water.

Vorticity modeling

The Froude law will not serve as a basis for scaling planetary fluid motions unless long, free gravity waves are involved, as they might be, for example, in the study of global tides and where the effects of rotation on wave velocities have to be taken into account. Ordinary wind waves and swell are too small to influence the circulation directly. The Reynolds law also falls short of planetary requirements, because the fluid motion in this case is not predominantly controlled by forces due to internal friction. For these and other reasons, ordinary modeling techniques seem inadequate to deal directly with planetary problems, but appropriate alternatives exist.

One of the most promising points of view is that provided by the vorticity theory of the wind-driven circulation. In terms of this conception the desired effects arise without regard for heat or dynamic topography, and a homogeneous fluid can be used in the model. Earth rotation and the stress of zonal winds can be provided by straightforward methods, but the meridional variation of the Coriolis parameter over a quadrant cannot be supplied through geometric similarity,[14] since there is at present no means by which a radial gravitational field of uniform strength can be produced or simulated in the laboratory.

The only apparent means for simulating the meridional variation of the Coriolis parameter through a latitude range of 90° in a uniform fluid is to employ the vorticity changes accompanying stretching and shrinking of fluid columns undergoing meridional motion in the presence of rotation. Physically, this can be accomplished quite simply by building a model of the earth in a flat-bottomed rotating tank in which the depth of water varies radially owing to the fact that the free surface of the water is necessarily paraboloidal. As vertical columns of water (parallel with the axis of rotation) move toward the center of the tank (poleward), they are squeezed vertically, extended horizontally, and thereby acquire anticyclonic vorticity relative to the tank. Vertical columns moving toward the rim (equatorward) will be stretched vertically, and in shrinking horizontally will acquire cyclonic vorticity relative to the tank. Were the relative vorticity so produced in transferring a column in either direction between the equator and the pole over the corresponding curve of the earth, the planetary vorticity tendency of the earth would be simulated in the model* (Fig. 10-7).

For a model in a shallow flat-bottomed tank the rotation rate necessary to produce the proper equatorial depth is a function of the size of the model and the central depth, as given by the equation for the free surface of a rotating fluid:

$$D_r - D_{90} = \frac{\Omega^2 r^2}{2g},$$
(10-19)

$$\Omega^2 = \frac{2gD_{90}(e-1)}{R^2},$$
(10-20)

where R is the radius at the equator and D_{90} is the depth at the center of

---

* The problem of simulating the planetary vorticity of the earth through the effects of horizontal divergence and curvature of the tank bottom, as when paraboloids are employed, has been examined analytically by A. Faller and the author (1959)[14] and by P. Raethjen, 1958, *Arch. Met.*, Wien, A., 10: 178–193.

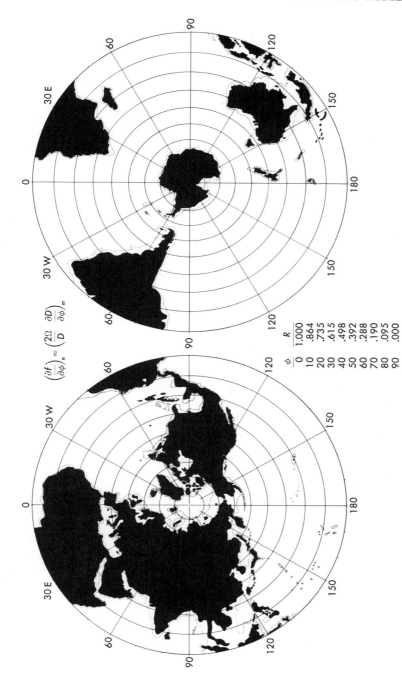

$$\left(\frac{\partial f}{\partial \phi}\right)_e \approx \left(\frac{2\Omega}{D}\,\frac{\partial D}{\partial \phi}\right)_m$$

ϕ	R
0	1.000
10	.864
20	.735
30	.615
40	.498
50	.392
60	.288
70	.190
80	.095
90	.000

Fig. 10–7. Hemispheric projections of coast lines and 100-meter isobaths for plane vorticity models. [From Chapter 1, *Physics and Chemistry of the Earth*, Vol. 2, 1957, L. H. Ahrens, *et al.*, editors, London: Pergamon.]

rotation. If the latitude $\phi = 0$ is taken to be the rim of the model, the depth at the rim must be eD_{90}.

The spacings of the latitude circles relative to the equatorial radius, R, are (for a plane model) independent of all other parameters, and are given by

$$\frac{r_\phi}{R} = \sqrt{\frac{e^{(1-\sin \phi)} - 1}{e - 1}}. \qquad (10\text{-}21)$$

Table 10–3 gives the relative positions of the latitude circles for a plane model. Given this spacing of latitude circles, the geography of the oceans can be plotted as shown in Fig. 10–7 and the continents carved out to the necessary depths as in Fig. 10–8. This scheme has been applied successfully in two recent experiments which concerned the wind-driven ocean circulation of both the Northern and Southern Hemispheres.

TABLE 10–3

ϕ	0°	10°	20°	30°	40°	50°	60°	70°	80°	90°
r_ϕ/R	1.000	0.864	0.735	0.615	0.498	0.392	0.288	0.190	0.095	0.000

The most essential feature of a rotating model is that the tank turn smoothly and steadily around a precisely vertical axis. The rotation provided by simply mounting a tank on a vertical spindle is fundamentally unsatisfactory. It is better in many ways to adapt the simple method originally used by Taylor (1921)[#] in which the model is floated in a stationary tank and driven by three water jets or through a fluid clutch.[15] See Fig. 10–9.

In order for planetary models to be housed, the horizontal scale must necessarily be on the order 10^{-7}. Mixing in these extremely small-scale models is usually excessive. The coefficient of eddy diffusivity in the oceans is said to reach values ranging from 10^4 to 10^8 cm^2/sec for large-scale phenomena. In a planetary model the diffusivity ratio is

$$A_r = \frac{A_m}{A_p} = \frac{L_r^2}{T_r} = \frac{(10^{-7})^2}{10^{-4}} = 10^{-10}. \qquad (10\text{-}22)$$

But since $A_m = 2 \times 10^{-5}$ cm^2/sec at the least, due to molecular diffusion, the turbulence in a planetary model must be negligible if diffusion is to be kept in scale. Toward this end a decrease in time scale or an increase in the length scale is desirable, but T_r is effectively limited to the range 10^{-4} to 10^{-5}, and L_r cannot be carried farther than 10^{-6} in even the most formidable installations.

Fig. 10–8. A flat-bottomed tank eight feet in diameter for vorticity models of wind-driven circulation. [From Chapter 1, *Physics and Chemistry of the Earth*, Vol. 2, 1957, L. H. Ahrens, *et al.*, editors, London: Pergamon.]

The theory of the vorticity model indicates nothing directly with regard to vertical scale. Experimentally a model of the surface-layer circulation of the oceans (above the main thermocline) can be satisfactorily produced in homogeneous water when the bottom topography of the oceans is represented only to the depth of the 200-meter isobath. This depth reaches the shoulder of most continental shelves and suggests the subsurface bulk of ocean islands and island chains, but excludes the deeper features such as swells and ridges on the ocean floor. It is found that when the mid-Atlantic Ridge, for example, is modeled in the North Atlantic, the wind-driven flow across it is strongly affected. This is undoubtedly due to the fact that between the upper Ekman friction layer (about 1 mm thick) and the bottom Ekman friction layer (also about 1 mm thick) the flow is essentially two-

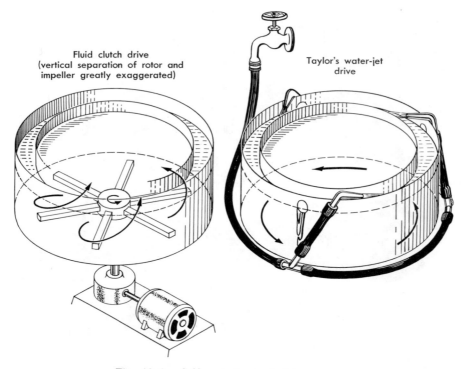

Fluid clutch drive
(vertical separation of rotor and
impeller greatly exaggerated)

Taylor's water-jet
drive

Fig. 10–9.   Self-centering tank drive systems.

dimensional. Therefore, in the presence of rotation any obstacle that protrudes above the bottom friction layer will act as though it extended through the full depth of the fluid even though it may not actually do so.

With these facts in mind, one may consider as significant to the vertical scale the preservation of the ratio of the thickness of the friction layer to the total depth in the oceans and in the model. The Ekman layer in the oceans is on the order of 100 meters thick, and the average total depth is about 4000 meters, so that their ratio is approximately 1:40. The mean depth of water in the model should therefore be in the neighborhood of 4 cm. This has proved to be a convenient and apparently suitable depth.

The vorticity model is driven by the torque of the wind stress over its water surface. Thus the direction of air flow is of no consequence, but it is of great importance, in order to accord with nature, that the inflection points in the mean zonal wind field be properly placed in latitude and that the ratio of torques between the trades and westerlies and between the westerlies and polar easterlies be reproduced. In order for the ratio of zonal wind torques to be preserved, one has either to know the value of the wind-stress coefficient and compute the required wind speed or to adjust these values in some empirical fashion.

The lowest wind speeds of suitable strength are obtained when the directions of air motion reproduce nature: from east to west in the trades and polar easterlies, and from west to east in the westerlies. The rotation of the model through the motionless air in the laboratory provides a ready-made easterly trade wind which need only be shaped by gentle air streams from a system of nozzles so directed as to produce a maximum trade at latitude 15°. Westerly winds must be produced by an air blast having sufficient speed to overtake the model, with maximum wind speed near latitude 45° in the Northern Hemisphere and near −50° in the Southern, and with inflection points at 30° and 60° and −30° and −68°, respectively, according to recent evidence. Through careful adjustment of the west-wind nozzles, a core of air will remain at rest over the polar regions of the model, and provide a suitable anticyclonic torque as the model rotates.

While the zonal wind speed ratios are preserved, the actual wind speeds employed in the model experiments are adjusted through their effect on the ocean circulation. To be in scale, the velocities of the ocean currents should bear the same ratio to the tangential velocity of the model at its equator as do the velocities of the actual ocean currents to the tangential velocity of the earth at its equator. This is a definition of the *kinematic Rossby number*, which for the western boundary currents has values in the vicinity of 0.3×10^{-2}. Proper adjustment of the Rossby number establishes the velocity scale of the model, and also has a sensitive influence on the pattern of the ocean currents. The results of experiments with models of both the Northern and Southern Hemispheres resemble the steady-state circulations of the world ocean remarkably well, especially in the Northern Hemisphere. This suggests that the winds, together with the effects of earth rotation and curvature, are the predominant influences governing the surface-layer circulations. In both Northern and Southern Hemisphere models, it has been observed that barotropic readjustment of the circulation to simulated seasonal changes in the applied wind field is complete within a matter of some 12 to 14 revolutions, or model "days." The response of the natural oceans to seasonal change in the atmospheric circulation is both smaller and slower owing to their baroclinicity, which serves as a stabilizing influence. The technique for studying the baroclinic circulation of oceans in laboratory models has not yet been worked out.

Among the interesting details shown by the rotating models of oceans are the distributions of dyes placed in the surface and bottom Ekman layers (Fig. 10–10). An initially uniform coating of slightly buoyant dye in the surface layer of the model ocean will gradually accumulate near the subtropical convergences to delineate the central water mass in each ocean. Heavier dye placed in the bottom layer will be swept out of these regions in the bottom Ekman layer to accumulate as coastal water masses and to fortify the subpolar water masses. On the western sides of the oceans the narrow boundary currents flow between the central and coastal water masses, as they do in nature, and upon turning eastward under the westerly winds these currents also flow eastward along the boundary between the central and subpolar water masses. In the Northern Hemisphere model both the Gulf Stream and Kuroshio develop meanders of a sort which, although probably maintained by quite different forces from those in nature, closely resemble their large-scale counterparts, discussed in Chapter 11.

STUDY QUESTIONS

1. From the shallow-water wave equation, derive suitable expressions for the scales of velocity, time, horizontal length, and depth (a) for an undistorted Froude model, (b) for a vertically distorted Froude model.

2. In a three-meter-square Froude model of an estuary measuring 40 km square and having a mean depth of 5 meters, calculate the vertical exaggeration required to hold the time scale at $T_r = 1/1440$.

3. What rate of rotation would be required to reproduce the Coriolis effects in the model of question 2? Could the $\sin \phi_m = 1$ assumption be justified?

4. What scales of time and vertical length could be chosen to accord with a rotation rate of 2π rad/min in the model of question 2?

5. In a vorticity model of constant normal depth, how much of the variation with latitude of the Coriolis parameter can be accommodated by the curvature of a paraboloid terminated by its latus rectum? To simulate the total variation of the Coriolis parameter, what different polar and equatorial depths would be required if the rim is to correspond with the equator and the vertex with a pole?

6. If the depth of frictional influence is determined by $D = \sqrt{\mu/\rho f}$ and $\mu = 10^{-2}$ for the dynamic viscosity of water in models, what is the thickness of the Ekman layers in a model rotating at 10π rad/min? Sketch the upper and lower Ekman spirals for a Southern Hemisphere model.

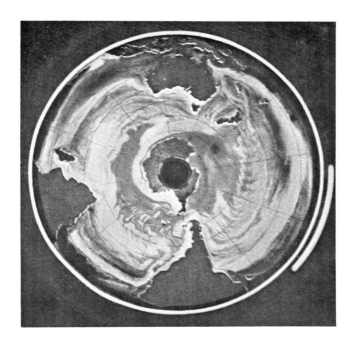

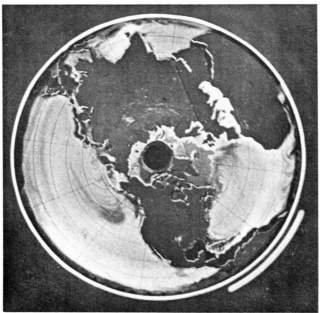

Fig. 10–10. Photographs of the dye distributions produced by the zonal wind pattern shown in Fig. 6–11 and in the eight-foot tank shown in Fig. 10–8. [From *Fortune*, November, 1959.]

REFERENCES

1. M. Rattray, Jr., and J. H. Lincoln, 1955, *Trans. Amer. geophys. Un.*, 36: 251–261.

2. W. S. von Arx, 1954, *Prof. Pap. U.S. Geol. Surv.*, 260-B:265–273.

3. H. Rouse, 1938, *Fluid Mechanics for Hydraulic Engineers*, New York: McGraw-Hill; 1946, *Elementary Mechanics of Fluids*, New York: Wiley.

4. J. C. Stevens, *et al.*, 1942, *Hydraulic Models*, Manual of Engineering Practice No. 25, New York: American Society of Civil Engineers.

5. R. Bonnefille, 1957, *Mémoires & Travaux de la S. H. F.*, 2: 154–161.

6. S. Ishiguro, "A Method of Analysis for Long-wave Phenomena in the Ocean, Using Electronic Network Models, Part II," *Phil. Trans.*, A., in preparation.

7. F. M. Exner, 1925, *Dynamische Meteorologie*, 2nd ed., Wien: Julius Springer.

8. C.-G. Rossby, 1926, *Mon. Weath. Rev.*, Wash. 54: 237–240.

9. D. Fultz, 1951, pp. 1235–1248, *Compendium of Meteorology*, Boston: American Meteorological Society.

10. R. R. Long, 1951, *J. Met.*, 8: 207–221.

11. A. J. Faller, 1956, *J. Met.*, 13: 1–4.

12. R. Hide, 1953, pp. 101–116, *Fluid Models in Geophysics*, Robert R. Long, editor, Washington, D.C.: U. S. Government Printing Office.

13. W. S. von Arx, 1952, *Tellus*, 4: 311–318; 1957, Ch. 1, pp. 1–29, *Physics and Chemistry of the Earth*, Vol. 2, L. H. Ahrens, F. Press, K. Rankama, and S. K. Runcorn, editors, London: Pergamon Press.

14. A. J. Faller and W. S. von Arx, 1959, pp. 53–70, *Proceedings of the Seventh Hydraulics Conference June 16–18, 1958*, Studies in Engineering Bulletin 39, A. Toch and G. R. Schneider, editors, Iowa City: State University of Iowa; A. A. Dmitriev, 1956, *Izv. Akad. Nauk SSSR, Ser. Geofiz.*, 3: 320–326.

15. W. S. von Arx, 1957, Ch. 1, pp. 1–29, *Physics and Chemistry of the Earth*, Vol. 2, L. H. Ahrens, F. Press, K. Rankama, and S. K. Runcorn, editors, London: Pergamon Press.

SUPPLEMENTARY READING

Langhaar, H. L., 1951, *Dimensional Analysis and Theory of Models*, New York: Wiley.

Prandtl, L., 1952, *Essentials of Fluid Dynamics*, translated by W. M. Deans, New York: Hafner.

Rouse, H., and S. Ince, 1957, *History of Hydraulics*, Iowa City: Iowa Institute of Hydraulic Research, State University of Iowa.

Wilson, E. B., Jr., 1952, *An Introduction to Scientific Research*, New York: McGraw-Hill.

CHAPTER 11

The Gulf Stream Problem

Having come this far in a study of the oceans and having become aware of the physical significance of western boundary currents, it seems appropriate to conclude this book with some discussion of the Gulf Stream, the most familiar example of such currents. East coast oceanographers tend to become preoccupied with the Gulf Stream from time to time, but few have confined their efforts to it exclusively. This is perhaps partly because of the powerful buffeting that its seas can give a ship and partly because of the enigmatic quality of its problems. But the intriguing presence of the Gulf Stream cannot long be ignored. Every so often a new idea or instrumental development will inspire a burst of optimism and a new research effort is launched. This chapter gives an account of some of the recent ones.

Geography of the Gulf Stream system

Prior to 1920 the Gulf Stream was thought to be well named because of its apparent origins in the Gulf of Mexico. It has been found, however, that the Gulf of Mexico contributes almost nothing to the volume of water involved in the current. For this reason both Nielsen (1925)[1] and Wüst (1924)# suggested that the old name, Florida Current, be used for the portion between the Gulf of Mexico and Cape Hatteras. Iselin suggested in 1933[2] and 1936# that for scientific purposes the name "Gulf Stream" be used to refer to part of a system of currents in the North Atlantic, and that this system be regarded as containing three main subdivisions. This suggestion has gained wide acceptance, but with the passage of time certain modifications have developed as follows:

The *Gulf Stream system* includes the complex of currents, having relation to the waters passing through the Straits of Florida, that extends as various branches, eddies, and closely coupled countercurrents along the east coast of North America onward toward the eastern North Atlantic. The *Florida Current* portion of the Gulf Stream system includes the water moving toward the Atlantic through the Straits of Florida. The *Gulf Stream* portion of the system extends from the Atlantic end of the Florida Straits to the vicinity of the Flemish Cap. The *North Atlantic Current* is that portion of the Gulf

312

Fig. 11–1.   Sargassum.   [Photograph by J. Hahn.]

Stream system lying to the eastward of Flemish Cap which has a continuous or quasi-continuous baroclinic structure connected at mid-depth with
that of the Gulf Stream off Newfoundland.

The Gulf Stream system forms the western and northern boundary of
the waters of the Sargasso Sea. Krümmel in 1891[#] defined the limits of this
legended area as that characterized by the distribution of Sargasso weed,
a brown alga which floats near the sea surface in small clumps or extensive
pads (Fig. 11–1).   Sargassum is scattered throughout a large oval area
delimited roughly by the Bermuda-Azores high-pressure cell. The eastern
and southern margins of the Sargasso Sea are rather difficult to place, but
the hydrographic limits of the western and northern boundaries of this area
are situated at the seaward margin of the Gulf Stream system.

A change in water type occurs between the Sargasso Sea proper and the
Gulf Stream, because the latter provides both rapid and relatively direct

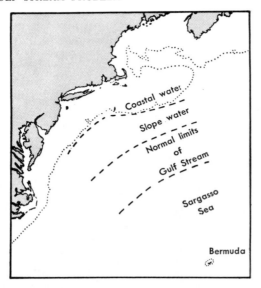

Fig. 11–2.   Surface water provinces between the New England coast and Bermuda (after Iselin, 1936).

importation of water from low latitudes.   Gulf Stream water is slightly warmer but slightly less salt than the average for the surface layers of the Sargasso Sea at corresponding latitudes.

Between the continental margin of the Gulf Stream and the shoulder of the continental shelf is a region which Huntsman (1924)[3] called *slope water* (Fig. 11–2).   Slope water extends from Cape Hatteras to the Grand Banks and has characteristics which are similar to but do not coincide quite exactly with the *T-S* characteristics of the central Atlantic water mass below 900 meters.   The slope water region is one of ephemeral, weak currents often running parallel with the Gulf Stream but not always in the same direction.

The coastal water band, described by Bigelow in 1915,[4] lies in the zone between the 100-fathom curve and the coast line from Cape Hatteras to the Maritime Provinces.   The circulation of this continental shelf region is dominated by tides and local winds.   The water on the continental shelf is a variable composite of the discharge of rivers and ground water into patches of ocean water which migrate inward across the shelf from the open sea.

Superficial appearances

In a ship steaming from Cape Cod toward Bermuda, the banded distribution of surface salinity, temperature, and water velocity roughly parallel

Fig. 11–3.   Slick patterns characteristic of slope water as seen from the air.

with the coast is so often met that it provides direct clues to the ship's position.   Inshore the temperature may range from about 5°C in midwinter to about 18°C in late summer, increase seaward, and rise abruptly to a maximum often near 28°C in the Gulf Stream.   In a corresponding way the salinity increases with distance offshore from about 32 $^0/_{00}$ to 34 $^0/_{00}$ against the coast to values which may exceed 36.5 $^0/_{00}$ in the Gulf Stream and Sargasso Sea.   The seaward gradients of salinity and temperature are broken by numerous patchy irregularities.

In the coastal water on the continental shelf, both sea temperature and sea color are correlated with the season of the year.   With the flowering of phytoplankton in late spring and again in early fall, the sea color changes from a dull to a clearer green.   Currents on the continental shelf have a systematic motion that is weak compared with the flow in the rotary tides.

Beyond the 200-meter isobath at the brink of the abyssal slope of the continental shelf, the regime of slope water begins.   Currents in this region are weak in comparison with tidal currents on the continental shelf, and possess an unsystematic pattern which is not resolved until the ship reaches the vicinity of the Gulf Stream.   In calm weather the slope-water region is banded by slicks (Fig. 11–3).   Entry into slope water is revealed by an increase in salinity and a patchy but otherwise steadily rising temperature in the wind-stirred layer.

Fig. 11–4.   Aerial view of the change of sea state often developed at the continental edge of the Gulf Stream.

With progress toward the inshore edge of the Gulf Stream, a narrow streamer of cool and relatively fresh water may be crossed before the water velocity suddenly increases along with the water temperature and salinity. This frontlike transition may be marked by a change in water color from the bluish or grayish green of slope water to the deep ultramarine of the Gulf Stream.   There may also be an associated change in sea state and increasing cloud cover (Figs. 11–4 and 11–7).   Owing to the effects of wave refraction in the shear zone of the current, the sea state will increase if the local wind has a component which opposes the current, and will generally decrease if a component of the local wind follows the current.

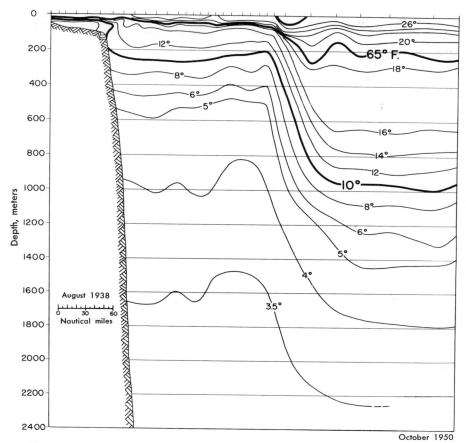

Fig. 11–5. A typical temperature section across the Gulf Stream (after Fuglister).

At the same time that the surface temperature, salinity, and velocity increase at the Gulf Stream front, the isotherms slope (about 10^{-2}) beneath the warm water of the current. It is often considered that when the 65°F (18°C) isotherm passes through the 200-meter level, the edge of the main current has been penetrated and the ship is nearing the core of maximum current.* Near the core there is on many occasions a very noticeable set accumulated on the navigational plot.   When electromagnetic current-

---

* It is easy to be misled by surface evidence alone. All surface indications may suggest that the ship has entered the Gulf Stream when indeed it has only traversed a branch or an eddylike structure associated with the current. For this reason ultimate reliance is placed on subsurface measurements to indicate where the main part of the current is to be found.

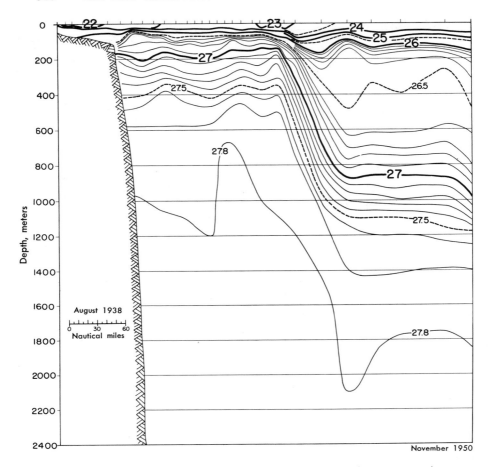

Fig. 11–6. A typical σ_t section across the Gulf Stream (after Fuglister).

measuring apparatus is aboard, one can observe, over a distance of 10 to 15 nautical miles, a rapid increase of water velocity from almost nil in slope water to as much as 5 or even 6 knots in midstream.

Seaward of the current maximum, which is always on the left of center, the change in velocity with distance at the surface is much more gradual on the right-hand side of the current (Fig. 11–8). In this region there is a tendency for the surface temperature to reach a maximum and fall, and for the deep isotherms to rise slightly in the counterflow sometimes found on the Sargasso Sea side of the Gulf Stream. The countercurrent, when present, will usually be crossed some 40 to 60 nautical miles seaward of the Gulf Stream front. The width of the countercurrent is roughly half that of the

Fig. 11–7. A band of cumulus cloud is sometimes excited in the lower air as it is warmed along the continental edge of the Gulf Stream.

Gulf Stream itself. The countercurrent may have speeds as great as 1.5 knots and is to be distinguished from the water motions encountered on the rest of the traverse to Bermuda mainly by its systematic direction of flow.

Upon reaching the Sargasso Sea area one enters an environment where sea temperatures and salinities and the arrangement of water properties are relatively uneventful. In this area sunshine is intense and cloudiness generally consists of isolated cumulus, except in winter when the strato-form clouds associated with frontal disturbances moving off the North American continent may extend eastward beyond Bermuda.

The Sargasso Sea is a maritime desert where evaporation exceeds pre-cipitation. Here the Ekman transports due to the planetary wind fields in temperate and tropical latitudes are thought to converge and thus produce a gigantic accumulation of slow-moving, warm, saline water which floats as a broad but rather thin lens in hydrostatic equilibrium on the cold but fresher water below. The outward pressure of the water in this lens can be considered to be balanced by the Coriolis forces accompanying the geo-strophic motions around its perimeter. Were an effect of this kind to be absent, Sargasso water would flow outward and reach the coast, causing sea level to rise somewhat less than one meter. Thus the Gulf Stream

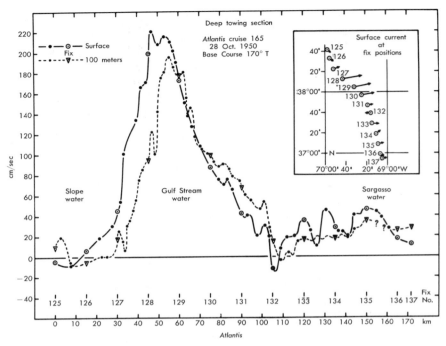

Fig. 11–8.   Transverse profile of GEK velocities observed in the Gulf Stream at the surface and 100-meter levels on 28 October 1950. [From W. S. von Arx, 1952, *Tellus*, 4(3).]

system acts as a dynamic barrier.   But owing to its constantly changing structure and velocity, there are occasional "leaks" associated with strong transverse components of flow.   Eddies cast off on the left of the current have been observed to move patches of clear blue ocean water and sargassum into coastal water and sometimes to the shore line.   Those cast off on the right of the current form cold pools in the Sargasso Sea (Fig. 4–12).

Origins of Gulf Stream water

The surface water of the Gulf Stream is mainly derived from the southern part of the north Equatorial Current.   Water driven by the trades moves slowly across the equatorial North Atlantic, through the Antilles into the Caribbean Sea, and onward through the Yucatan Channel into the southeastern portion of the Gulf of Mexico before entering the Straits of Florida and the Gulf Stream system.   The remarkable part of this progress is that the water is little modified in its transit across the Caribbean, and passes along at least two routes through the eastern Gulf of Mexico without appreciable mixing.   On some occasions the flow from the Yucatan Channel proceeds almost due north to the latitude of St. Petersburg before it turns southeastward and rejoins the Florida Current. Another portion of the entering flow through the Yucatan Channel may turn sharply around Cape San Antonio (westernmost Cuba) and make its way directly to the Straits of Florida. On occasion the flow into the east-central part of the Gulf of Mexico is absent altogether.

In addition to the surface water from the north Equatorial Current, the Gulf Stream is fed by a part of the south Equatorial Current, which appears to be split against the easternmost promontory of South America. At this point the Brazil Current branches southward and the Guiana Current flows along the northern coast of Brazil and the Guianas, past Trinidad, and into the Caribbean Sea. The velocity of the Guiana Current is considerably greater than that of the north Equatorial Current and may exceed 2 knots off Trinidad. There is a sharp contrast between the clear blue color of this water and the muddy outflow from the Amazon and Orinoco Rivers.

The water derived from the south Equatorial Current by the Guiana route is not sufficiently different from that supplied by the north Equatorial Current to form any conspicuous interface as it passes through the Yucatan Channel. However, water contributed from deeper layers of the South Atlantic—Antarctic Intermediate water—can be readily distinguished from the surface-layer contributions. Antarctic Intermediate water is formed in a region of heavy precipitation and has therefore a distinctly lower salinity than the surface water conditioned in the tropics. It also follows a different route. Antarctic Intermediate water lies too deep (600 to 800 meters) to make its way in quantity through the Antilles arc, and instead apparently finds its way into the Gulf Stream by the outside route adjoining the current at the exit of the Straits of Florida. Iselin (1936)[#] suggested that water below the 8°C isotherm which has joined the Gulf Stream in this fashion may contribute as much as 15% of the total transport.

The surface water feeding the Gulf Stream by way of the Caribbean tends to be low in oxygen. Water reaching the Gulf Stream by the outside route, on the other hand, tends to contain more dissolved oxygen. Richards and Redfield (1955)[5] have studied the dissolved oxygen concentration of the Gulf Stream in the vicinity of the Blake Plateau, finding that although the water volume contributed to the upper levels of the Gulf Stream by way of the Antilles varied irregularly in the period 1950 to 1953, it was not zero. But in spite of indirect evidence of flow, no persistent Antilles Current has been observed.

The temperature-salinity correlation of Gulf Stream water is different from that of the Sargasso Sea. In the vicinity of Cape Hatteras, Gulf Stream surface water has a salinity of 36.0 $^0/_{00}$ and a temperature of 26°C or higher. At 200 meters the temperature falls to 20°C but the salinity rises to 36.6 $^0/_{00}$ or slightly above. In the depths between 200 and 1000 meters there is a steady decrease of both temperature and salinity to values near 6°C and 35 $^0/_{00}$ at the greater depth. Sargasso Sea water, on the other hand, is usually somewhat cooler than Gulf Stream water and somewhat more salt at the surface, having characteristic values of 20°C or more, depending on the season, and a salinity near 36.6 $^0/_{00}$.

The temperature and salinity of Sargasso Sea water decrease almost linearly with depth to the level of the 10°C isotherm, usually found near the 800-meter level; from there on the water decreases in temperature more rapidly than in salinity through the main thermocline layer. Between the surface and the upper part of the main thermocline there is a large volume of water having a temperature near 18°C. The origin and maintenance of this extensive water type is discussed in a recent paper by Worthington (1959).[6]

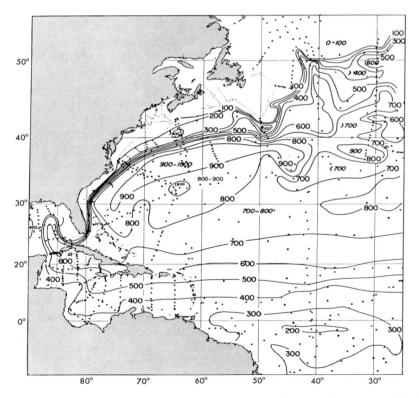

Fig. 11–9.   Chart showing the depth of the 10° isotherm in the western North Atlantic (after Iselin, 1936).

Main thermocline layer

The descent of the isotherms beneath the Gulf Stream marks the interface between the warm, saline water of the central Atlantic water mass as it rests on the cool, somewhat fresher water beneath. This temperature contrast is both conspicuous and widespread and provides a region of very strong vertical stability ordinarily referred to as the main thermocline layer of the North Atlantic. Because of its significance as an indicator of the depth of the main thermocline layer, Iselin's (1936)[#] chart of the 10°C isotherm for the western North Atlantic, shown in Fig. 11–9, gives an illuminating description of the topography of this interface. The 10°C isothermal surface reaches a maximum depth between 800 and 900 meters under the Sargasso Sea, may often rise to the 200-meter level in slope water, and actually reach the surface in the vicinity of Nova Scotia and Newfoundland.

The crowding of isotherms in the main thermocline may involve a change of temperature as great as 15°C in 500 or 600 meters of depth. The maximum slope observed for isotherms in the region of intense baroclinicity along the Gulf Stream front may be in the order of 1%.

Structure of the current

The Gulf Stream system is probably composed of a complex of filamentary flows, eddies, and meanders. These are confined within diverging limits along the western perimeter of the Sargasso Sea (Fig. 11–10). Near the Florida Straits and again at Cape Hatteras these limits may be as little as 60 nautical miles apart. Off Newfoundland, the total width of the system may be as much as 300 nautical miles.

Figure 11–11 shows the temperature and salinity structure of a section through the Gulf Stream made between Chesapeake Bay and Bermuda, and Fig. 11–12 shows these same properties between Nova Scotia and Bermuda. The marked descent of both the isothermal and isohaline surfaces is related to a change in density supporting the pressure gradient with which the Gulf Stream is theoretically in balance. The Chesapeake Bay sections show a single slope in these properties, while the Nova Scotia

The sections of Figs. 11–11 and 11–12 are plotted with a vertical exaggeration of 370 in the first 2000 meters of depth and 148 below 2000 meters where the changes in water properties are less abrupt. In this and in the earlier work of A. D. Bache (1844),[#] vertical exaggeration causes the slope of the isotherms to appear to be nearly vertical. Bache used the term "cold wall" for the thermal interface between Gulf Stream water and slope water. This terminology has been criticized by Fuglister and Worthington (1951) because of its mixed connotations. In its place the meteorological term "front" has been used, since this defines a discontinuity of both velocity and fluid properties. See F. C. Fuglister and L. V. Worthington, 1951, *Tellus*, 3: 1–14.

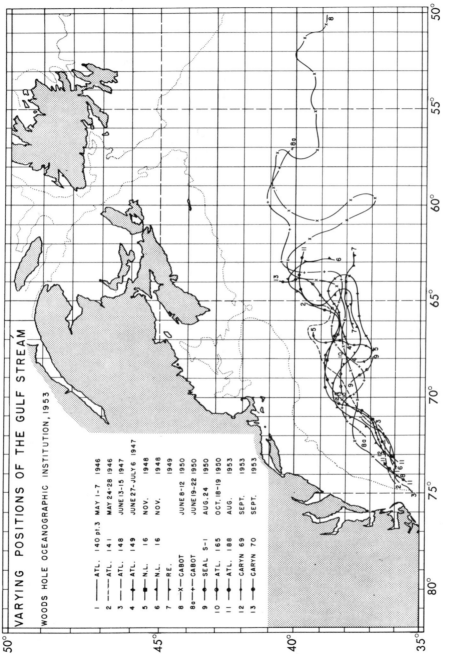

Fig. 11–10. Varying positions of the inshore margin of the Gulf Stream from 1946 to 1953. [From F. C. Fuglister and L. V. Worthington, 1951, *Tellus*, 3(1).]

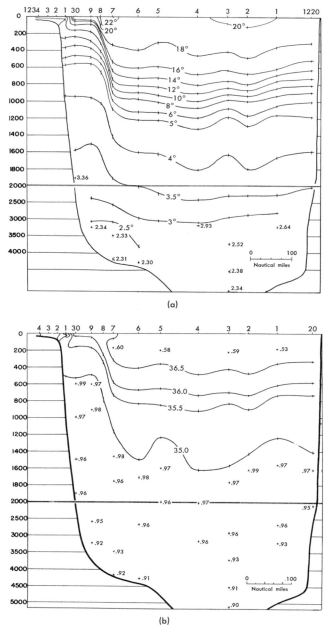

(a)

(b)

Fig. 11–11. (a) Temperature section and (b) salinity section for Chesapeake Bay–Bermuda during 17–23 April 1932 (after Iselin, 1936).

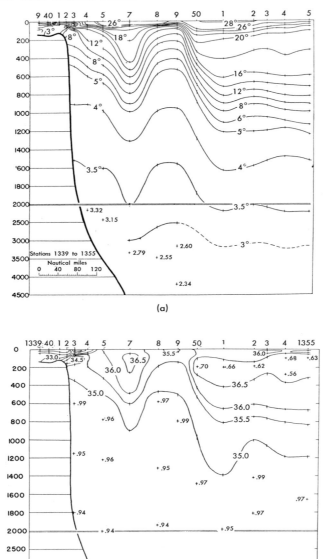

Fig. 11–12. (a) Temperature section and (b) salinity section for Nova Scotia-Bermuda during 14–20 August 1932 (after Iselin, 1936).

sections generally show a double one, which suggests that the current may have branched or developed a parallel companion current. Within the limits of the complex, however, there is very often one main filament which has a set of very characteristic properties.

The main current has an effective width (measured between 10 cm/sec isokinetics) which ranges from about 50 km in cyclonic bends to about 75 km in anticyclonic bends. The maximum surface temperature is usually encountered to the right of center and the maximum velocity is found to the left of center. The transverse profile of surface velocities is variable from an asymmetrically developed single peak similar to that shown in Fig. 11–8 to a broadened crest or even a double-peaked profile on occasions where the current structure is in the process of branching or becoming multiple. When these processes are completed, however, the characteristically asymmetrical and sharply peaked profile is usually found in each of the current filaments. The horizontal shear on either side of the current maximum has values which are generally not far removed from the local value of the Coriolis parameter.

Theoretically the anticyclonic shear on the right of the current maximum should have an upper limit equal to the Coriolis parameter, about $1.0 \times 10^{-4}\ \text{sec}^{-1}$ in the latitudes of New England. The averages of corrected shear on the right-hand side of the current maximum in these latitudes range between 0.5×10^{-4} and $0.8 \times 10^{-4}\ \text{sec}^{-1}$, so that it seems probable that the theoretical limit is seldom approached. The lower limit for the left-hand shear zone is more often approached, but this shear may occasionally be very much greater than the local value of the Coriolis parameter.

The velocity of the Gulf Stream is usually greatest at or very near the surface, where it may range from 100 cm/sec to values approaching 300 cm/sec. Unless local winds have mixed the surface momentum downward, the velocity will decrease regularly with depth to about the 500-meter level. At this level there is often a slight inflection or a secondary maximum below which the velocity again decreases with depth, reaching values in the order 1 to 10 cm/sec in depths between the 1500- to 2000-meter levels. There is some uncertainty about this because of the relatively few direct measurements of the current velocities in the portion of the current downstream from Cape Hatteras. The structure is better known in the vicinity of Florida, Georgia, and the Carolinas. The first systematic observations of the subsurface velocities in the Gulf Stream in this region are those by Pillsbury, begun in 1885 (1890)[#] with a mechanical current meter of his own design. Among the direct measurements downstream from Cape Hatteras are those recently made by Malkus and Johnson (1954)[7] using a ram-pressure recorder called the "bathypitotmeter,"[8] and the Swallow float observations made in 1960 by L. V. Worthington and W. G. Metcalf.

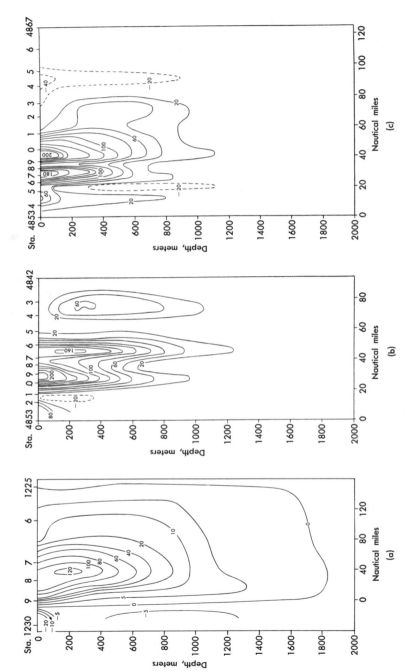

Fig. 11–13. (a) Velocity section (in cm/sec) across the Gulf Stream off Chesapeake Bay during 20–22 April 1932. (b) Velocity profile (in cm/sec) across the Gulf Stream south of Cape Cod during October 1950. (c) Velocity profile (in cm/sec) obtained on the return crossing of the Gulf Stream following that of (b). [Parts (b) and (c) from L. V. Worthington, 1954, *Tellus*, 6(2).]

But the method of dynamic sections has provided the bulk of present information on the structure of the Gulf Stream (Fig. 11–13).

A number of dynamic measurements of the velocity field through a vertical plane are available, but those made by Worthington in 1950[9] are probably the least generalized because of the extraordinarily close spacing of stations. Occupation of oceanographic stations in a current as swift as the Gulf Stream is not a task for a novice. To hold position, the ship must stem the flow of the surface water at speeds up to 5 knots. This maneuver keeps the wire angle small enough for the messenger to slide freely, trip the water bottles, and permit deep casts to be made; but at the same time it introduces an uncertainty in station position because the ship may move as much as a mile or more during the cast. Since the shallow and deep casts are made separately, there may be a difference of as much as a mile between the shallow and deep casts at what is nominally the same oceanographic station. When the interval between stations is as short as ten miles, an average error in position of one mile can amount to an average error of $\pm 20\%$ in the computed velocity and transport between consecutive stations. Possible curvatures of flow must be considered, as well as errors associated with the choice of a suitable level of no motion. Nevertheless, all that is known of the subsurface characteristics within the current has been obtained by dynamic sections. Estimates of the longitudinal change in the thickness of the current have been obtained from electromagnetic and radio navigational techniques.

It will be recalled from Chapter 9 that electrical thickness can be established for broad currents (which is not quite the nature of the Gulf Stream) from the expression $d = (1 - 1/k)\mathcal{Z}$ [Eq. 9–22]. The levels discerned by

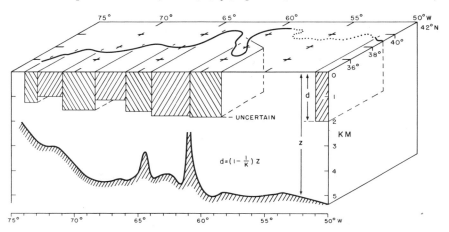

Fig. 11–14. The electrical thickness, d, of the Gulf Stream during 6–14 June 1950, and the average charted depth, z, plotted against longitude. (See Fig. 11–17.)

TABLE 11-1

ELECTRICAL THICKNESS OF PORTIONS OF THE GULF STREAM OBSERVED
DURING THE MULTIPLE-SHIP SURVEY OF 1950 (Fig. 11-17)

Position, west longitude	Description, see Fig. 11-14	$\overline{K}$	$\overline{z}$, meters	$\overline{d}$, meters	Total observations
73-74° (0)	Approach to first anticyclonic bend	1.56	3460	1240	27
74-70.5° (A)	First anticylconic bend	1.38	3610	1000	49
70.5-68.5° (A')	First cyclonic bend	1.50	4800	1590	18
68.5-66° (B)	Second anticyclonic bend	1.36	4850	1290	25
66-64° (B')	Second cyclonic bend	1.55	4750	1690	31
64-61° (C)	Third anticyclonic bend	1.53	5100	1760	54

this relation agree with that of the 25 cm/sec isokinetic in Worthington's very detailed dynamical section made across the Gulf Stream in October-November, 1950,[9] and seem generally to reflect the position of about the 4° isotherm and 35 $^0/_{00}$ isohaline shown in diagrams by Iselin (1936, 1940),[#] indicating a relationship to the level below which the circulation begins to be weak. The data in Table 11-1, considered in this sense, indicate that the electrical thickness of the current varies systematically with the sense of curvature, and that the current also extends to greater depths downstream (Figs. 11-14 and 11-17).

Seasonal changes

Much of our present knowledge of the variability of the Gulf Stream has been obtained from dynamic sections made serially in space or time. In 1936[#] Iselin reported the results of a series of sections he made from Nova Scotia and Chesapeake Bay to Bermuda. In 1940[#] he published the results of additional sections between Montauk Point and Bermuda. All these sections were made in an attempt to discover seasonal variations in the transport and structure of the current. From such studies, together with measurements of the difference in sea level revealed by tide gauges at Charleston

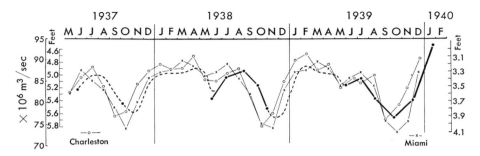

Fig. 11–15. Thirteen transport values from the Montauk Point-Bermuda section compared with the tide-gauge record (monthly means) from Miami, Florida, and Charleston, South Carolina, for the same period (after Iselin, 1940).

and Bermuda, Iselin (1940)[#] showed a variation in the transport of the Gulf Stream which is roughly related to the season of the year. Monthly departures from annual means were as great as 6 to 10 $\times$ 10^6 m^3/sec, with a minimum in October and a maximum distributed between January and July (Fig. 11–15). Montgomery (1938)[10] and Hela (1952)[11] computed seasonal fluctuations in the Florida Current from a comparison of tide-gauge records of sea-surface levels along and across the current system, finding a maximum gradient in July, a minimum in October, and secondary maxima and minima in January and April. Fuglister (1951)[12] studied the variability of speed of the Gulf Stream using data gleaned from ships' logs. He also found evidence for a minimum in October or November and a maximum which is predominantly in July, and companion peaks in the early spring. Studies of the surface velocity do not reflect changes of transport unless assumptions are made concerning the vertical profile of velocity.

The transport of the Gulf Stream varies with the place as well as the time of measurement. The flows through the Yucatan Channel and the Straits of Florida are usually regarded as being alike, but estimates of the transport differ. Sverdrup (1942)[13] places this transport at 26 $\times$ 10^6 m^3/sec, Montgomery (1941)[14] between the limits 26 $\times$ 10^6 and 30 $\times$ 10^6 m^3/sec, and Parr (1937)[#] between 30 $\times$ 10^6 and 34 $\times$ 10^6 m^3/sec. The mass transport through the Straits of Florida is also known to swing through the range from about 20 $\times$ 10^6 to 40 $\times$ 10^6 m^3/sec. The transport in the Gulf Stream off Charleston, North Carolina, and Montauk, Long Island, obtained by Iselin (1940)[#] ranges from 76.4 $\times$ 10^6 to 93.5 $\times$ 10^6 m^3/sec, with reference to the 2000-decibar surface, and the mean of 15 sections made between June 1937 and January 1940 is 82.5 $\times$ 10^6 m^3/sec. This requires that the volume transport of the Gulf Stream increase by some 1.8 fold after passing through the Straits of Florida. The mechanism for this addition of mass is unknown.

Sverdrup (1942)[#] assumed that $12 \times 10^6 \text{ m}^3/\text{sec}$ are contributed to the flow of the Gulf Stream by the Antilles Current. Neither geostrophic nor nongeostrophic methods give certain evidence of the existence of this current, and yet some flow from this source and from others along the coast as far north as Cape Hatteras must occur in order for the baroclinic volume transport of the Gulf Stream to increase as observations at Hatteras show that it does. The temperature and salinity of water passing through the Straits of Florida and that in the deeper layers of the Gulf Stream in the Hatteras region are quite different. From this it has been suggested that some augmentation of the volume transport of the current may take place in the vicinity of the Blake Plateau and be virtually complete by the time the current has moved free of the coast beyond Cape Hatteras. The augmentation may be discontinuous.

Geostrophic estimates of the volume transport of the current depend critically on a suitable choice of the reference level of no motion on which the dynamic topography is based. The level of no motion has been assumed to coincide with the level of least vertical shear, situated between the 1500- and 2000-meter levels. Recently this choice has been vindicated in one small region of the Gulf Stream through a series of direct observations with the Swallow float.[15]

Synoptic oceanography

The Gulf Stream changes its position and therefore cannot be charted once and for all. But since the Gulf Stream is composed of a collection of well-known water types, it is possible to estimate the structure of the current from a rapid examination of the temperature field alone.

In 1938 a very simple device, the bathythermograph, was devised by A. F. Spilhaus[16] (later improved by W. M. Ewing, J. L. Worzel, and A. C. Vine) to measure the temperature of sea water as a continuous function of pressure. A bathythermograph in its present form (Fig. 11–16) can be lowered to depths below the 200-meter level from a ship moving at speeds of 10 to 12 knots. This is sufficient to reach the seasonally undisturbed water of the Gulf Stream. Because of the rapidity with which lowerings can be made from a moving ship, it is possible to measure the distribution of temperature in the water column at intervals of one-half hour or even less and thus to generate shallow temperature sections through complicated thermal structures.* Many thermal studies of the structure of the Gulf Stream have been made from moving ships by Fuglister and his colleagues to generate what amount to synoptic maps of small areas of the Gulf Stream

---

* Instrumental developments by Richardson and Hubbard (1959) now permit this to be done on a practically continuous basis. See W. S. Richardson and C. J. Hubbard, 1959, *Deep-Sea Res.* 6: 239–244.

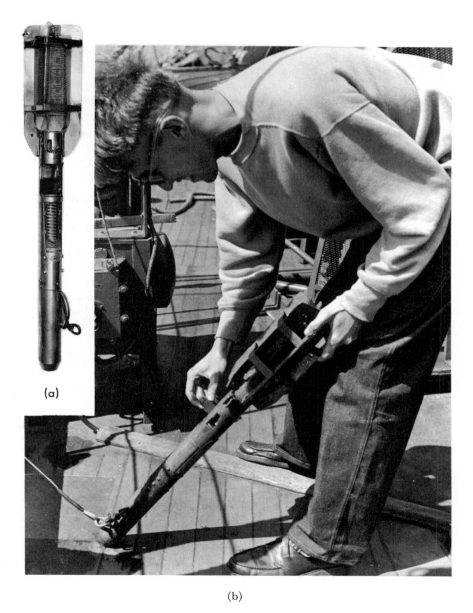

(a)

(b)

Fig. 11–16.   (a) Cutaway view of a bathythermograph (BT), showing bourdon tube, stylus, slide holder, and pressure sylphon. [Photograph by G. Boltz.]   (b) A bathythermograph being loaded. [Photograph by J. Hahn.]

structure, principally in those portions off New England. The results of these investigations proved so interesting that in June, 1950, six ships were assembled in a cooperative effort to map the current for a period of three weeks between Cape Hatteras and the offing of Nova Scotia. In this multiple-ship survey the bathythermograph, Loran, and electromagnetic methods of current measurement were used simultaneously, with the result that in the period from the 6th to the 23rd of June, 1950, a series of observations were made of the Gulf Stream which provided a synoptic picture of its behavior. Some results of this survey are given in a paper by Fuglister and Worthington (1951).[#]

The results of the "current" surveys by Fuglister agree very well with the observations of currents obtained by Loran dead reckoning and the electromagnetic method, so that it is felt that in this area of the ocean, at least, the temperature field is a reliable indication of the velocity field where the currents are strong.

In other areas where the T-S correlation is well enough known, LaFond (1949, *Trans. Amer. geophys. Un.*, 30: 231–237) has used the bathythermograph as a means of making dynamic computations. Stommel (1947, *J. Mar. Res.*, 6: 85–92) suggests that bathythermograms may also be used to improve the results of usual dynamic computations by removing the errors of interpolation between individual measurements near the surface.

Direct utilization of these observations has yielded a synoptic plot of the temperature and surface velocity of the Gulf Stream, and of the meander pattern and its changes during the period of observations. A generalized chart of the changing trend of the Gulf Stream during this time is shown in Fig. 11–17. Fuglister and Worthington (1951)[#] have indicated that the trend of the surface isotherms and the direction of currents observed electromagnetically are in good agreement. If the temperature pattern is considered identical with the flow pattern, the rates of meander motion are in the range 10 to 15 mi/day. However, a recent four-ship survey of the same area from April through mid-June, 1960, indicates that the synoptic pattern of Gulf Stream meanders can also change very slowly.

Between the Straits of Florida and Cape Hatteras, meanders do not develop conspicuous amplitudes. Where the excursions of the current are limited by the proximity of the coast, as off the Carolina bays, the amplitude of meanders ranges from 10 to 30 nautical miles. Downstream from Cape Hatteras, where the current is free of continental influences, meanders may develop amplitudes as great as 300 nautical miles, in extreme cases. The wavelength of meanders seems to range from about 80 to 250 nautical miles.

Gulf Stream meanders may degenerate into eddies, much as meandering rivers develop oxbow lakes. These developments have been noted during both the 1950 and 1960 multiple-ship surveys of the current off New England. On the 1950 survey, one of the cyclonic bends was observed to pinch

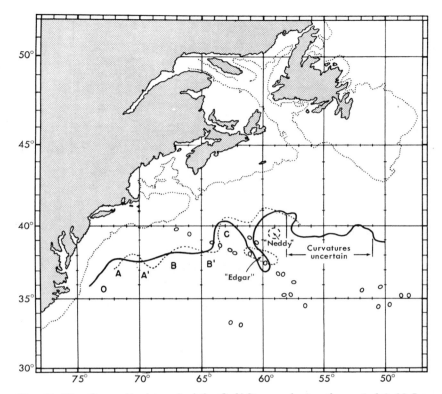

Fig. 11–17. Generalized trend of the Gulf Stream during the period 8–22 June 1950 (Operation CABOT). Seamounts are indicated by ovals. [Base chart drawn from data supplied by U.S. Navy Hydrographic Office.]

off and separate itself as an independent eddy on the Sargasso Sea side of the main current structure. This is the first instance where an eddy, so often mentioned in the oceanographic literature, was actually mapped in detail. In one earlier case an eddy of this type was circumnavigated by *Atlantis*. The highest Gulf Stream velocities on record were observed on that occasion (about 6 knots). Figures 11–18 through 11–20 show some details of the structure of the Gulf Stream observed during the multiple-ship survey of 1950, and include a detailed map of the eddy before the time of its separation from the main current. A sketch of the development of this feature, prepared by Fuglister, is shown in Fig. 11–21.

On the shoreward side of the same meander pattern, Ford, Longard, and Banks (1952)[17] found a narrow and presumably shallow ribbon of cool water having a somewhat lower salinity than that which is characteristic for slope water. This feature was observed during the 1960 survey to extend downstream as far as the Grand Banks, and at other times as far upstream as the Carolina Bays.[18] The ribbon is probably discontinuous.

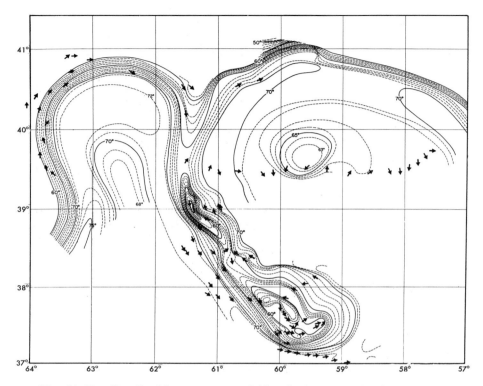

Fig. 11–18. Details of the temperature field at the 200-meter depth in the feature labeled "Edgar" in Fig. 11–17. GEK vectors are superimposed. [From F. C. Fuglister and L. V. Worthington, 1951, *Tellus*, (3)1.]

Multiple currents

Fuglister (1951)[19] has said: "In nearly all discussions of the Gulf Stream and of the North Atlantic Current the terms countercurrent and eddy are frequently employed.[20] Many temperature and salinity profiles across the currents indicate the presence of countercurrents and/or eddies and certain theoretical considerations require that they exist.[21] On the other hand charts showing the average distribution of temperature and salinity, and charts showing the currents in the western North Atlantic do not indicate that countercurrents or eddies are permanent features of the system."[22]

The narrow filaments that are so often observed to compose the Gulf Stream change position or appear and disappear to such an extent that they are lost in the data-averaging processes used to produce mean charts of the circulation. Even when given adequate numbers of data, features that have a characteristic longevity less than the period of averaging will neces-

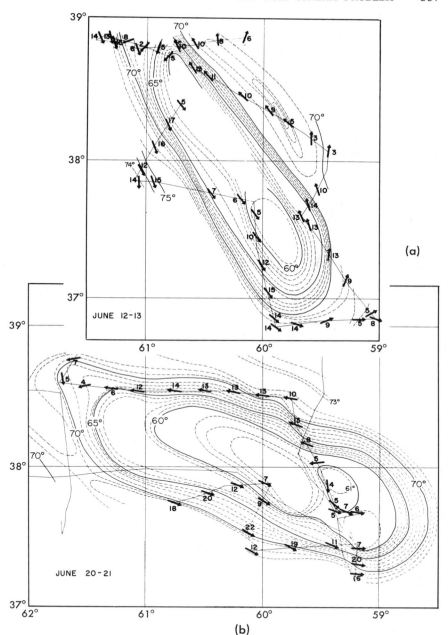

Fig. 11–19. "Edgar" during (a) the first and (b) the last periods of Operation
CABOT, showing average temperature of the upper 200-meter layer. Current
arrows from the GEK speed are in decimeters per second. [From F. C. Fuglister and
L. V. Worthington, 1951, *Tellus*, 3(1).]

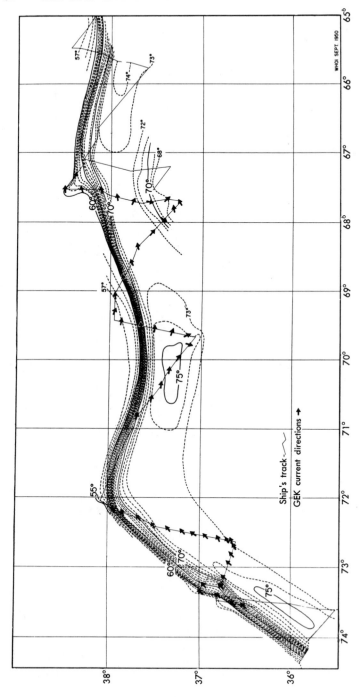

Fig. 11–20. Trend of the current and isotherms in the upper 200-meter layer of the Gulf Stream on 8 June 1950. [From F. C. Fuglister and L. V. Worthington, 1951, *Tellus* 3(1).]

sarily vanish from the final picture. Average charts therefore have a built-in filter whose working depends on the time span as well as on the spatial interval between sampling points.

Recognizing this problem, Fuglister (1951)# studied the results of three sections made concurrently and at right angles to the North American coast between Cape Hatteras and the Grand Banks, and from generalizations based on these drew the idealized chart shown in Fig. 11–22. It is believed that, if they are real, these streamers may be of a quasi-permanent character and that this is a much more valid interpretation of the current system in this area than the older concept that the Gulf Stream is a single tortuous current.

Rapid changes in the Gulf Stream

At present there are two methods by which the fluctuations of the position and shape of the Gulf Stream can be examined: the first is to cross and re-cross the current several times each day in a ship sailing a line fixed in space so that, with time, the current structure passes under the keel of the observing ship; the second is to fly repeatedly along the length of the current so that a sufficient length is observed in one day to establish a synoptic chart of its surface structure. Both of these methods have been employed since 1953.

Serial sections made from a ship become particularly valuable in a Loran service area. Through a comparison of the dead reckoning with the actual position of the ship as observed by Loran, it is possible to estimate the surface currents and with electromagnetic observations to estimate the mass transport. With the aid of the GEK the ship may be steered so as to make good a given geographic line by turning the bow upstream an amount just sufficient to provide a component of forward speed equal to the set of the current. Loran is used to confirm the cumulative accuracy of such corrections and permits measurements of the width of the current and changes in its structural geometry to an accuracy of about ±1 nautical mile. These observations, together with bathythermograms to measure the vertical structure of temperature to a depth exceeding 200 meters and a surface thermograph and salinity meter, provide data of a quasi-continuous nature developed in time and one dimension of space. Such studies can be made from a ship moving at the highest speed that weather conditions will permit. Some results of one such survey are described in a paper by von Arx, et al., (1955)[23] and in a thesis by T. F. Webster (1961).[24] Coordinated observations were made from the air.

The principal value of an airplane is its high mobility. When its course can be guided by visual and radiometric observations of the structure of the sea surface, it is possible to track the Gulf Stream front. The position of the front can be determined partly by visible indications (Fig. 11–23).

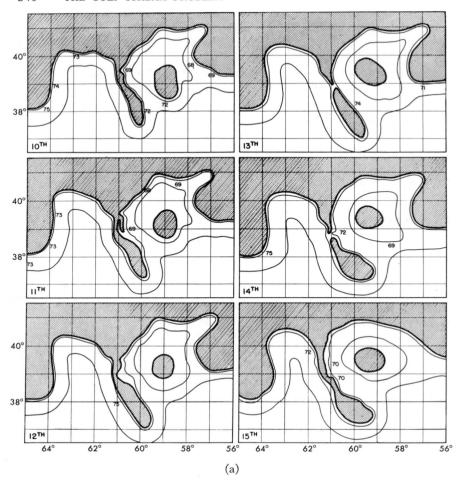

(a)

Fig. 11–21. (a) and (b). The evolution of "Edgar" during 10–22 June 1950. [From F. C. Fuglister and L. V. Worthington, 1951, *Tellus*, 3(1).]   Part (b) on opposite page.

These include the contrast in water color between Gulf Stream and slope water, the change in sea state across the frontal outcrop, the accumulated sargassum and other flotsam on the frontal convergence, and, in calm weather, characteristic patterns of slicks which tend to develop in slope water and to lie parallel with the front. Cloud base over the Gulf Stream is normally between 1500 and 2000 feet, so that it is necessary to fly below this level to observe the sea surface. By Loran it is possible to fix the position of significant indications that the frontal outcrop has been crossed. An additional clue to such events is a sudden change in the thermal emission of

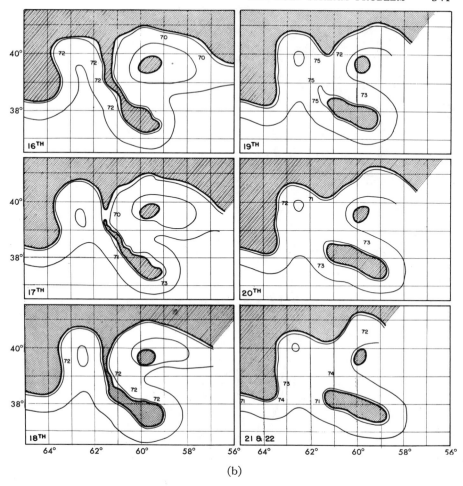

(b)

the sea surface, and to some extent that of the water vapor in the intervening air, detected by the airborne radiation thermometer.[25]

By these means it is possible to follow the frontal outcrop from the air, flying along the current when its edge is clearly visible, and zigzagging across the expected position of the edge using the radiation thermometer as a guide when visual indications are poor.

The relationship between the frontal position detected from ships and from aircraft has been studied. It is found that the frontal outcrop is closely associated with the steepest horizontal gradients of temperature at depths of 100 to 200 meters during the winter months. In late spring and summer the solar warming of the surface layer tends to blur this relationship or even obscure it altogether.

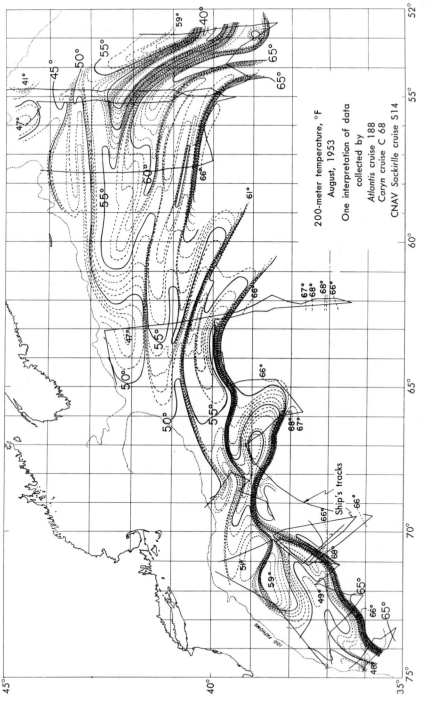

Fig. 11–22. Fuglister's multiple-current interpretation of the data collected simultaneously on three widely separated ship tracks. [From F. C. Fuglister, 1955, *Deep-Sea Res.*, 2.]

200-meter temperature, °F
August, 1953
One interpretation of data
collected by
Atlantis cruise 188
Caryn cruise C 68
CNAV *Sackirlle* cruise S14

Fig. 11–23.   The continental edge of the Gulf Stream as seen from the air (1500 feet) after passing tankers have been pumping water ballast.

Figure 11–24 shows the frontal outcrop of the Gulf Stream observed from an altitude of 1500 feet in February 1953. The frontal outcrop is composed of overlapping segments. In following each segment, it was often found that the temperature contrast across the front and the change of sea state and sea color coincident with the temperature contrast grew stronger and then weaker as observations were pursued downstream, until at last none of these clues was sufficiently pronounced to direct the airplane further. Course was then changed 90° left to renew contact with the frontal outcrop. The fresh contact was usually strong, so it can be inferred that the filament extended upstream some distance. Visual observations from the airplane indicated a finer overlapping structure in the frontal outcrop, which interrupted its continuity at intervals of the order of 10 km. When horizontal visibility was good, it was possible to fly across the smaller irregularities without losing track of the larger-scale features.

The method of serial sections from ships, augmented by longitudinal surveys of the Gulf Stream from the air, shows promise of revealing information on the short-period fluctuations of the current in the context of its large-scale geometry. For example, by these methods it has been shown that the rise and fall of local tides moves the Gulf Stream bodily from side to side. Tidal oscillations of the position of the current maximum are observed to have an amplitude as great as 10 nautical miles and a mean amplitude near 5 miles. The oscillation of the position of the current maximum associated with meandering has an amplitude of from 10 to 30 miles in the region of Onslow Bay, North Carolina, where the lateral excursions of the current are limited by the proximity of the coast.

The characteristic frequencies with which the current executes major swings from side to side are observed as one, two, four, and perhaps six cycles per lunar month. The position of the current maximum, relative to a fixed point such as Cape Hatteras, tends to oscillate offshore for a period of days, after which it approaches the coastline rapidly, covering a distance of as many as 35 miles in 4 days and then retreating again to its mean position in the ensuing 3 or 4 days. It is very much as though the Gulf Stream "bounced" off the continental shelf. At the same time that the core of the current approaches the coast, the field of sloping isotherms moves shoreward but is often left behind when the current maximum regains its normal position. On other occasions the interface between slope water and Gulf Stream water may move shoreward from the position of the current maximum without appreciable change in the latter's position, whereupon the steep slope of the isotherms will either move offshore to the position of the current maximum or regenerate itself anew in the more normal configuration.

It remains to be learned whether or not this motion is related to the meander pattern downstream from Cape Hatteras or, indeed, if meanders

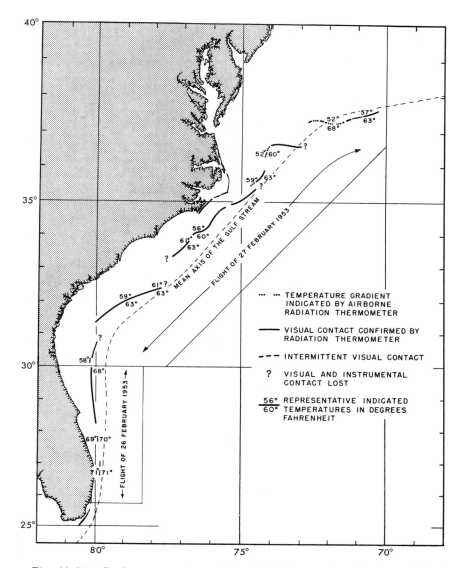

Fig. 11–24. Surface expression of the Gulf Stream front between Miami, Florida, and the meridian 70°W (flight altitude, 1500 feet). [From W. S. von Arx, D. F. Bumpus, and W. S. Richardson, 1955, *Deep-Sea Res.*, 3.]

progress, regress, or merely oscillate around some mean position. Observations of the Kuroshio collected by M. Uda suggest that the latter possibility may ultimately turn out to be the case.

Tidal modulation of flow rates

Pillsbury (1890)[#] was among the first to detect diurnal fluctuations in the flow of the Florida Current and Gulf Stream by direct measurements in the Straits of Florida and off Cape Hatteras. Parr (1937)[#] reported similar fluctuations and lateral motions of the isotherms and isopleths of salinity in the Straits of Florida, which he related to changes in the surface velocity observed with a taffrail log. Murray (1952)[26] and Wagner and Chew (1953)[27] have reported from GEK observations short-period fluctuations in the surface current near Miami. Using moored electrodes on the telegraph cable between Key West, Florida, and Havana, Cuba, Wertheim has obtained a long series of electromagnetic measurements of the fluctuations of volume transport through the Straits of Florida.[28] These reveal a daily variation in the flow of the Florida Current which is believed to be at least partly real. Two facts emerge from these studies: first, that, as shown in Fig. 11–25 (Wertheim 1954),[28] there is a close correlation in the diurnal variation of emf (interpreted as transport) between a linear combination of the tide at Miami displaced by two hours and inverted to represent the motion of the wave upstream and the Tampico tide taken directly and displaced by six hours. Such seemingly capricious juggling of tidal signatures can be given physical justification in terms of water depth and the equation for shallow-water waves.

In addition to the tidal variations in the Florida Current there is an irregular variation of transport based on 24-hour averages taken so as to eliminate the tidal influences. These irregularities show a wide range of variation (by a factor of two) in the course of a month (December, 1952, for example) and other oscillations having a similar slope but lesser amplitude irregularly distributed during other times of the interval August, 1952, through July, 1953. Stommel (1958, pp. 141–143, *The Gulf Stream*, Berkeley, California: University of California Press) attempted to interpret these in relation to the five-day mean pressure differences between 20° and 30° north latitude, averaged across the North Atlantic, as an indication of the strength of the trade winds. He found, however, that while there were sustained periods during which the Bermuda-Azores high weakened, these periods of light winds corresponded with high discharge rates in the Florida Current. More than this, the sustained period of weakened trades *preceded* the periods of high flow rates in the Florida Current by about 30 days.

The astronomical tide in the Gulf of Mexico is predominantly diurnal, being interrupted at fortnightly intervals by a short succession of weak

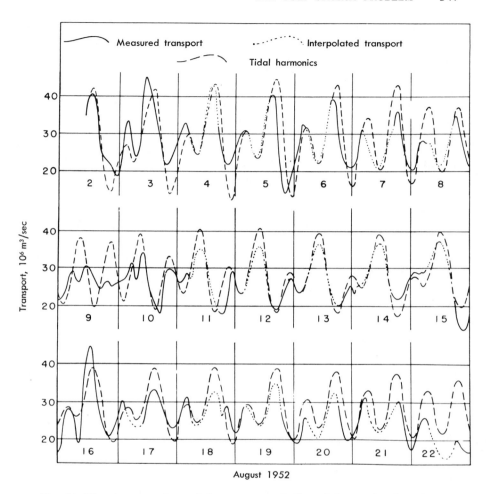

Fig. 11–25. A comparison of the transports indicated by motional emf across the Florida Straits with certain tidal harmonics characteristic of the Gulf of Mexico. [From G. K. Wertheim, 1954, *Trans. AGU*, 35(6): 878, 881.]

semidiurnal tidal seiches. The diurnal tide in the Gulf of Mexico is essentially in phase at all points around the perimeter, which requires that there be a diurnal change in the volume of water in the Gulf when this tidal species predominates. Since the flow in the Straits of Florida is unidirectional, the increase in volume required for a rise in the diurnal tide must be accomplished through a change in the rate of outflow through the Straits of Florida if the rate of inflow through the Yucatan Channel is actually constant.

To examine these effects quantitatively, it may be said that the amplitude of the average diurnal tide in the Gulf of Mexico is about 0.4 meter and its area 1.3×10^{12} square meters. Hence the volume that must enter to supply this rise in the diurnal tide is approximately 0.5×10^{12} cubic meters. If the average rate of inflow through the Yucatan Channel is 30×10^6 m^3/sec, the outflow through the Straits of Florida must be less than this figure in order for the tidal rise to occur. Assuming that the rate of rise is linear during a 12-hour period, approximately 10×10^6 m^3/sec must be contributed to effect the tidal rise; hence the flow through the Florida Straits must be this amount short of average, namely, about 20×10^6 m^3/sec during the rise of the diurnal tide. On the ebb, the Straits of Florida must transport about 40×10^6 m^3/sec. Since the cross-sectional area of the Straits of Florida is not materially changed by the rise and fall of the diurnal tide, the mean velocity must change by something like a factor of two during the course of one diurnal tidal cycle. This requirement is not inconsistent with the daily range of transport measured on the Key West-Havana cable.

A note on methods

In the study of the Gulf Stream we find all the essentials of the problem facing those who concern themselves with the study of the natural world: how can potentially simple and understandable processes be separated from the general complex of natural phenomena? Dr. Warren Weaver once pointed out:

Physical nature . . . seems to be on the whole very *loosely coupled*. That is to say, excellently workable approximations result from studying physical nature bit by bit, two or three variables at a time, and treating these bits as isolated. Furthermore, a large number of the broadly applicable laws are, to a useful approximation, linear, if not directly in the relevant variables, then in nothing worse than their second time derivatives. And finally, a large fraction of physical phenomena (meteorology is sometimes an important exception) exhibit *stability;* perturbations tend to fade out, and great consequences do not result from very small causes.[29]

The oceans and atmosphere form a system in which probably many different physical mechanisms play interacting roles. For this reason it is hardly possible to study the oceanic circulation out of context with the atmosphere, or the general atmospheric circulation without regard for both the oceans and the continents. But we also realize that, in general, geophysical feedback systems vary widely in size and frequency and must be degenerative. For example, the instabilities which form the meander pattern of the Gulf Stream do not grow to such a size that the general circulation pattern of the North Atlantic is greatly changed. Ordinary cyclones, and even hurricanes in their turn, do not modify the Gulf Stream circulation noticeably, while less conspicuous general climatic variations do. From this it may be reasonable to say that we have to contend with a spectrum of motional regimes, and that these do not interact very readily unless their natural periods are matched fairly closely. If this be true, effects on the general ocean circulation must be produced by widespread causes of sufficient duration to match approximately the natural period of response of the oceans as a whole, and small-scale high-frequency effects must be associated with causes of similar small scale and short period.

We know that the oceans exhibit two kinds of responses to external forces. Barotropic responses are effective through the full depth of the water column, but the motional readjustment is complete in a few pendulum days. If the external change is sustained, baroclinic responses will, in time, cancel barotropic effects through a readjustment of the fields of density and motion within the water. To the extent that the baroclinic responses cannot keep step with external changes of even seasonal frequency, it is to be expected that barotropic effects will predominate where these are of subseasonal or even subannual duration.

Except for some means to measure the geopotential of the sea surface or that of an arbitrarily chosen isobaric surface, the method of dynamic sections is adequate for studies of baroclinic processes in the ocean. New observational techniques are needed which embrace not only the larger scales of motion but those ordinarily classed as turbulence if studies of transient, barotropic processes are to progress as rapidly.

REFERENCES

1. J. N. Nielsen, 1925, *Geogr. Tidsskr.*, 28: 49–59.

2. C. O'D. Iselin, 1933, *Trans. Amer. geophys. Un.*, 14: 226–231.

3. A. G. Huntsman, 1924, pp. 274–290, *Handbook of the British Association for the Advancement of Science*, University of Toronto Press.

4. H. B. Bigelow, 1915, *Bull. Mus. Comp. Zool. Harv.*, 59 (4): 359 pp.

5. F. A. Richards and A. C. Redfield, 1955, *Deep-Sea Res.*, 2: 182–199.

6. L. V. Worthington, 1959, *Deep-Sea Res.*, 5: 297–305.

7. W. Malkus and K. Johnson, 1954, *Atlantis Cruise 198 and Caryn Cruise 78, A Drift Study of the Gulf Stream*, Technical Report, Woods Hole Oceanographic Institution, Reference No. 54–67, unpublished manuscript.

8. W. V. R. Malkus, 1953, *J. Mar. Res.*, 12: 51–59.

9. L. V. Worthington, 1954, *Tellus*, 6: 116–123.

10. R. B. Montgomery, 1938, *J. Mar. Res.*, 1: 165–185.

11. I. Hela, 1952, *Bull. Mar. Sci.*, 1: 241–248.

12. F. C. Fuglister, 1951, *J. Mar. Res.*, 10: 119–127.

13. H. U. Sverdrup, 1942, *Oceanography for Meteorologists*, New York: Prentice-Hall.

14. R. B. Montgomery, 1941, *J. Mar. Res.*, 4: 32–37.

15. J. C. Swallow and L. V. Worthington, 1957, *Nature, London,* 179: 1183–1184.

16. A. F. Spilhaus, 1938, *J. Mar. Res.,* 1: 95–100.

17. W. L. Ford, J. R. Longard, and R. E. Banks, 1952, *J. Mar. Res.,* 11: 281–293.

18. P. E. Church, 1937, *Un. géod. géophys. int., Ass. Océanogr. phys. Publ. Sci.,* 4; S. L. Strack, 1953, *Surface Temperature Gradients as Indicators of the Position of the Gulf Stream,* Woods Hole Oceanographic Institution, Reference No. 53–53, unpublished manuscript.

19. F. C. Fuglister, 1951, *Tellus,* 3: 230.

20. C. O'D. Iselin, 1936, *Pap. phys. Oceanogr.,* 4(4): 101 pp.; 1940, *Pap. phys. Oceanogr.,* 8(1): 40 pp; H. U. Sverdrup, *et at.,* 1942, *The Oceans,* New York: Prentice-Hall.

21. C.-G. Rossby, 1936, *Pap. phys. Oceanogr.,* 5(1): 43 pp.

22. G. Wüst and A. Defant, 1936, *Atlas zur Schichtung und Zirkulation des Atlantischen Ozeans, Deutsche Atlantische Expedition "Meteor" 1925–1927,* Wiss. Erg., Bd. VI, Atlas, Berlin; F. C. Fuglister, 1947, *Pap. phys. Oceanogr.,* 10(2): 25 pp.

23. W. S. von Arx, D. F. Bumpus and W. S. Richardson, 1955, *Deep-Sea Res.,* 3: 46–65.

24. T. F. Webster, 1961, *Deep-Sea Res.,* 7: (in press) and *Tellus* (in press).

25. W. S. Richardson and C. H. Wilkins, 1958, *Deep-Sea Res.,* 5: 62–71.

26. K. M. Murray, 1952, *Bull. Mar. sci.,* 2: 360–375.

27. L. P. Wagner and F. Chew, 1953, *Some Results of the Florida Current Survey,* University of Miami Technical Report 53–9, unpublished manuscript.

28. G. K. Wertheim, 1954, *Trans. Amer. geophys. Un.,* 35:872–882.

29. W. Weaver, 1955, *Science,* 122:1256.

SUPPLEMENTARY READING

PIERCE, J. A., A. A. MCKENZIE, and R. H. WOODWARD, editors, 1948, *Loran,* Massachusetts Institute of Technology Radiation Laboratory Series, L. N. Ridenour, editor-in-chief, No. 4, New York: McGraw-Hill.

SPILHAUS, A. F., 1938, "A Bathythermograph," *J. Mar. Res.,* 1: 95–100.

STOMMEL, H., 1950, "The Gulf Stream, A Brief History of Ideas Concerning Its Cause," *Sci. Mon.,* N. Y. 70: 242–253.

STOMMEL, H., 1958, *The Gulf Stream,* Berkeley, California: University of California Press.

WARREN, B.A., 1963, "Topographic Influences on the Path of the Gulf Stream," *Tellus* 15(2): 167–183.

WEBSTER, F., 1961, "A Description of Gulf Stream Meanders off Onslow Bay," *Deep-Sea Res.,* 8: 130–143.

WEBSTER, F., 1961, "The Effect of Meanders on the Kinetic Energy Balance of the Gulf Stream," *Tellus* 12(3): 392–401.

Appendix A

SOME EVENTS THAT HAVE INFLUENCED THOUGHT IN THE MARINE SCIENCES

with Ann J. Martin

Original sources which were accessible to the writers are cited in this appendix. For some of the older items it was necessary to rely on secondary sources, which include the reprinted collections contained in *Source Books in the History of the Sciences*, McGraw-Hill series, Gregory D. Walcott, general editor. To avoid redundancy, full reference to these books is given below, and further references to them are made simply by title.

A Source Book in Astronomy, 1929, H. Shapley and H. E. Howarth, New York: McGraw-Hill.

A Source Book in Chemistry, 1952, H. M. Leicester and H. S. Klickstein, New York: McGraw-Hill.

A Source Book in Greek Science, 1948, M. R. Cohen and I. E. Drabkin, New York: McGraw-Hill.

A Source Book in Mathematics, 1929, D. E. Smith, New York: McGraw-Hill (Dover edition 1959).

A Source Book in Physics, 1935, W. F. Magie, New York: McGraw-Hill.

Some of the events listed in this appendix are not peculiar to oceanography. They are included nevertheless to give occasional reminders of past intellectual climates and of developments that have influenced contemporary thoughts and practices in the marine sciences.

640?– THALES OF MILETUS (of the Ionian school of philosophers) thought
546 that all things are derived from the single element *water* and that the
B.C. solid earth was afloat on the world-encircling flat ocean.

> G. Sarton, 1952, pp. 169–173, *A History of Science*, Vol. 1, Cambridge: Harvard University Press.

611–   ANAXIMANDER made what may be the first of the Greek maps of the
547   known world.
B.C.

> *Ibid.*, p. 175; W. A. Heidel, 1921, "Anaximander's Book, the Earliest Known
> Geographical Treatise," *Proc. Amer. Acad. Arts Sci.*, 56: 237–288.

384–   ARISTOTLE, one of the founders of the inductive method and the
322   accepted authority on scientific matters for one thousand years, pub-
B.C.   lished a series of treatises on natural and human philosophy which
       are encyclopedic records of the learning of his time. Among these are
       Meteorologica, Mechanica, and *Problemata*.

> G. Sarton, *op. cit.*, 476–485, 509–511, 513–519; *A Source Book in Greek Science*,
> pp. 189–194, 245–247, 374–375, 378–379, 384–385, 387–388, 391–392 *passim*.

3rd c. ARISTARCHUS OF SAMOS anticipated the Copernican solar system.
B.C.

> G. Sarton, 1959, *A History of Science*, Vol. 2, Cambridge: Harvard University
> Press, pp. 56–58; *A Source Book in Greek Science*, pp. 107–109.

3rd c. ERATOSTHENES OF ALEXANDRIA estimated the circumference of the
B.C.   earth to be 252,000 stades from simultaneous observations of the solar
       altitude taken from places at different latitudes. This estimate is in
       very fair agreement with modern values.

> A. Diller, 1949, "The Ancient Measurements of the Earth," *Isis*, 40: 6–9.

287?– ARCHIMEDES formulated the laws of hydrostatics. He also combined
212   numbers with experiments and gave the principle of the level.
B.C.

> G. Sarton, 1959, *op. cit.*, pp. 69–80; *A Source Book in Greek Science*, pp. 186–
> 189, 235–238; T. L. Heath, editor, 1953, *The Works of Archimedes*, Reissue of
> 1897 edition, including 1812 supplement, New York: Dover.

160?– HIPPARCHUS (OF NICAEA) attempted to apply astronomical methods
125?   to the problem of mapping the positions of places on the earth's sur-
B.C.   face. He also invented spherical trigonometry, discovered the preces-
       sion of the equinoxes, and classified according to magnitudes the
       brightness of naked-eye stars.

> G. Sarton, 1959, *op. cit.*, pp. 284–288, 298–302, 415.

64?–   STRABO (STRABON) wrote (*ca.* 20 A.D.) his *Geography*, in which he
21?   summarized the works of the Greek mathematical geographers to
B.C.   obtain the total size of the earth, estimated the inhabited fraction,
       and gave a description of the climatic zones of the globe.

> G. Sarton, *ibid.*, pp. 418–424; *A Source Book in Greek Science*, pp. 153–159.

150   Ptolemy (Claudius Ptolemaeus) was among the last of the Alex-
(*ca.*)   andrians and of the early scholars to be concerned with an objective
        study of natural phenomena.   He wrote the *Almagest*, which contains
        an explanation of the precession of the equinoxes and the Ptolemaic
        solar system, and a catalogue of fixed stars.   In his *Geography* he
        described a factual map of the known world.

> *Encyclopaedia Britannica*, 1886, 9th ed., Vol. 20, pp. 87–96, New York:
> Scribners.

With the decline of ancient civilization and the progressive incineration
of the great Library of Alexandria, partially at the hands of Christian Bishop
Theophilus (*ca.* 390 A.D.) and finally as a result of the Moslem conquest of
Egypt (640 A.D.), it remained for the Persian and Arabian scholars to as-
semble, record, and to continue to enlarge scientific knowledge.   The pall of
the Dark and Middle Ages lay over western Europe until the beginning of
the twelfth century, when translations of Arabic records into Latin and
various western tongues began to appear.   Ptolemy's *Almagest* was translated
into Latin by an anonymous scholar in about the year 1160.

1291   Ugolino Vivaldi and his brother, Vadino, Genoese explorers, ven-
        tured beyond the Straits of Gibraltar in search of a route to India but
        did not return.

> D. B. Durand, 1936, "Precursor of Columbus?" *Geogr. Rev.*, 26: 525–526.

1360   European trade routes well established on land and sea.
(*ca.*)

1452–   Leonardo da Vinci discussed the principle of continuity in connec-
1519   tion with incompressible fluid flow.

> Leonardo da Vinci, 1938, "The Nature of Water," pp. 11–131, *The Note-
> books of Leonardo da Vinci*, translated and arranged by Edward MacCurdy,
> Vol. 2, New York: Reynal & Hitchcock.

1500   Pedro Alvares Cabral discovered what is now Brazil, before
        rounding the Cape of Good Hope on the route established three
        years earlier by Vasco da Gama.

> A. Fontoura da Costa, 1939, "The Discovery of Brazil in 1500," *Int. hydr.
> Rev.*, 16(1): 147–151.

1513   Juan Ponce de Léon gave the first description of the Florida Current.

> L. D. Scisco, 1913, "The Track of Ponce de Leon in 1513," *Bull. Amer.
> Geogr. Soc.*, 45: 721–735.

1515   PETER MARTYR conjectured about the physical origins of the Gulf
Stream.

P. Martyr, 1912, pp. 346–348 (Third Decade, Book VI), *De Orbo Novo, The
Eight Decades of Peter Martyr D'Anghera*, translated by Francis Augustus
MacNutt, Vol. 1, New York: G. P. Putnam's Sons.

1521   FERDINAND MAGELLAN attempted possibly the first deep-sea sounding
in the Pacific Ocean with hand lines, going to a depth of about 1200
feet, but failed to reach bottom, which is now known to be 12,000
feet.

H. S. Bailey, Jr., 1953, "The Voyage of the 'Challenger'," *Sci. Amer.*, 188:
88–94.

1543   NICOLAUS COPERNICUS published *de Revolutionibus Orbium Coelestium*, in
which he proposed the heliocentric concept of the solar system.

W. Norlind, 1953, "Copernicus and Luther: A Critical Study," *Isis*, 44:
273–276; H. Dingle, 1943, "Nicolaus Copernicus, 1473–1543," *Endeavour*, 2:
136–141; *A Source Book in Astronomy*, pp. 1–12.

1567   WILLIAM BOURNE gave an account of the log and line for measuring
a ship's way in a volume entitled *An Almanack and Prognostication for
iii Yeres with Serten Rules of Navigation*, which later appeared as a
manual called *A Regiment for the Sea*. This was published in 1574
and was widely used by English and Dutch sailors of the time.

E. G. R. Taylor, 1957, pp. 201–205, *The Haven-Finding Art*, New York:
Abelard-Schuman; G. D. Zerfass, 1952, "What's Our Speed? The Evolution of Ship-Logs," *Navigation*, 3: 232–239.

1569   GERARDUS MERCATOR (GERHARD KREMER) devised a nautical chart
distorted in such a way that the meridional and zonal scales were
equal at each point. This kind of chart facilitates both dead reckoning and the problems of rhumb line sailing.

C. H. Deetz and O. S. Adams, 1944, pp. 103–144, *Elements of Map Projection*,
5th ed., U.S. Coast and Geodetic Survey Special Publication No. 68; J. W.
van Nouhuys, 1933, "Mercator's World Atlas 'ad usum Navigantium',"
Int. hydr. Rev., 10(2): 237–241.

1586   SIMON STEVIN (STEVINUS) established the "law of equilibrium of a
body on an inclined plane" and made the first use of the parallelogram of forces.

A Source Book in Physics, pp. 23–27; G. Sarton, 1934, "Simon Stevin of Bruges
(1548–1620)," *Isis*, 21: 241–303.

1600   SIR WILLIAM GILBERT, court physician to Queen Elizabeth and James I, published his work *De Magnete* describing the dipole field of the earth, which he likened to a magnetized sphere that he called the "terrella" (little earth). He proposed that the magnetic dip needle be used as a means for measuring latitude at sea on overcast days.

A Source Book in Physics, pp. 387–393; R. Suter, 1950, "Dr. William Gilbert of Colchester," *Sci. Mon., N.Y.*, 70: 254–261.

1604   JOHANN KEPLER (KEPPLER) explained the faint light on the dark side of the moon as a second reflection of light from the sunlit hemisphere of the earth.

A. Danjon, 1954, p. 726, *The Earth as a Planet* (The Solar System, Vol. 2), G. P. Kuiper, editor, Chicago: University of Chicago Press.

1605   WILLEBRORD SNELL VAN ROIJEN (SNELLIUS), a Dutch physicist, described the loxodrome for rhumb line navigation, and in about 1621 established the law of optical refraction.

G. Sarton, 1934, "Simon Stevin of Bruges (1548–1620)," *Isis*, 21: 278–279; J. A. Vollgraff, 1936, "Snellius' notes on the reflection and refraction of rays," *Osiris*, 1: 718–725.

1609   GALILEO GALILEI employed the telescopic optical system to open a new era in observational astronomy, and eventually in navigation.

A Source Book in Astronomy, pp. 41–52; E. Rosen, 1951, "Galileo and the Telescope," *Sci. Mon., N.Y.*, 72: 180–182.

1609   JOHANN KEPLER (KEPPLER) summarized the planetary observations of Tycho Brahe (1546–1601) in three laws of planetary motion. The first and second were published in *De Motibus Stellae Martis* (Commentaries on the Motion of Mars) and the third in *De Harmonice Mundi* (Harmony of the World) in 1619.

A Source Book in Astronomy, pp. 29–40.

1610   GALILEO GALILEI estimated the distance between the earth and the moon to be sixty earth-diameters, or twice the true distance.

E. Rosen, 1952, "Galileo on the Distance Between the Earth and the Moon," *Isis*, 43: 344–348.

1629   RENÉ DESCARTES developed analytic geometry, and may have been one of the first to consider the property of inertia and the principle of conservation of momentum.

A. C. Crombie, 1959, "Descartes," *Sci. Amer.*, 201: 160–173, 218; R. Descartes, *Geometry*, translated by D. E. Smith and M. L. Latham, New York: Dover (1954); *A Source Book in Mathematics*, pp. 397–402.

1638   GALILEO GALILEI published his *Dialoghi della Nuove Scienze*, in which are established the principles of accelerated motion which laid the foundations of mechanics. He was among the first to make a clear distinction between *mass* and *weight*.

Galileo, *Dialogues Concerning Two New Sciences*, translated by Henry Crew and Alfonso de Salvio, New York: Dover.

1661   CHRISTIAN HUYGHENS (HUGENIUS) completed the first timekeeper specifically intended for the purpose of finding longitude at sea—a marine pendulum clock.

W. V. Cannenburg, 1936, "The Marine Clock of Christian Huyghens," *Int. Hydr. Rev.*, 13(2): 162–165; S. L. Chapin, 1952, "A Survey of the Efforts to Determine Longitude at Sea, 1660–1760. Part III: A Perfect Timekeeper," *Navigation*, 3: 296–303.

1663   BLAISE PASCAL discussed the hydrostatic paradox and the axiom of hydrostatics (that the pressure at each point in a fluid at rest is transmitted equally in all directions) in his *Traité de l'Equilibre des Liqueurs*, published posthumously.

A Source Book in Physics, pp. 75–80.

1663   ISAAC S. VOSSIUS (VOSS) suggested that the North Atlantic circulation consisted of a general clockwise motion.

I. Vossius, 1663, *De motu marium et ventorum*, The Hague: Adrian Vlacq.

1665   ROBERT HOOKE adopted Boyle's suggestion that the freezing point of water be used as the zero point on a scale of temperature, and specified that the water be distilled.

M. K. Barnett, 1956, "The Development of Thermometry and the Temperature Concept," *Osiris*, 12: 293.

1666   ROBERT HOOKE made the first step toward an efficient instrument for measuring the angles between the moon and fixed stars by the method of reflecting coincidences.

E. G. R. Taylor, 1957, p. 252, *The Haven-Finding Art*, New York: Abelard-Schuman; G. D. Zerfass, 1952, "Two Mirrors: The Story of the Invention of the Sextant," *Navigation*, 3: 131–137.

1673   CHRISTIAN HUYGHENS (HUGENIUS) published *Horologium Oscillatorium*, in which are discussed his method for the pendulum control of clocks and some theorems on the conical pendulum.

A Source Book in Physics, pp. 27–30; C. A. Crommelin, 1950, "The Clocks of Christian Huyghens," *Endeavour*, 9: 64–69.

1686   EDMOND HALLEY attempted a systematic study of the primary wind systems and their relation to the main ocean currents.

E. Halley, 1686, "An Historical Account of the Trade Winds, and Monsoons, observable in the Seas between and near the Tropicks, with an attempt to assign the Phisical cause of the said Winds," *Phil. Trans.*, 16: 153–168.

1686   GOTTFRIED WILHELM LEIBNITZ (LEIBNIZ), codiscoverer with Newton of the infinitesimal calculus, established the principles of work against gravity and of the release of potential into kinetic energy in free fall in *Acta Eruditorum*.

A Source Book in Physics, pp. 52–55.

1687   SIR ISAAC NEWTON, in his *Principia Mathematica* put forward principles of physical mechanics, and in connection with astronomical problems stated the principle of universal gravitation.   The latter provided physical insight into Kepler's laws of planetary motion and the origins and properties of the tide-generating force.

Sir Isaac Newton, 1848, *Newton's Principia*, translated by Andrew Motte, New York: Daniel Adee; *A Source Book in Physics*, pp. 30–46; *A Source Book in Astronomy*, pp. 74–93; H. Crew, 1942, "Sir Isaac Newton, 1642–1727," *Sci. Mon., N.Y.*, 55: 279–284; I. B. Cohen, 1955, "Isaac Newton," *Sci. Amer.*, 193: 73–80.

1687   PIERRE VARIGNON developed the principles of the moments of force and formalized the use of the parallelogram of forces introduced by Stevin in 1586.

A Source Book in Physics, pp. 46–49.

1697   WILLIAM DAMPIER, buccaneer and explorer, published "A Discourse of the Winds," in which the steady trades were distinguished geographically from the more variable westerlies.

John Masefield, editor, 1906, "A Discourse of the Trade-winds, Breezes, Storms, Seasons of the Year, Tides and Currents of the Torrid Zone throughout the World," pp. 229–321, *Dampier's Voyages*, by Captain William Dampier, Vol. 2, Part 3, London: E. Grant Richards; W. C. D. Dampier-Wetham, 1929, "William Dampier, Geographer," *Geogr. J.*, 74: 478–480.

1698   EDMOND HALLEY, later Astronomer Royal, made one of the first purely scientific voyages to measure the longitude of places and study the variation of the compass.

E. Halley, 1714, "Some Remarks on the Variations of the Magnetical Compass published in the Memoirs of the Royal Academy of Sciences, with regard to the General Chart of those Variations made by E. Halley; as also concerning the true Longitude of the Magellan Streights," *Phil. Trans.*, 28: 165–168; N. T. Bobrovnikoff, 1942, "Edmond Halley, 1656–1742," *Sci. Mon., N.Y.*, 55: 439–446; E. Bullard, 1956, "Edmond Halley (1656–1741)," *Endeavour*, 15: 189–199.

1699   SIR ISAAC NEWTON showed to the Royal Society a reflecting octant
(*ca.*) he had devised, with which angles between celestial bodies could be measured by the method of reflecting coincidences.

I. Newton, 1742, "A true Copy of a Paper found, in the Hand Writing of Sir Isaac Newton, among the Papers of the late Dr. Halley, containing a Description of an Instrument for observing the Moon's Distance from the Fixt Stars at Sea," *Phil. Trans.*, 42: 155–156; I. B. Cohen, 1955, "Isaac Newton," *Sci. Amer.*, 193: 73–80; E. G. R. Taylor, 1957, p. 252, *The Haven-Finding Art*, New York: Abelard-Schuman.

1714   QUEEN ANNE authorized a public reward to any person who might invent a practicable method for determining a ship's longitude in the open sea. In this connection Sir Isaac Newton said, "That, for determining the Longitude at Sea, there have been several Projects, true in the Theory, but difficult to execute: One is, by a Watch to keep time exactly: But, by reason of the Motion of a Ship, the Variation of Heat and Cold, Wet and Dry, and the Difference of Gravity in different Latitudes, such a Watch hath not yet been made."

R. T. Gould, 1923, p. 13, *The Marine Chronometer*, London: J. D. Potter.

1715   EDMOND HALLEY suggested that the age of the ocean might be esti-
.mated by measuring the rate at which salt is carried into the sea by rivers.

E. Halley, 1715, "A Short Account of the Cause of the Saltness of the Ocean, and of the several Lakes that emit no Rivers; with a Proposal, by help thereof, to discover the Age of the World," *Phil. Trans.*, 29: 296–300; N. T. Bobrovnikoff, 1942, "Edmond Halley, 1656–1742," *Sci. Mon., N.Y.*, 55: 439–446.

1724   GABRIEL DANIEL FAHRENHEIT devised the thermometric scale now bearing his name, in which the interval between the melting point of ice and the boiling point of water is divided into 180 equal parts.

D. G. Fahrenheit, 1724, "Experimenta circa gradum caloris liquorum nonnullorum ebullientium instituta," *Phil. Trans.*, 33: 1–3, and the translation in *A Source Book in Physics*, pp. 131–133.

1725 COUNT LUIGI FERDINANDO MARSIGLI published a monograph on the natural history of the seas consisting of five parts: the bottom topography of the seas, the waters of the sea, the motions of the waters, plants in the sea, and animals of the sea. He reported measurements of the tides, and seasonal temperature changes of the air and water to depths as great as 120 fathoms. He also drew certain conclusions in which he states, in effect: Research into the nature of the sea is very difficult, and this perhaps has been the reason why many learned men—because they could not see an end to such an undertaking— were of the opinion that it was useless to make a beginning.

Original work, in French, published in Amsterdam, 1725. Dutch edition, *Natuurkundige Beschryving der Zeën*, The Hague, 1786.

1731 JOHN HADLEY, elder brother of George Hadley, devised a reflecting octant which employed a fixed horizon mirror and a movable index mirror similar to the design of modern sextants.

J. Hadley, 1731, "The Description of a new Instrument for taking Angles," *Phil. Trans.*, 37: 147–157.

1732 HENRI PITOT described a mechanism to measure speed of flow from the difference between the ram and static pressures at a pair of orifices.

H. Pitot, 1732, "Description d'un machine pour mesurer la vitesse des eaux et le sillage des vaisseaux," *Histoire de l'Academié des Sciences;* H. Rouse and S. Ince, 1957, pp. 114–116, *History of Hydraulics*, State University of Iowa: Iowa Institute of Hydraulic Research.

1735– PIERRE BOUGUER made measurements of the astro-geodetic arc in
1739 Peru which, with other evidence obtained by the Venerable John Henry Pratt, Archdeacon of Calcutta, in India, 1854, and by Pierre-Louis Moreau de Maupertuis in Lapland, 1736, led to the theory of isostasy.

S. L. Chapin, 1952, "Expeditions of the French Academy of Sciences, 1735," *Navigation*, 3: 120–122; H. B. Glass, 1955, "Maupertuis, a Forgotten Genius," *Sci. Amer.*, 193: 100–110; J. H. Pratt, 1855, "On the Attraction of the Himalaya Mountains, and of the elevated regions beyond them, upon the Plumbline in India," *Phil. Trans.*, 145: 53–100.

1735 GEORGE HADLEY proposed a theory based on the conservation of angular momentum to explain the existence of trade winds. This meteorological feature has become known as the "Hadley cell."

G. Hadley, 1735, "Concerning the Cause of the General Trade-Winds," *Phil. Trans.*, 39: 58–62.

1736   PIERRE-LOUIS MOREAU DE MAUPERTUIS directed astro-geodetic measurements to determine the linear dimensions of a degree in Lapland,
and in 1738 published the results in *La Figure de la Terre*.

H. B. Glass, 1955, "Maupertuis, a Forgotten Genius," *Sci. Amer.*, 193:
100–110.

1738   DANIEL BERNOULLI published his *Hydrodynamica*, which deals with the
statics and dynamics of fluids. He also stated his theorem on the
conservation of energy in fluids.

D. Bernoulli, 1738, *Hydrodynamica, sive de viribus et motibus fluidorum com-
mentarii*, Argentorati.

1740   DANIEL BERNOULLI made the first study of the equilibrium tide.

D. Bernoulli, 1740, "Traité sur le flux et reflux de la mer," *Pièces qui ont
remporté le prix de l'Académie en 1740*, Paris, p. 55.

1740   LEONHARD EULER showed that the significant component of the tide-
generating force is that which is tangent to the earth.

L. Euler, 1740, "Inquisitio physica in causam fluxus ac refluxus maris,"
Pièces qui ont remporté le prix de l'Acad. Sci., Paris.

1742   ANDERS CELSIUS, of Uppsala, Sweden, devised the centigrade thermometric scale, in which the interval between the freezing point and
boiling point of water at atmospheric pressure is divided into 100
equal parts.

A. Celsius, 1742, "Observationer on twänne beständiga Grader pa en Thermometer," *K. svenska VetenskAkad. Handl.*, 3: 171–180; N. V. E. Norden-
mark, 1952, "Anders Celsius (1701–1744)," pp. 66–73, *Swedish Men of Science
1650–1950*, Sten Lindroth, editor, translated by Burnett Anderson, Stockholm: The Swedish Institute/Almqvist & Wiksell.

1743   JEAN LE ROND D'ALEMBERT distinguished between momentum and
kinetic energy in his *Traité de Dynamique*.

A Source Book in Physics, pp. 55–58.

1743   ALEXIS CLAUDE CLAIRAULT (CLAIRAUT) wrote the formula for the
gravitational acceleration to be expected at the ocean surface on the
assumption that the interior of the earth is in hydrostatic equilibrium.

A. C. Clairaut, 1743, *Théorie de la figure de la terre, tirée des principes de l'hydrostatic*, Paris.

1747   NILS GISSLER showed the statistical effect on sea level of an atmospheric pressure change.

N. Gissler, 1747, "Anledning at finna Hafvets affall för vissa är," *K. svenska VetenskAkad. Handl.*, 8: 142–149.

1748   JAMES BRADLEY published his discovery, from observations started in 1727, that the principal period of nutation for the earth, due to the inclination of the moon's orbit, is 18.6 years. He had earlier (1729) announced his discovery of the aberration of starlight.

J. Bradley, 1729, "A Letter from the Reverend Mr. James Bradley Savilian Professor of Astronomy at Oxford, and F.R.S. to Dr. Edmond Halley Astronom. Reg. &c. giving an Account of a new discovered Motion of the Fix'd Stars," *Phil. Trans.*, 35: 637–661, and reprint in *A Source Book in Astronomy*, pp. 103–108; J. Bradley, 1748, "A Letter to the Right honourable George Earl of Macclesfield concerning an apparent Motion observed in some of the fixed Stars," *Phil. Trans.*, 45: 1–43, and reprint in *A Source Book in Astronomy*, pp. 108–112; G. Sarton, 1931, "Discovery of the aberration of light," *Isis*, 16: 233–265; G. Sarton, 1932, "Discovery of the main nutation of the earth's axis," *Isis*, 17: 333–383.

1749   PIERRE BOUGUER, member of the French astro-geodetic expedition to Peru, published *La Figure de la Terre*, in which the gravity anomalies of the Andes were related to a deficiency of mass in the earth's crust.

P. Bouguer, 1749, *La figure de la terre, déterminée par les observations de Messieurs Bouguer, & de la Condamine, envoyés par ordre du Roy au Pérou, pour observer aux environs de l'équateur. Avec une relation abregée de ce voyage, qui contient la description du pays dans lequel les opérations ont été faites*, Paris: C.-A. Jombert.

1752   PHILIPPE BUACHE used bathymetric charts to show that there is continuity between the topographic features of the land and sea floor.

P. Buache, 1753, *Considérations géographiques et physiques sur les nouvelles découvertes au nord de la Grand Mer, appellée vulgairement le Mer du Sud, Avec des Cartes qui y sont relatives*, Paris: Ballard.

1754   IMMANUEL KANT pointed out that the tide-generating forces of the moon might act through the oceans to produce a secular retardation of the rotation of the earth. The attendant acceleration in the orbital motion of the moon had been suggested by Halley in 1695.

A Source Book in Astronomy, pp. 124–125; E. Halley, 1695, "Some Account of the Ancient State of the City of Palmyra, with short Remarks upon the Inscriptions found there," *Phil. Trans.*, 19: 174.

1755   LEONHARD EULER proposed a means for the discussion of the mechanics of fluid motion in purely analytical terms, including the Lagrangian method of 1788.

L. Euler, 1755, "Principes généraux de l'état d'équilibre des fluides," "Principes généraux du mouvement des fluides," "Continuation des recherches sur la théorie du mouvement des fluides," *Histoire de l'Académie de Berlin*.

1759   JOHN HARRISON completed the first satisfactory marine timekeeper, a copy of which was used on Cook's last voyage of 1776. "Cook, the most exact and least enthusiastic of men, had nothing but praise for it . . ."

R. T. Gould, 1923, pp. 54 and 72, *The Marine Chronometer*, London: J. D. Potter.

1762   JOHN CANTON found that water is somewhat compressible.

J. Canton, 1762, "Experiments to prove that Water is not incompressible," *Phil. Trans.*, 52 (2): 640–643.

1770   BENJAMIN FRANKLIN as Postmaster General authorized Captain Folger of Nantucket to draw a map of the Gulf Stream, and arranged its publication for the improvement of mail service from England.

B. Franklin, 1786, "A Letter from Dr. Benjamin Franklin, to Mr. Alphonsus le Roy, Member of several Academies, at Paris. Containing sundry Maritime Observations," *Trans. Amer. phil. Soc.*, 2: 314–315.

1772   JOHANN HEINRICH LAMBERT devised a conformal conic projection which has become standard for charting coastal geography.

C. H. Deetz and O. S. Adams, 1944, p. 79, *Elements of Map Projection*, 5th ed., U.S. Coast and Geodetic Survey Special Publication No. 68.

1774   THE REVEREND NEVIL MASKELYNE made the first attempt, proposed by him in 1772, to determine the mass of the earth from consideration of its density as revealed by astro-geodetic arcs measured near mountains.

The Reverend Nevil Maskelyne, 1775, "An Account of Observations made on the Mountain Schehallien for finding its Attraction," *Phil. Trans.*, 65: 500–542, and reprint in *A Source Book in Astronomy*, pp. 133–139.

1775   MARQUIS PIERRE SIMON DE LAPLACE wrote, in connection with the study of the tides, hydrodynamic equations of motion which include the horizontal components of the Coriolis acceleration. This antedates the work of Coriolis (1835).

P. S. Laplace, 1775, "Recherches sur plusieurs points du système du monde," *Mém. Acad. R. Sci.* (Paris), p. 75; A. T. Doodson and H. D. Warburg, 1941, pp. 3–4, 189–190, 193, *Admiralty Manual of Tides*, London: H. M. Stationery Office.

1780   HORACE BÉNEDICT DE SAUSSURE made one of the first serial measurements of the thermal gradient in the Mediterranean, using thermometers lagged in wax.

B. Helland-Hansen, 1912, p. 50, "The Ocean Waters. An Introduction to Physical Oceanography. I. General Part (Methods)," *Int. Rev. Hydrobiol.*, Hydrogr. Suppl., 1 (2): 84 pp.

1781   CHARLES BLAGDEN reported a series of measurements of surface temperature he had made along the North American coast, and suggested that the thermometer might be used as an aid to navigation.

C. Blagden, 1781, "On the Heat of the Water in the Gulf-stream," *Phil. Trans.*, 71: 334–344.

1781   JOSEPH-LOUIS LAGRANGE introduced the concept of velocity potential and made the first use of the stream function in the analysis of fluid motion.

M. de la Grange, "Mémoire sur la Théorie du mouvement des fluides," *Nouv. Mem. Acad. R. Berl.*, for 1781, 1: 151–198.

1782   JAMES SIX described the thermometer he had devised to register "the greatest degree of heat and cold which happened in the observer's absence," which has become known as Six's maximum and minimum thermometer. He gave this reason for the invention: "The sultry heat of the summer's days, and freezing cold of the winter's nights, which is commonly most severe at a late unseasonable hour, render it very unpleasant to be abroad in the open air, although it is absolutely necessary for the thermometer to be placed in such a situation."

J. Six, 1782, "Account of an improved Thermometer," *Phil. Trans.*, 72: 72–81.

1783   THOMAS EARNSHAW developed his chronometer escapement to essentially its modern form. Earnshaw's chronometer No. 1503 made the voyage of the *Bounty* with Captain William Bligh in 1791.

R. T. Gould, 1923, pp. 116 and 120, *The Marine Chronometer*, London: J. D. Potter.

1785   JAMES HUTTON first set forth his views disputing the cataclysmic doctrine of the world's origin in a paper read to the Royal Society of Edinburgh, which later formed the first chapter of his two-volume *Theory of the Earth* (1795). "No powers are to be employed that are not natural to the globe, no action to be admitted except those of which we know the principle."

J. Hutton, 1788, "Theory of the Earth; or an Investigation of the Laws observable in the Composition, Dissolution, and Restoration of Land upon the Globe," *Trans. roy. Soc. Edinb.*, 1(2): 209–304; M. MacGregor, 1947, "James Hutton, the founder of modern geology: 1726–97," *Endeavour*, 6: 109–111; the quotation above is from W.C.D. Dampier-Wetham, 1930, pp. 289–290, *A History of Science*, 4th ed., 1949, New York: Macmillan.

1788   CHARLES BLAGDEN found that the presence of dissolved substances in water lowers its freezing point.

C. Blagden, 1788, "Experiments on the cooling of Water below its freezing Point," *Phil. Trans.*, 78: 129–130; and "Experiments on the Effect of various Substances in lowering the Point of Congelation in Water," *Phil. Trans.*, 78: 277–312.

1788   JOSEPH-LOUIS LAGRANGE published his treatise *Mécanique Analytique*, in which the approach to problems in mechanics could be expressed in equations offering only mathematical rather than the diagrammatic difficulties encountered by Newton. Two such methods were originally proposed by Euler in 1755, but the analysis referred to a fixed point is now named the Eulerian method, and where reference is taken to the particle itself, the analysis is called the Lagrangian method.

A Source Book in Physics, pp. 61–65.

1798   HENRY CAVENDISH derived a mean density for the earth of 5.48 gm/cm^3 from his measurements of the constant of gravitation.

H. Cavendish, 1798, "Experiments to determine the Density of the Earth," *Phil. Trans.*, 88: 509–522; *A Source Book in Physics*, pp. 106–111.

1802   NATHANIEL BOWDITCH published the first edition of his *New American Practical Navigator*.

1802   FRANZ JOSEPH V. GERSTNER published the first theory of surface waves in deep water.

G. G. Stokes, 1880, p. 219, *Mathematical and Physical Papers*, Vol. 1, Cambridge University Press.

1802   JOHN PLAYFAIR published *Illustrations of the Huttonian Theory of the Earth*, the first of his arguments in favor of that theory of land forms expounding the uniformitarian doctrine, as opposed to the catastrophic doctrine so much in vogue before his time.

Facsimile reprint of Edinburgh first edition, with an introduction by George W. White, published 1956, Urbana: University of Illinois Press.

1803   LUKE HOWARD, an English pharmacist, made the first successful attempt to classify cloud forms, using the Latin names which are still accepted.

L. Howard, 1803, "On the Modification of Clouds, and on the Principles of their Production, Suspension, and Destruction; being the Substance of an Essay read before the Askesian Society in the Session 1802–3," *Phil. Mag.*, 16: 99–100.

1804   ADAM JOHANN V. KRUSENSTERN made use of Six's maximum and minimum thermometers, invented in 1782, for measurements of sea-water temperature at depth to 125 fathoms.

Captain A. J. von Krusenstern, 1813, pp. 8, 203, *Voyage Round the World in the Years, 1803, 1804, 1805 & 1806, by order of His Imperial Majesty Alexander the First, on Board the Ships Nadeshda and Neva, under the Command of Captain A. J. von Krusenstern of the Imperial Navy*, translated by Richard Belgrave Hoppner, Vol. 1, London: John Murray.

1805   ADRIEN-MARIE LEGENDRE published the first statement of the method of least squares, in an appendix to his *Nouvelles méthodes pour la détermination des orbites des comètes*. The principle was later firmly established by Karl Friedrich Gauss in his *Theoria Motus* (1809), a treatise on celestial mechanics.

For Legendre see *A Source Book in Mathematics* pp. 576–579; for Gauss see *A Source Book in Astronomy*, pp. 183–195.

1806   ADMIRAL SIR FRANCIS BEAUFORT, later Hydrographer of the British Navy, devised a scale from 0–12 for reporting the estimated wind force at sea. A similar scale is often used for the local sea state associated with Beaufort wind speeds.

L. G. Garbett, 1926, "Admiral Sir Francis Beaufort and the Beaufort scales of wind and weather," *Quart. J. R. met. Soc.*, 52: 161–168.

1814   FRIEDRICH WILHELM HEINRICH ALEXANDER V. HUMBOLDT wrote that water is cold at great depths in the tropics, and explained this as a consequence of the sinking and outflow of water conditioned in the polar regions. The five-year voyage on which he made the observations began in 1799.

A. de Humboldt and A. Bonpland, 1822, p. 64, *Personal Narrative of Travels to the Equinoctial Regions of the New Continent, during the years 1799–1804*, translated by H. M. Williams, 3d ed., Vol. 1, London: Longman, Hurst, Rees, Orme, and Brown; G. Wüst, 1959, pp. 90–104, *Alexander von Humboldt*, J. H. Schultze, editor, Berlin: Walter de Gruyter & Co.

1818   SIR JOHN ROSS made use of the self-registering thermometer for measuring subsurface temperatures in the ocean to a depth of 1000 fathoms.

> J. Ross, 1819, Appendix, pp. cxxxiii–cxxxiv, *A Voyage of Discovery, Made under the Orders of the Admiralty, in His Majesty's Ships Isabella and Alexander, for the Purpose of Exploring Baffin's Bay, and Inquiring into the Probability of a North-West Passage*, London: John Murray.

1822   JEAN BAPTISTE JOSEPH FOURIER published his *Théorie Analytique de la Chaleur*, in which he proposed physical dimensions for the measurement of heat and a statement of the theorem on heat conduction which now bears his name.

> *A Source Book in Physics*, pp. 175–178.

1823   CLAUDE LOUIS MARIE HENRY NAVIER extended the Eulerian equations of fluid motion by postulating that, in molecular momentum exchange, the forces of mutual repulsion between molecules are directly proportional to their velocities and inversely related to their spacing.

> Navier, 1823, "Mémoire sur les lois du mouvement des fluides," *Mém. Acad. Sci., Paris*, 6: 389–440.

1828   J. R. MERIAN published his theory of seiches.

> J. R. Merian, 1828, "Über die Bewegung tropfbarer Flüssigkeiten in Gefässen," Basle.

1831   MICHAEL FARADAY discovered magnetically induced electric currents and in 1832 suggested that electromagnetic induction accompanying sea-water motion through the magnetic field of the earth might yield measurable signals.

> M. Faraday, 1832, "The Bakerian Lecture. Experimental Researches in Electricity. Second Series," *Phil. Trans.*, 1832, 131–139, and reprint in M. Faraday, 1839, pp. 7–16, *Experimental Researches in Electricity*, Vol. 1, London: Richard and John Edward Taylor; M. Faraday, 1832, "The Bakerian Lecture. Experimental Researches in Electricity. Second Series. Terrestrial Magneto-electric Induction," *Phil. Trans.*, 1832, 163–177, and reprint in *Experimental Researches in Electricity, op. cit.*, 42–57; H. Kondo, 1953, "Michael Faraday," *Sci. Amer.*, 189: 90–98.

1831   WILLIAM C. REDFIELD deduced the rotary character of hurricanes from a study of ships' logs, plotted the paths of hurricane motion in the western North Atlantic, and made contributions to the study of tides and of the prevailing currents of the ocean and atmosphere.

W. C. Redfield, 1831, "Remarks on the prevailing storms of the Atlantic coast, of the North American States," *Amer. J. Sci.*, 20: 17–51.

1832   JAMES RENNELL published (posthumously) summaries of data on file at the British Admiralty office in which he distinguished clearly between drift currents and stream currents in the oceans.

J. Rennell, 1832, *An Investigation of the Currents of the Atlantic Ocean*, London: Published for Lady Rodd by J. G. & F. Rivington.

1833   WILLIAM WHEWELL proposed his progressive wave theory of the ocean tide and introduced the concept of cotidal lines.

Reverend W. Whewell, 1833, "Essay towards a First Approximation to a Map of Cotidal Lines," *Phil. Trans.*, 1833, 147–236.

1835   GASPARD GUSTAVE DE CORIOLIS published his basic paper, "Mémoire sur les équations du mouvement relatif des systèmes de corps," in which he studied the accelerations in rotating systems, with particular reference to fluid motions on a rotating earth.

G. Coriolis, 1835, *J. Éc. Roy. polyt.*, *Paris*, 15: 142–154.

1837   CÉSAR MANSUÈTE DESPRETZ showed that cooling sea water does not reach its greatest density above its freezing point.

C. Despretz, 1837, "Recherches sur le maximum de densité des liquides," *C. R. Acad. Sci.*, *Paris*, 4: 124–130.

1837   CLAUDE SERVAIS MATHIAS POUILLET devised the pyrheliometer, which was later improved by Charles Greeley Abbot of the Smithsonian Institution and used in his studies of the "solar constant" in 1905.

C. S. M. Pouillet, 1838, "Mémoire sur la chaleur solaire, sur les pouvoirs rayonnants et absorbants de l'air atmosphérique, et sur la température de l'espace," *C. R. Acad. Soc.*, *Paris*, 7: 24–65; C. G. Abbot and F. E. Fowle, Jr., 1908, *Ann. astrophys. Obs. Smithson. Instn*, 2: 39–44; C. G. Abbot, F. E. Fowle, and L. B. Aldrich, 1908, *Ann. astrophys. Obs. Smithson. Instn*, 3: 47–60.

1839   GEORGE GREEN derived the formula for the phase speed of waves in terms of wavelength.

G. Green, 1839, "Note on the Motion of Waves in Canals," *Trans. Camb. phil. Soc.*, 7: 87–95.

1839   SIR JAMES CLARK ROSS led the British Antarctic expedition (1839–1843) and made soundings as deep as 2677 fathoms some 450 miles west of the Cape of Good Hope with a line "three thousand and six hundred fathoms, or rather more than four miles in length, fitted with swivels to prevent it unlaying in its descent, and strong enough to support a weight of seventy-six pounds."

Captain Sir James Clark Ross, 1847, pp. 26–32, *A Voyage of Discovery and Research in the Southern and Antarctic Regions, during the Years 1839–43*, Vol. 1, London: John Murray.

1840   EDWARD FORBES began his studies of the fauna of the Aegean Sea with the Naturalist's Dredge, and came to the conclusion that marine animals were distributed in zones of depth and that no life existed below about 300 fathoms, the "azoic zone." This hypothesis has been abundantly refuted.

E. Forbes, 1859, *The Natural History of the European Seas*, R. Godwin-Austen, editor, London: John Van Voorst.

1840   SIR CHARLES WHEATSTONE proposed the first telegraphic connection between England and France across the Straits of Dover before a committee of the House of Commons. The first cable was laid by John Watkins Brett and Jacob Brett ten years later.

G. R. M. Garratt, 1950, *One Hundred Years of Submarine Cables*, Ministry of Education, Science Museum, London: H. M. Stationery Office.

1842   GEORGES AIMÉ first used his invention of a crude form of the reversing thermometer. The instrument came into general use only after it was re-devised by Negretti and Zambra in 1874.

Aimé, 1845, "Mémoire sur les Températures de la Méditerranée," *Ann. Chim. (Phys.)*, 3, 15: 5–34; H. Negretti and J. W. Zambra, 1874, "On a New Deep-sea Thermometer," *Proc. roy. Soc.*, 22: 238–241, and reprint in *Phil. Mag.*, 4, 48: 306–309; H. Negretti and J. W. Zambra, 1874, "On a New Deep Sea and Recording Thermometer," *Quart. J. R. met. Soc.*, 2: 188–191.

1842   JULIUS ROBERT MAYER, a ship's surgeon, advanced a theory of the mechanical nature of heat and its equivalence to work, established by Joule (1843), which led him to the concept of the conservation of energy later formalized by Helmholtz.

J. R. Mayer, 1842, "Bemerkungen über die Kräfte der unbelebten Natur," *Liebigs Ann.*, 42: 233–240, and English reprints by G. C. Foster, "Remarks on the Forces of Inorganic Nature," 1862, *Phil. Mag.*, 4, 24: 371–377, and in *A Source Book in Physics*, pp. 197–203; J. P. Joule, 1843, "On the Calorific Effects of Magneto-Electricity, and on the Mechanical Value of Heat," *Phil. Mag.*, 3, 23: 263–276, 347–355, 435–443.

1843   CHRISTIAN DOPPLER published "Ueber das farbige Lichte der Doppel-sterne," in which he discussed "the way in which the velocity of a moving source of light should affect the frequency as perceived by an observer."

E. N. da C. Andrade, 1959, "Doppler and the Doppler Effect," *Endeavour*, 18: 14–19.

1843   BARRÉ ADHÉMAR JEAN CLAUDE DE SAINT-VENANT derived the Navier-Stokes equations on the single assumption that fluid resistance to slipping is exercised in the very direction of the slipping. He also introduced the term "celerity" to distinguish between the velocity of wave propagation with respect to the fluid and the fluid velocity itself.

Barre de Saint-Venant, 1843, "Note à joindre au Mémoire sur la dynamique des fluides, présenté le 14 avril 1834," *C. R. Acad. Sci., Paris*, 17: 1240–1243.

1843   THOMAS HUBBARD SUMNER published instructions on the use of the principle of the line of position (a short segment of an equal altitude circle for a given celestial body) which he had first employed in 1837.

R. S. Richardson, 1943, "Captain Thomas Hubbard Sumner, 1807–1876," *Publ. astr. Soc. Pacif.*, 55: 136–144, and reprint in *Navigation*, 1: 35–40, 1946.

1844   ALEXANDER DALLAS BACHE, Benjamin Franklin's great-grandson, directed the survey of the Gulf Stream, completed in 1860, by the U.S. Coast and Geodetic Survey, to which he was appointed super-intendent in 1843.

A. D. Bache, 1860, "Lecture on the Gulf Stream," *Proc. Amer. Ass. Adv. Sci.* 14: xlix–lxxii; *Amer. J. Sci.*, 2, 30: 313–329.

1845   SIR GEORGE BIDDELL AIRY reasoned that since there is near agree-ment between the calculated and observed ellipticity of the earth, the earth must have once been in a fluid or semifluid state.

A Source Book in Astronomy, pp. 202–207.

1845   SIR GEORGE BIDDELL AIRY published his article "Tides and Waves," in which, among other things, he gave a general explanation of the tides of the Bay of Fundy.

Encyclopaedia Metropolitana.

1845   GEORGE GABRIEL STOKES showed that for two-dimensional motion the circulation around the perimeter of an area is equal to the integral of vorticity over the enclosed surface (Stokes' theorem).

G. G. Stokes, 1849, "On the Theories of the Internal Friction of Fluids in Motion, and of the Equilibrium and Motion of Elastic Solids," *Trans. Camb. phil. Soc.*, 8: 287–319, and reprint in G. G. Stokes, 1880, p. 81 ff., *Mathematical and Physical Papers*, Vol. 1, Cambridge University Press; G. G. Stokes, 1905, "Smith's Prize Examination Papers, February 1854, Problem 8," p. 320, *Mathematical and Physical Papers*, Vol. 5, Cambridge University Press.

1847   HERMANN LUDWIG FERDINAND V. HELMHOLTZ presented a paper, "Ueber die Erhaltung der Kraft," before the Physical Society of Berlin, in which he established the modern concept of the conservation of energy.

A *Source Book in Physics*, pp. 212–220; A. C. Crombie, 1958, "Helmholtz," *Sci. Amer.*, 198: 94–102.

1847   GEORGE GABRIEL STOKES studied the form and propagation of deep water waves on a homogeneous incompressible frictionless fluid of uniform depth.

G. G. Stokes, 1849, "On the Theory of Oscillatory Waves," *Trans. Camb. phil. Soc.*, 8: 441–445, and reprint in G. G. Stokes, 1880, pp. 197–229, *Mathematical and Physical Papers*, Vol. 1, Cambridge University Press.

1848   JAMES CLARK ROSS, sent to search for Sir John Franklin, observed, while his ships were ice-bound for 47 days in Port Leopold, that except for storm surge effects the mean sea level is altered inversely as the height of the mercury column of a barometer and in proportion to the ratio of the densities of the two liquids.

J. C. Ross, 1854, "On the Effect of the Pressure of the Atmosphere on the Mean Level of the Ocean," *Phil. Trans.*, 144: 285–296.

1849   JOSEPH HENRY, best known for his researches in electromagnetism, began the collection of meteorological data by telegraph while acting as the first secretary of the Smithsonian Institution.

G. B. Goode, editor, 1897, p. 146, *The Smithsonian Institution 1846–1896*, Washington, D.C.: The Institution.

1849   GEORGE GABRIEL STOKES showed that Clairaut's assumption of hydrostatic equilibrium (1743) was unnecessary to his conclusions concerning the variation with latitude of the acceleration due to gravity, and that the same results obtain by "supposing that it [the earth] consists of nearly spherical strata of equal density."

G. G. Stokes, 1849, "On the Variation of Gravity at the Surface of the Earth," *Trans. Camb. phil. Soc.*, 8: 672–695, and reprint in G. G. Stokes, 1883, pp. 131–171, *Mathematical and Physical Papers*, Vol. 2, Cambridge University Press.

1850   RUDOLPH JULIUS EMMANUEL CLAUSIUS, one of the founders of the kinetic theory of gases, stated the second law of thermodynamics and named the concept of entropy in 1865.

R. Clausius, 1850, "Ueber die bewegende Kraft der Wärme und die Gesetze, welche sich daraus für die Wärmelehre selbst ableiten lassen," *Ann. Phys.*, *Lpz.*, 79(3, 19): 368–397, 500–524, and English translation in *Phil. Mag.*, 4, 2: 1–21, 102–119, and in *A Source Book in Physics*, pp. 228–233; R. Clausius, 1865, "Ueber verschiedene für die Anwendung bequeme Formen der Hauptgleichungen der mechanischen Wärmetheorie," *Ann. Phys.*, *Lpz.*, 125(5, 5): 353–400, and English translation in *A Source Book in Physics*, pp. 234–236.

1850   GEORGE GABRIEL STOKES formulated what is known today as "Stokes' law" for the terminal velocity of fall of a sphere in a viscous fluid.

G. G. Stokes, 1856, "On the Effect of the Internal Friction of Fluids on the Motion of Pendulums. Section IV," *Trans. Camb. phil. Soc.*, 9(2): 48–57, and reprint in G. G. Stokes, 1901, pp. 55–67, *Mathematical and Physical Papers*, Vol. 3, Cambridge University Press.

1851   JEAN BERNARD LÉON FOUCAULT first observed the apparent rotation of the plane of oscillation of a pendulum, caused by the rotation of the earth. He also invented the gyroscope in 1852 in connection with his studies of the absolute rotation of the earth.

L. Foucault, 1851, "Démonstration physique du mouvement de la terre au moyen du pendule," *C. R. Acad. Sci.*, *Paris*, 32: 135–138; L. Foucault, 1852, "Sur une nouvelle demonstration experimentale du mouvement de la terre fondée sur la fixité du plan de rotation," *C. R. Acad. Sci.*, *Paris*, 35: 421–424; L. Foucault, 1852, "Sur les phenomenes d'orientation des corps tournants entraînes par un axe fixe à la surface de la Terre-Nouveaux signes sensibles du mouvement diurne," *C. R. Acad. Sci.*, *Paris*, 35: 424–427.

1853   JAMES H. COFFIN described a tropical storm: ". . . in the northern hemisphere, a wind arriving from its mean direction always finds the point of maximum pressure on its left, and the minimum on its right." This became known as Buys-Ballot's law after it was popularized by the chief of the Dutch Meteorological Services (1854–1889), Christoph Heinrich Diedrich Buys-Ballot.

J. H. Coffin, 1853, "An Investigation of the Storm Curve, deduced from the Relation existing between the Direction of the Wind, and the Rise and Fall of the Barometer," *Proc. Amer. Ass. Adv. Sci.*, Cleveland: 83–101.

1855   MATTHEW FONTAINE MAURY published *The Physical Geography of the Sea* as a companion explanation to the wind and current charts he constructed from a study of ships' logbooks.

Published by Harper in 1855, 8th and last American edition in 1861; C. L. Lewis, 1927, *Matthew Fontaine Maury*, Annapolis: U.S. Naval Institute.

1856   OTWAY H. BERRYMAN made a line of soundings from Newfoundland
       to Ireland from the U.S.S. *Arctic* to verify the existence of the sub-
       marine ridge on which it was proposed to lay a telegraph cable. The
       first successful Atlantic cable was in operation from August to Octo-
       ber, 1858, and a permanent connection was established in 1866.

       "Geographical Notices. No. III," 1858, *Amer. J. Sci.*, 2, 26: 219–223; G. R.
       M. Garratt, 1950, *One Hundred Years of Submarine Cables*, Ministry of Educa-
       tion, Science Museum, London: H. M. Stationery Office.

1856   WILLIAM FERREL described the effects of the earth's rotation on the
       distribution of ocean currents caused by the wind, and derived, in
       1874, the equation relating the barometric gradient of pressure to
       the velocity of the wind.

       W. Ferrel, 1856, "An Essay on the Winds and the Currents of the Ocean,"
       Nashville J. Med. Surg., 11(4 and 5), and reprint in *Prof. Pap. Signal Service*,
       12: 7–19, 1882; W. Ferrel, 1874, "Relation between the Barometric Gradient
       and the Velocity of Wind," *Amer. J. Sci.*, 8, and reprint in *Prof. Pap. Signal
       Service*, 12: 39–49, 1882.

1857   SIR WILLIAM THOMSON (LORD KELVIN) gave an expression for the
       adiabatic temperature gradient in the deep sea.

       W. Thomson, 1857, "On the Alterations of Temperature accompanying
       Changes of Pressure in Fluids," *Proc. roy. Soc.*, A, 8: 566–569, and reprints in
       Phil. Mag., 4, 15: 540–542, 1858, and W. Thomson, 1890, pp. 236–239,
       Mathematical and Physical Papers, Vol. 3, Cambridge University Press.

1858   WILLIAM FERREL discussed the nature of inertial motion.

       W. Ferrel, 1858, "The influence of the earth's rotation upon the relative
       motion of bodies near its surface," *Astronom. J.*, 5: 97–100, 113–114.

1858   HERMANN LUDWIG FERDINAND V. HELMHOLTZ proposed an analysis
       of vortex motions and the interaction of couples.

       H. Helmholtz, 1858, "Ueber Integrale der hydrodynamischen Gleichungen,
       welche den Wirlbelbewegungen entsprechen," *J. reine angew. Math.*, 55:
       25–55, and reprint in H. Helmholtz, 1882, pp. 101–134, *Wissenschaftliche
       Abhandlungen*, Vol. 1, Leipzig: Johann Ambrosius Barth; English translations
       in "On the Integrals of the Hydrodynamical Equations, which express
       Vortex-motion," 1867, *Phil. Mag.*, 4, Suppl. 33: 485–510, and Cleveland
       Abbe, translator, 1893, "The Mechanics of the Earth's Atmosphere, A Col-
       lection of Translations," *Smithson. misc. Coll.*, 34(10): 31–57.

1865   JOHANN GEORG FORCHHAMMER made chemical analyses of sea water
       from many localities and found that the ratios of concentration of the
       constituent ions show only slight variation.

G. Forchhammer, 1865, "On the Composition of Sea-water in the different parts of the Ocean," *Phil. Trans.*, 155: 246–262.

1866   NATHANIEL BOWDITCH. His *American Practical Navigator* was transferred to the United States Hydrographic Office for publication after his death in 1838.

American Practical Navigator (Bowditch), 1958, U.S. Navy Hydrographic Office Publication No. 9.

1867   SIR WILLIAM THOMSON (LORD KELVIN) gave his theorem on the constancy of the circulation around a closed curve moving with an incompressible, frictionless fluid.

W. Thomson, 1869, "On Vortex Motion," *Trans. roy. Soc. Edinb.*, 25: 217–260, and reprint in W. Thomson, 1910, pp. 13–66, *Mathematical and Physical Papers*, Vol. 4, Cambridge University Press.

1868   SIR CHARLES WYVILLE THOMSON and WILLIAM B. CARPENTER carried out oceanographic work on board H.M.S. *Lightning*, the results of which led to the *Challenger* expedition.

C. Wyville Thomson, 1874, pp. 49–81, *The Depths of the Sea*, 2d ed., London: Macmillan.

1870   JAMES CROLL gave an estimate of poleward heat flux provided by ocean currents.

J. Croll, 1870, "On Ocean-currents. Part I. Ocean-currents in relation to the Distribution of Heat over the Globe," *Phil. Mag.*, 4, 39: 82–84.

1872   H.M.S. *Challenger* was commissioned by the British Admiralty for the Royal Society of London to make a global survey of the physical, chemical, and biological properties of the oceans (1872–1876). This expedition produced the first extensive and authoritative survey of the world ocean, the results of which have been published in a succession of some 50 volumes, under the general editorship of Sir John Murray.

Sir John Murray, editor, *Report on the Scientific Results of the Voyage of H.M.S. Challenger during the Years 1873–76*, H.M. Stationery Office; J. W. Hedgpeth, 1946, "The Voyage of the *Challenger*," *Sci. Mon., N.Y.*, 63: 194–202; H. S. Bailey, Jr., 1953, "The Voyage of the 'Challenger'," *Sci. Amer.*, 188: 88–94.

1872   SIR WILLIAM THOMSON (LORD KELVIN) conceived the tide predictor with the help of "Mr. Tower." The first was built by Messrs. Roberst and Légé during the winter of 1872–1873.

W. Thomson, 1882, "The Tide Gauge, Tidal Harmonic Analyser, and Tide Predictor," *Min. Proc. Instn. civ. Engrs.*, March 1, 1882, and reprint in W. Thomson, 1911, pp. 285–296, *Mathematical and Physical Papers*, Vol. 6, Cambridge University Press.

1874   JAMES CROLL, who had earlier maintained that direct thermodynamic processes could not maintain the ocean circulation, suggested the importance of wind stress as the intermediate process by which the sun's heat is converted to motion in the sea.

J. Croll, 1874, "On Ocean-currents. Part III. On the Physical Cause of Ocean-currents," *Phil. Mag.*, 4, 47: 168–190.

1875   JAMES CROLL made formal inquiry into the suggestion made by J. Adhémar in 1832 that variations in the climate of the earth during the geologic past could be accounted for by the influence of regular changes in the earth's orbit. He showed that the distribution of insolation could be altered sufficiently, when the perihelion passage coincided with the summer solstice in one hemisphere, to encourage glaciation in the other.

J. Croll, 1875, *Climate and Time in their Geological Relations . . .*, New York: Appleton.

1875   COMMANDER ADOLPHE-LAURENT-ANATOLE MARCQ DE BLONDE DE SAINT-HILAIRE, later Admiral Saint-Hilaire, devised a method for finding the line of position corresponding to the observed altitude and direction of a celestial body relative to some assumed dead-reckoning position.

W. A. Mason, 1939, "Marcq Saint-Hilaire—Father of the New Navigation," *Proc. U.S. Nav. Inst.*, 65: 1171–1176.

1876   SIMON NEWCOMB raised objections to Croll's hypothesis (1875) on grounds of quantitative insufficiency.

S. Newcomb, 1876, "Review of Croll's Climate and Time with especial reference to the Physical Theories of Climate maintained therein," *Amer. J. Sci.*, 3, 11: 263–273.

1876   GEORGE GABRIEL STOKES was first to point out that the phase velocity of deep water waves is twice their group velocity.

G. G. Stokes, 1905, "Smith's Prize Examination Paper, Wednesday, February 2, 1876, Problem 11," p. 362, *Mathematical and Physical Papers*, Vol. 5, Cambridge University Press.

1877   ALEXANDER AGASSIZ directed the survey by the *Blake*, a United States

Coast Survey steamer, in the Caribbean Sea and Gulf of Mexico and substituted steel wire rope for dredging in place of hemp. In 1888 he published a general account of the results.

A. Agassiz, 1888, "The Cruises of the United States Coast and Geodetic Survey Steamer 'Blake' in the Gulf of Mexico, in the Caribbean Sea, and along the Atlantic Coast of the United States from 1877 to 1880," *Bulletin of the Museum of Comparative Zoology*, Vol. 14 and 15.

1879 HORACE LAMB published his *Treatise on the Mathematical Theory of Fluid Motion*. The title was changed in the 1895 edition to *Hydrodynamics* and has so remained through many subsequent editions.

H. Lamb, 1932, *Hydrodynamics*, 6th ed., New York: Dover.

1879 JOSEF STEFAN concluded that the total energy radiated by a body is proportional to the fourth power of the absolute temperature ($R = CT^4$). From Maxwell's electromagnetic theory of light (1873) Boltzmann showed that Stefan's law is strictly true only for black bodies.

J. Stefan, 1879, "Über die Beziehung zwischen der Wärmestrahlung und der Temperatur," *S.B. Acad. Wiss. Wien*, 2, 79: 391–428; L. Boltzmann, 1884, "Ueber eine von Hrn. Bartoli entdeckte Beziehung der Wärmestrahlung zum zweiten Hauptsatze," *Ann. Phys., Lpz.*, N.F. 22: 31–39; L. Boltzmann, 1884, "Ableitung des Stefan'schen Gesetzes, betreffend die Abhängigkeit der Wärmestrahlung von der Temperatur aus der electromagnetischen Lichttheorie," *Ann. Phys., Lpz.*, N.F. 22: 291–294.

1879 SIR WILLIAM THOMSON (LORD KELVIN) devised the tidal harmonic analyser, based on the disk-globe-cylinder integrator invented by his brother, James Thomson, in 1876–1878.

W. Thomson, 1882, "The Tide Gauge, Tidal Harmonic Analyser, and Tide Predictor," *Min. Proc. Instn. civ. Engrs.*, March 1, 1882, and reprint in W. Thomson, 1911, pp. 280–285, *Mathematical and Physical Papers*, Vol. 6, Cambridge University Press.

1879 SIR WILLIAM THOMSON (LORD KELVIN) and PETER GUTHRIE TAIT
(and   published *A Treatise on Natural Philosophy*, in two parts, which repre-
1883) sented an authoritative summary of the current theories and concepts concerning the earth's properties up to that time, together with a discussion of their own distinguished contributions. Published by Cambridge University Press and Macmillan.

1880 WILLIAM FERREL designed a tide prediction machine of 19 components for the U.S. Coast and Geodetic Survey, which indicated both the heights and times of maxima and minima. In 1910 Rollin A.

Harris extended the capacity of this machine to integrate 37 components.

W. Ferrel, "Description of a maxima and minima tide-predicting machine," *Annual Report of the Superintendent, U.S. Coast and Geodetic Survey, for 1883*, Appendix 10: 253–272; *Annual Report of the Superintendent, U.S. Coast and Geodetic Survey, for 1910*, pp. 69–70.

1880   SIR JOHN MURRAY and T. H. TIZARD re-examined the Faroe Channel, discovering the bottom feature (they named Wyville Thomson Ridge) which separates the Arctic Basin from the North Atlantic Basin.

T. H. Tizard and J. Murray, 1882, "Exploration of the Faroe Channel, during the Summer of 1880, in H.M.'s hired ship 'Knight Errant,'" *Proc. roy. Soc. Edinb.*, 11: 638–677.

1881   SAMUEL PIERPONT LANGLEY, third secretary of the Smithsonian Institution, developed the bolometer and fundamental methods for measuring radiant heat as a function of wavelength in the solar spectrum.

S. P. Langley, 1881, "The Bolometer and Radiant Energy," *Proc. Amer. Acad. Arts and Sci.*, 16: 342–358.

1882   FRANÇOIS MARIE RAOULT established the law of the depression of the freezing point by dissolved substances in solvents.

F. M. Raoult, 1882, "Loi générale de congélation des dissolvants," *C.R. Acad. Sci., Paris*, 95: 1030–1033; English translation in *A Source Book in Chemistry*, pp. 471–473.

1882   SIR WILLIAM THOMSON (LORD KELVIN) showed the existence of both semidiurnal and diurnal variations in the barometric pressure which he ascribed to atmospheric dilation due to solar heating rather than to gravitational forces coupled with mechanical oscillations of the atmosphere as a whole.

Sir William Thomson, 1882, "On the Thermodynamic Acceleration of the Earth's Rotation," *Proc. roy. Soc. Edinb.*, 11: 398–400, and reprint in Sir William Thomson, 1890, pp. 343–344, *Mathematical and Physical Papers*, Vol. 3, Cambridge University Press.

1883   SAMUEL PIERPONT LANGLEY published his "Researches on Solar Heat," the starting point of geophysical studies of atmospheric transparency and the radiation balance of the earth.

S. P. Langley, 1883, "Researches on Solar Heat and Its Absorption by the Earth's Atmosphere. A Report of the Mount Whitney Expedition," *Prof. Pap. Signal Service*, 15: 242 pp.; *A Source Book in Astronomy*, pp. 345–349.

1883   OSBORNE REYNOLDS published "An Experimental Investigation of the Circumstances which determine whether the Motion of Water shall be Direct or Sinuous, and of the Law of Resistance in Parallel Channels," which contains a formulation now known as the Reynolds number.

O. Reynolds, 1883, *Phil. Trans.*, 174: 935–982, and reprint in O. Reynolds, 1901, pp. 51–105, *Papers on Mechanical and Physical Subjects*, Vol. 2, Cambridge University Press.

1884   WILHELM DITTMAR made analyses of 77 water samples, considered to be representative of all the oceans, which had been collected on the *Challenger* expedition. These showed that the ratios of the halides, sulfate, bicarbonate, magnesium, calcium, potassium, and strontium ions remained essentially constant, so that by measuring one, usually chloride, it is possible to estimate the total salt content of water from the open sea.

W. Dittmar, 1884, p. 203, *Report on the Scientific Results of the Voyage of H.M.S. Challenger during the Years 1873–76*, Sir John Murray, editor, Physics and Chemistry, Vol. 1, H.M. Stationery Office.

1884   ULRICH FRIEDRICH VETTIN made what was probably the first experimental model of atmospheric circulation.

U. F. Vettin, 1884, "Experimentelle Darstellung von Luftbewegungen unter dem Einflusse von Temperatur-Unterscnieden und Rotations-Impulsen," *Met. Z.*, 1: 227–230, 271–276.

1885   FRANÇOIS ALPHONSE FOREL suggested that the submarine trenches in the deltas of Lake Constance and Lake Geneva were eroded by the streams of cold high-density water entering these lakes from Alpine glaciers. In 1887 he suggested that, in addition to the effect of lowered temperature, the density of these river discharges is increased by their burden of sediment.

F.-A. Forel, 1885, "Les ravins sous-lacustres des fleuves glaciaires," *C.R. Acad. Sci., Paris*, 101: 725–728; 1887, "Le ravin sous-lacustre du Rhone dans le lac Leman," *Bull. Soc. Vaud. Sci. Nat.*, 23: 85–107.

1885   HENRIK MOHN gave a formula for computing the horizontal velocities of ocean currents at right angles to the measured slope of isobaric surfaces 18 years before Helland-Hansen and Sandström (1903) derived the same formula from V. Bjerknes' theorem on circulation (1898).

H. U. Sverdrup, M. W. Johnson, and R. H. Fleming, 1942, p. 460, *The Oceans*, New York: Prentice-Hall.

1885   ALBERT I, PRINCE OF MONACO, began systematic oceanographic studies of the Mediterranean and North Atlantic aboard his yachts *Hirondelle, Princesse Alice, Hirondelle II* and *Princesse Alice II.* In addition to publishing many of the results privately, he used his wealth to found and endow the Oceanographic Museum in Monaco and the Oceanographical Institute in Paris.

Sir William A. Herdman, 1923, "The Prince of Monaco and the Oceanographic Museum," pp. 119–133, *Founders of Oceanography and their Work,* London: Edward Arnold.

1886   STEPAN OSIPOVICH MAKAROFF led an expedition (1886–1889) around the world in the Russian steamer *Vitiaz,* making observations of the temperature and specific gravity of water, particularly in the North Pacific. He later served on the committee of the International Council for the Exploration of the Sea which led to the publication (1901) of Knudsen's *Hydrographical Tables.*

S. O. Makaroff, 1894, *Le Vitiaz et l'Ocean Pacifique,* St. Petersburg; O. Krümmel, 1893, "Russische Arbeiten zur Ozeanographie des Nordpacifischen Ozeans," *Petermanns geogr. Mitt.,* 39: 85–88.

1887   HEINRICH RUDOLPH HERTZ found experimental confirmation of Maxwell's electromagnetic theory, and thus laid the foundations for radio communications and radio navigation.

H. Hertz, 1887, "Ueber sehr schnelle electrische Schwingungen," *Ann. Phys., Lpz.,* N.F., 31: 421–448; P. and E. Morrison, 1957, "Heinrich Hertz," *Sci. Amer.,* 197: 98–106.

1889   CLARENCE EDWARD DUTTON coined and introduced the term "isostasy" to scientific usage in his paper "On Some of the Greater Problems of Physical Geology," delivered before the Philosophical Society of Washington on April 27 of that year.

C. E. Dutton, 1892, *Bull. phil. Soc. Wash.,* 11: 51–64.

1889   PETER GUTHRIE TAIT made an extended series of measurements of the compressibility of sea water.

P. G. Tait, 1889, "Report on some of the Physical Properties of Fresh Water and of Sea Water," *Report on the Scientific Results of the Voyage of H.M.S. Challenger during the Years 1873–76,* Sir John Murray, editor, Physics and Chemistry, Vol. 2, H.M. Stationery Office.

1890   JOHN ELLIOTT PILLSBURY published his observations of the velocity and structure of the Gulf Stream made from the steamer *Blake,* beginning in 1885 at anchor stations in the Florida Straits and in the waters off Cape Hatteras and the passages of the Windward Islands.

J. E. Pillsbury, 1890, "The Gulf Stream. A Description of the Methods Employed in the Investigation, and the Results of the Research," *Annual Report of the Superintendent, U.S. Coast and Geodetic Survey for 1890*, Appendix 10: 538–579.

1891 SETH CARLO CHANDLER published his discovery of a small variation in latitude with a half-period of 222 days and a range of about 0.7 sec of arc. This polar motion was later shown to contain two periods, 12 months and 14 months, which have since been associated with geophysical causes.

S. C. Chandler, 1891, "On the Variation of Latitude," *Astronom. J.*, 11: 59–61, and reprint in *A Source Book in Astronomy*, pp. 377–379.

1891 OTTO KRÜMMEL published a description of the central Atlantic water mass.

O. Krümmel, 1891, "Die nordatlantische Sargassosee," *Petermanns geogr. Mitt.*, 37: 129–141.

1892 SIMON NEWCOMB calculated that the elasticity of the earth could lengthen the period of Eulerian nutation sufficiently to account for the 14-month (427 days) motion discovered by Chandler, 1891. Oppolzer computed the period for a rigid earth to be 10 months (305 days).

S. Newcomb, 1892, "On the Dynamics of the Earth's Rotation, with respect to the Periodic Variations of Latitude," *Mon. Not. R. astr. Soc.*, 52: 336–341.

1892 LORD RAYLEIGH (JOHN WILLIAM STRUTT) generalized the principles of dimensional analysis as a logical procedure.

Lord Rayleigh, 1892, "On the Question of the Stability of the Flow of Fluids," *Phil. Mag.*, 5, 34: 59–70.

1892 JAMES THOMSON, in his paper "On the Grand Currents of Atmospheric Circulation" conceived the experimental scheme and the results that might be obtained from rotating models of the atmosphere.

J. Thomson, 1892, *Phil. Trans.*, A, 183: 653–684.

1893 FRIDTJOF NANSEN started his remarkable drift across the North Polar Sea in the *Fram*. During this time he made oceanographic soundings and observations which showed that the North Pole was covered by deep sea.

F. Nansen, 1897, pp. 631–638, *Farthest North*, Vol. 2, Westminster: Archibald Constable.

1894 CHARLES VERNON BOYS improved the design of the Cavendish balance and redetermined the gravitational constant, obtaining $G = 6.6576 \times 10^{-8}$ (cgs).

C. V. Boys, 1895, "On the Newtonian Constant of Gravitation," *Phil. Trans.*, A, 186: 1–72.

1894 OSBORNE REYNOLDS read his basic paper, "On the Dynamical Theory of Incompressible Viscous Fluids and the Determination of the Criterion," in which he distinguished between turbulent and mean motion components of fluid flow and gave the critical Reynolds number above which a small disturbance in a given flow must increase in amplitude with time.

O. Reynolds, 1895, *Phil. Trans.*, A, 186: 123–164, and reprint in O. Reynolds, 1901, pp. 535–577, *Papers on Mechanical and Physical Subjects*, Vol. 2, Cambridge University Press.

1895 HORACE LAMB described the effects of bottom topography on fluid motion—ideas which were further developed by Sverdrup in 1941.

Sir Horace Lamb, 1895, Ch. 8, Art. 193, *Hydrodynamics*, London: Cambridge University Press; H. U. Sverdrup, 1941, "The influence of bottom topography on ocean currents," *Applied Mechanics*, Theodore von Kármán Anniversary Volume: 66–75.

1896 VALENTIN JOSEPH BOUSSINESQ proposed a formula for the wind stress on water and introduced the concept of eddy coefficients of viscosity.

J. Boussinesq, 1896, "Expression du frottement extérieur l'écoulement tumultueux d'un fluide," *C. R. Acad. Sci., Paris*, 122: 1445–1451; J. Boussinesq, 1896, "Formules du coefficient des frottements intérieurs, dans l'écoulement tumultueux graduellement varié des liquides," *C. R. Acad. Sci., Paris*, 122: 1517–1523.

1898 VILHELM FRIMAN KOREN BJERKNES published his classical paper, "Ueber einen hydrodynamischen Fundamentalsatz und seine Anwendung besonders auf die Mechanik der Atmosphäre und des Weltmeeres"; the basis of the method of dynamic sections.

V. Bjerknes, 1898, *K. svenska VetenskAkad. Handl.*, 31(4): 35 pp.

1899 JOHN JOLY estimated the age of the ocean at 80 to 90 million years, a figure he changed in 1911 to 150 million, on the basis of the uniform annual accumulation of sodium in the ocean from the discharge of rivers.

J. Joly, 1911, "The Age of the Earth," *Phil. Mag.*, 6, 22: 357–380.

1899   LÉON PHILIPPE TEISSERENC DE BORT discovered the cessation of the
       lapse rate with altitude at the atmospheric tropopause, and called
       the stratosphere the "isothermal zone." In 1928 Albert Defant
       pointed out that the concepts of troposphere and stratosphere have a
       parallel in the structure of the oceans.

       L. Teisserenc de Bort, 1899, "Sur la température et ses variations dans
       l'atmosphère libre, d'après les observations de quatre-vingt-dix ballons-
       sondes," *C. R. Acad. Sci., Paris*, 129: 417–420.

1901   HENRI BÉNARD described the cellular processes of stable thermal con-
       vection in fluids heated from below.

       H. Bénard, 1901, "Les tourbillons cellulaires dans une nappe liquide trans-
       portant de la chaleur par convection en régime permanent," *Ann. Chim.
       (Phys.)*, 7, 23: 62–144.

1901   MARTIN KNUDSEN edited *Hydrographical Tables* for the conversion of
       chlorinity to salinity and σ_t.

       M. Knudsen, editor, 1931, *Hydrographical Tables*, according to the measurings
       of Carl Forch, J. P. Jacobsen, Martin Knudsen, and S. P. L. Sørensen, 2nd
       ed., Copenhagen: G. E. C., GAD, and London: Williams & Norgate.

1902   VAGN WALFRID EKMAN published a theory, prompted by Nansen's
       observation that polar ice drifts some 20° to 40° to the right of the
       wind, in which the earth's rotation accounts for deflections of 45°
       cum sole under certain idealized conditions.

       V. W. Ekman, 1902, "Om jordrotationens inverkan pa vindströmmar i
       hafvet," *Nyt. Mag. f. Naturvid.*, 40: 37–63.

1902   The International Council for the Exploration of the Sea was formed
       July 22 at Copenhagen after preliminary meetings (first prompted by
       the Swedish Hydrographic Commission) at Stockholm (1899) and
       later at Christiania (1901). Nations represented were Denmark,
       Germany, England, Finland, the Netherlands, Norway, Russia, and
       Sweden.

       P. P. C. Hoek, 1902/1903, "Report of Administration for the first year:
       22nd July 1902–21st July 1903," *Rapp. Cons. Explor. Mer*, 1: I–XXXIX.

1902   JOHAN SANDSTRÖM and BJØRN HELLAND-HANSEN made a careful study
       of the possibility of obtaining the field of relative motion in the sea
       from an exact knowledge of the mass distribution on the basis of the
       Bjerknes circulation theorem (1898).

       J. W. Sandström and B. Helland-Hansen, 1902, "Ueber die Berechnung
       von Meeresströmungen," *Rep. Norweg. Fish. Invest.*, 2(4): 43 pp.

1903   MARTIN KNUDSEN tabulated the equilibrium temperature for ice and sea water at different salinities.

M. Knudsen, 1903, "Gefrierpunkttabelle fuer Meerwasser," *Publ. Circ. Cons. Explor. Mer*, 5: 11–13.

1904   ROLLIN ARTHUR HARRIS proposed the amphidromic regime of tides.

R. A. Harris, 1904, "Manual of Tides, Part IVB. Cotidal Lines for the World," *Annual Report of the Superintendent, U.S. Coast and Geodetic Survey for 1904*, Appendix 5: 313–400.

1904   VAGN WALFRID EKMAN explained the phenomenon of dead water.

V. W. Ekman, 1904, *On Dead-Water: Being a description of the so-called phenomenon often hindering the headway and navigation of ships in Norwegian Fjords and elsewhere, and an experimental investigation of its causes etc.* (*The Norwegian North Polar Expedition 1893–1896. Scientific Results*, F. Nansen, editor, Vol. 5, no. 15), Christiania: Jacob Dybwad and London: Longmans, Green.

1905   PHILIP HERBERT COWELL detected, from a study of the eclipses recorded in the Almagest, an acceleration of the mean motion of the sun, presumably resulting from the action of the solar tide on the earth's oceans.

P. H. Cowell, 1905, "On the Secular Acceleration of the Earth's Orbital Motion," *Mon. Not. R. astr. Soc.*, 66: 3–5; "On the Ptolemaic Eclipses of the Moon recorded in the Almagest," *ibid.*, 5–7; "A tentative explanation of the apparent Secular Acceleration of the Earth's Orbital Motion," *ibid.*, 352–355.

1905   VAGN WALFRID EKMAN, Professor of Hydrodynamics, University of Lund, described the effects of a steady wind blowing on an infinite ocean of uniform eddy viscosity in the paper "On the influence of the Earth's Rotation on Ocean-Currents," in which he stated the concepts now known as the Ekman spiral and the Ekman transport.

V. W. Ekman, 1905, *Ark. Mat. Astr. Fys.*, 2(11): 52 pp.

1905   ERIK IVAR FREDHOLM obtained a solution for the Ekman layer under a transient wind.

Ibid., 16.

1905   *Galilee*, first research ship of the Carnegie Institution of Washington, began observations of the magnetic, electric, and chemical properties of the oceans on a world-wide basis.

The Work of the Carnegie and Suggestions for Future Scientific Cruises, Oceanography IV, Carnegie Institution of Washington, Publication 571, 1946.

1906   MAX MARGULES wrote the equations for the slope of an interface in
       the atmosphere which have been widely used in meteorology and
       were adapted to oceanographic use by Defant in 1929.

       M. Margules, 1906, "Über Temperaturschichtung in stationär bewegter und
       in ruhender Luft," *Met. Z.*, Hann-band: 243–254.

1906   ERNST RUPPIN showed that the specific conductance of sea water in-
       creases with both temperature and salinity.

       E. Ruppin, 1906, "Bestimmung der elektrischen Leitfähigkeit des Meer-
       wassers," *Wiss. Meeresuntersuch.*, Abt. Kiel, N. F., 9: 179–182.

1907   SIR GEORGE HOWARD DARWIN. The first volume of the collected
       Scientific Papers of this distinguished theoretician was published by
       Cambridge University Press. The reports include, among many
       other things, discussion of the body tides (1878), the lunar disturbance
       of gravity (1881), the equilibrium theory of a rotating liquid (1887),
       and the harmonic analysis of ocean tides (1883).

1908   JOHAN SANDSTRÖM, in agreement with Croll (1874), concluded that
       the ocean must be a very inefficient thermodynamic machine since the
       intake and output of heat occurs at the free surface which is every-
       where at essentially the same geopotential level.

       J. W. Sandström, 1908, "Dynamische Versuche mit Meerwasser," *Ann.
       Hydrogr., Berl.*, 36: 6–23; J. Croll, 1874, "On Ocean-currents. Part III. On
       the Physical Cause of Ocean-currents," *Phil. Mag.*, 4, 47: 94–122, 168–190.

1909   *Carnegie*, a nonmagnetic ship, began observations of the physical,
       chemical, and biological properties of the oceans on a world-wide
       basis under the auspices of the Carnegie Institution of Washington.
       Observations were continued until 1929 when the *Carnegie* burned at
       Apia, Samoa, while loading fuel for her nonmagnetic engine.

       The Work of the Carnegie and Suggestions for Future Scientific Cruises, Oceanog-
       raphy 4, Carnegie Institution of Washington, Publication 571; J. H. Paul,
       1932, *The Last Cruise of the Carnegie*, Baltimore: Williams and Wilkins.

1909   ANDRIJA MOHOROVIČIĆ detected a major discontinuity in the velocity
       of seismic waves beneath the outer layer of the earth's crust in central
       Europe. This feature, related to an estimated 10% increase of rock
       density, was later found to be world wide and at a shallower depth
       beneath oceans than under continents.

       A. Mohorovičić, 1910, "Das Beben vom 8. Okt. 1909," *Jb. met. Obs. Zagreb*,
       1909, 9(4).

1912   BJØRN HELLAND-HANSEN employed the concept of potential temperature in the discussion of the stability of deep water masses where the
temperature *in situ* may increase slightly with depth owing to adiabatic warming under compression.

> B. Helland-Hansen, 1912, "The Ocean Waters. An Introduction to Physical
> Oceanography. I. General Part (Methods)," *Int. Rev. Hydrobiol.*, Hydrogr.
> Suppl., 1(2): 51–53.

1912   FRIDTJOF NANSEN suggested that the North Atlantic bottom water is
formed at the surface in the Irminger Sea, where it sinks because of
cooling by conduction to the atmosphere and long-wave radiation to
space.

> F. Nansen, 1912, "Das Bodenwasser und die Abkühlung des Meeres," *Int.
> Rev. Hydrobiol.*, 5(1): 1–42.

1912   LEWIS FRY RICHARDSON, one month after the *Titanic* disaster, applied
for British patent No. 11,125 for an acoustic invention described as
an "Apparatus for Warning a Ship at Sea of its Nearness to Large
Objects Wholly or Partly under Water."

> British patent specification No. 11,125, available from Patent Office, 25
> Southampton Buildings, London W.C. 2, England.

1912   ALFRED WEGENER first proposed his displacement theory, or "continental drift hypothesis," in a lecture on January 6 to the Geological
Association of Frankfort-on-Main. In *The Origin of Continents and
Oceans*, first published in 1915, he expanded the theory which has
remained the center of vigorous discussion and considerable dis-
sention.

> A. Wegener, 1924, *The Origin of Continents and Oceans*, translated from 3rd
> German edition by J. G. A. Skerl, New York: E. P. Dutton.

1913   ARTHUR HOLLY COMPTON observed the change of relative vorticity
of water in a horizontal circular tube when the tube is suddenly
overturned, providing a nonastronomical measurement of latitude
and an additional experimental proof of the earth's rotation.

> A. H. Compton, 1913, "A Laboratory Method of Demonstrating the Earth's
> Rotation," *Science*, 37: 803–806; 1915, "A Determination of Latitude, Azi-
> muth, and the Length of the Day Independent of Astronomical Observa-
> tions," *Phys. Rev.*, 2, 5: 109–117.

1914   EDGAR BUCKINGHAM advanced the π theorem by means of which the
functional relations between physical quantities can be expressed in
dimensionally homogeneous terms.

E. Buckingham, 1914, "On Physically Similar Systems; Illustrations of the Use of Dimensionless Equations," *Phys. Rev.*, 2, 4: 345–376; 1914, "Physically Similar Systems," *J. Wash. Acad. Sci.*, 4: 347–353.

1914   VAGN WALFRID EKMAN computed the adiabatic lapse rate for the deep ocean.

V. W. Ekman, 1914, "Der adiabatische Temperaturgradient im Meere," *Ann. Hydrogr., Berl.*, 42: 340–344.

1914   L. LEIGH FERMOR suggested that isostatic compensation is accomplished by an exothermic transition of ultra-basic materials in the mantle to undifferentiated silicates at the base of the crust.

L. L. Fermor, 1914, "The Relationship of Isostasy, Earthquakes, and Vulcanicity to the Earth's Infra-Plutonic Shell," *Geol. Mag.*, 51: 65–67.

1914   International Ice Patrol was established as a result of the sinking of the *Titanic*. Thirteen maritime nations, using shipping lanes in the ice regions, met at the International Conference on the Safety of Life at Sea, and on January 20 signed an agreement to establish the patrol and pay its expenses.

F. B. Bassett, 1924, "International Ice Observation and Ice Patrol Service," *Int. hydr. Rev.*, 2(1): 129–130; "International Ice Observation and Ice Patrol Service in the North Atlantic Ocean," *Bull. U.S. Coast Guard*, 3: 3–5.

1914   ALBERT ABRAHAM MICHELSON and HENRY GORDON GALE used the interferometer to measure the body tide in the solid earth by the disturbance of the water level in two vertical tubes with a long horizontal connection underground.

A. A. Michelson and H. G. Gale, 1914, "Preliminary Results of Measurements of the Rigidity of the Earth," *Astrophys. J.*, 39: 105–138; 1919, "The Rigidity of the Earth," *Astrophys. J.*, 50: 330–345.

1915   GEOFFREY INGRAM TAYLOR found that the wind stress coefficient over Salisbury Plain had a mean value near 0.002. This value has been widely used for lack of better information.

G. I. Taylor, 1916, "Skin Friction of the Wind on the Earth," *Proc. roy. Soc.*, A, 92: 196–199.

1916   BJØRN HELLAND-HANSEN proposed the temperature-salinity diagram as a means for studying the distribution of these properties and the hydrostatic stability of water masses in the ocean.

B. Helland-Hansen, 1916, "Nogen hydrografiske metoder," *Forh. skand. naturf. Møte*, 16: 357–359.

1918   GEOFFREY INGRAM TAYLOR studied the frictional dissipation of tides in the Irish Sea and found that this area might account for about 2% of the total.

G. I. Taylor, 1919, "Tidal Friction in the Irish Sea," *Phil. Trans.*, A, 220: 1–33.

1920   MIECZYSLAW OXNER published "Manual Pratique de l'analyse de l'eau de mer," in which the procedures are given which implement Knudsen's (1901) *Hydrographical Tables.*

M. Oxner, 1920, *Bull. Comm. int. Explor. Méditerr.*, 3: 36 pp.

1920   LEWIS FRY RICHARDSON formulated a criterion which states that when turbulent kinetic energy is neither increasing nor decreasing, its production would exactly balance the loss in work done against stratification.

L. F. Richardson, 1920, "The Supply of Energy from and to Atmospheric Eddies," *Proc. roy. Soc.*, A, 97: 354–373.

1921   WILHELM BRENNECKE explained the origin of Antarctic bottom water as a consequence of cooling in high southern latitudes to produce water of the highest density in the open oceans.   (See Sverdrup, 1942.)

W. Brennecke, 1921, "Die Ozeanographischen Arbeiten der Deutschen Antarktischen Expedition 1911–1912," *Aus. d. Arch. dtsch. Seew.*, 39(1): 216 pp.

1921   GEOFFREY INGRAM TAYLOR called attention to the fact that in a rotating fluid the component velocities are independent of distance in the direction of the axis of rotation—the so-called Taylor Ink Walls.

G. I. Taylor, 1921, "Experiments with Rotating Fluids," *Proc. roy. Soc.*, A, 100: 114–121.

1922   LEWIS FRY RICHARDSON gave a practical means for the solution of the problem of numerical weather prediction based on the integration of the equations of motion of the atmosphere.

L. F. Richardson, 1922, *Weather Prediction by Numerical Process*, Cambridge University Press.

1923   FELIX MARIA EXNER showed the effects of rotation on the convection of a rotating fluid heated from below.

F. M. Exner, 1923, "Über die Bildung von Windhosen und Zyklonen," *S.B. Akad. Wiss. Wien*, IIa, 132: 1–16.

1923   SIR HAROLD JEFFREYS estimated the frictional dissipation in shallow seas over two days as equal to the tidal energy of the world ocean.

H. Jeffreys, 1923, "Tidal Dissipation of Energy," *Nature, Lond.*, 112: 622; 1920, "Tidal Friction in Shallow Seas," *Phil. Trans.*, A, 221: 239–264.

1923   FELIX ADOLF VENING MEINESZ, Professor of Geodesy, University of Utrecht, began an extensive survey of the earth's gravitational field at sea with a pendulum-type marine gravimeter of his own design.

F. A. Vening Meinesz, 1925, "The Determination of Gravity at Sea in a Submarine," *Geogr. J.*, 65: 501–521.

1924   SIR HAROLD JEFFREYS proposed the sheltering effect as part of the mechanism of the generation of wind waves on the sea surface.

H. Jeffreys, 1925, "On the Formation of Water Waves by Wind," *Proc. roy. Soc.*, A, 107: 189–206.

1924   SIR HAROLD JEFFREYS published his treatise *The Earth*, the first of a series of important discussions of the problems of terrestrial physics under that title.

Cambridge University Press and Macmillan; 4th ed., 1959.

1924   *Meteor* was commissioned by the Hydrographic Department of the German Navy for research in oceanography primarily in the Atlantic Ocean. The work of the men on this ship has provided one of the most thorough studies of an entire ocean ever made.

F. Spiess, 1932, *Das Forschungsschiff und seine Reise* (*Wissenschaftliche Ergebnisse der Deutschen Atlantischen Expedition auf dem Forschungs- und Vermessungsschiff "Meteor" 1925–1927*, A. Defant, editor, Bd. 1) Berlin: Walter de Gruyter.

1924   GEORGE WÜST showed a close correlation between geostrophic currents in the Florida Straits and Pillsbury's direct observations (1891), taking Pillsbury's zero velocity level as the level of no motion.

G. Wüst, 1924, "Florida- und Antillenstrom. Eine hydrodynamische Untersuchung," *Veröff. Inst. Meeresk. Univ. Berl.* N. F. A., 12: 48 pp.

1925   *Discovery* was commissioned by the Discovery Committee of the Colonial Office, Great Britain, for oceanographic research and biological reconnaissance of the whale fishery of the Southern Ocean.

Discovery Reports, Vol. 1, 1929, pp. 141–232, Cambridge University Press and Macmillan.

1925   FRIDTJOF NANSEN gave an early description of the Nansen bottle as
       it was used aboard the *Amauer Hansen*.

       B. Helland-Hansen and F. Nansen, 1925, "The Eastern North Atlantic,"
       Geofys. Publ., 4(2): 3–76.

1926   LUDWIG PRANDTL published the first results of experiments in the ro-
       tating laboratory at Göttingen.

       L. Prandtl, 1926, "Erste Erfahrungen mit dem rotierenden Laboratorium,"
       Naturwissenschaften, 14: 425–427.

1926   HARALD ULRIK SVERDRUP interpreted the behavior of tides observed
       during the drift of the *Maud* (1922–1924) across the North Siberian
       shelf as a consequence of the combined effects of friction and the
       deflecting force of earth rotation.

       H. U. Sverdrup, 1926/27, "Dynamic of Tides on the North Siberian Shelf.
       Results from the 'Maud' Expedition," *Geofys. Publ.*, 4(5): 75 pp.

1928   ALBERT DEFANT suggested that the ocean be considered to possess a
       troposphere and stratosphere by analogy with the atmosphere.

       A. Defant, 1928, "Die systematische Erforschung des Weltmeeres," *Z. Ges.
       Erdk. Berl.*, Sonderband: 459–505.

1929   JONAS EKMAN FJELDSTAD developed a theory for the behavior of a
       free wave in an estuary which contributes both damping and re-
       flection.

       J. E. Fjeldstad, 1929, *Contribution to the Dynamics of Free Progressive Tidal Waves
       (The Norwegian North Polar Expedition with the "Maud" 1918–1925, Scientific
       Results*, Vol. 4, no. 3), Bergen: Geofysisk Institutt.

1929   PETER LASAREFF, Director of the Geophysical Institute in Moscow,
       simulated the ocean circulation by means of trade winds blowing
       over a plaster model at rest.

       P. Lasareff, 1929, "Sur une méthode permettant de démontrer la dépendance
       des courants océaniques des vents alizés et sur le rôle des courants océaniques
       dans le changement du climat aux époques géologiques," *Beitr. Geophys.*, 21:
       215–233.

1929   *Willebrord Snellius* was commissioned in Holland for hydrographic
       survey work in the Netherlands East Indies.

       P. M. van Riel, 1938, "Programme of Research and Preparations," pp. 1–37,
       The Snellius-Expedition in the eastern part of the Netherlands East-Indies 1929–1930,
       Vol. 1, Leiden: E. J. Brill.

1930   *Atlantis*, a 142-ft auxiliary ketch especially built for the Woods Hole Oceanographic Institution, was launched December 31 from the ship-building yards in Copenhagen, Denmark, and sailed July 2, 1931, under the command of Columbus O'D. Iselin, later director of the institution.

H. B. Bigelow, 1933, "Reports of the Director," pp. 16–24, *The Woods Hole Oceanographic Institution Report for the Years 1930–1932, Woods Hole Oceanographic Institution Collected Reprints 1933.*

1932   ALBERT DEFANT developed a theory of free and forced internal waves between two layers, taking account of the effects of earth rotation.

A. Defant, 1932, *Die Gezeiten und inneren Gezeitenwellen des Atlantischen Ozeans* (*Wissenschaftliche Ergebnisse der Deutschen Atlantischen Expedition auf dem Forschungs- und Vermessungsschiff "Meteor" 1925–1927*, Vol. 7, Part 1), Berlin: Walter de Gruyter.

1932   KARL G. JANSKY discovered the radio-frequency emission from celestial objects, thereby establishing the basis for research in radio astronomy and certain applications of these principles to navigation.

K. G. Jansky, 1932, "Directional Studies of Atmospherics at High Frequencies," *Proc. Inst. Radio Engrs, N.Y.*, 20: 1920–1932; 1933, "Electrical Disturbances Apparently of Extraterrestrial Origin," *ibid.*, 21: 1387–1398; 1935, "A Note on the Source of Interstellar Interference," *ibid.*, 23: 1158–1163; G. C. Southworth, 1956, "Early History of Radio Astronomy," *Sci. Mon., N.Y.*, 82: 55–66.

1933   VILHELM FRIMAN KOREN BJERKNES, in collaboration with J. BJERKNES, H. SOLBERG, and T. BERGERON, published their treatise on geophysical fluid mechanics, *Physikalische Hydrodynamik*. Berlin: Julius Springer.

1933   JONAS EKMAN FJELDSTAD established a general theory for the motions of internal waves.

J. E. Fjeldstad, 1933, "Interne Wellen," *Geofys. Publ.*, 10(6): 35 pp.

1933   GEORGE RIDSDALE GOLDSBROUGH discussed hypothetical ocean circulations produced by evaporation and precipitation in opposite hemispheres on a rotating globe.

G. R. Goldsbrough, 1933, "Ocean Currents Produced by Evaporation and Precipitation," *Proc. roy. Soc.*, A, 141: 512–517.

1936   TORSTEN GUSTAFSON and BÖRJE KULLENBERG described observations made between 17 and 24 August 1933 of inertial motion in the Baltic Sea.

T. Gustafson and B. Kullenberg, 1936, "Untersuchungen von Trägheits-Strömungen in der Ostsee," *Svenska hydrogr.-biol. Komm. Skr.*, Ny, Hydro 13: 28 pp.

1936   COLUMBUS O'DONNELL ISELIN published an account of the circulation of the western North Atlantic based mainly on the evidence of dynamic sections.

C. O'D. Iselin, 1936, "A study of the circulation of the western North Atlantic," *Pap. phys. Oceanogr.*, 4(4): 101 pp.

1936   ANTON JAKHELLN discussed the equations for water transport in gradient currents.

A. Jakhelln, 1936, "The Water Transport of Gradient Currents," *Geofys. Publ.*, 2(11): 14 pp.

1936   CARL-GUSTAF A. ROSSBY borrowed the findings of Tollmien of the Göttingen school of hydrodynamicists in an attempt to liken the downstream increase of mass transport in the Gulf Stream to that of a turbulent wake stream.

C.-G. Rossby, 1936, "Dynamics of Steady Ocean Currents in the light of Experimental Fluid Mechanics," *Pap. phys. Oceanogr.*, 5(1): 43 pp.

1937   G. E. R. DEACON described the general hydrology of the Southern Ocean.

G. E. R. Deacon, 1937, "The hydrology of the Southern Ocean," pp. 1–124, *Discovery Reports*, Vol. 15, Cambridge University Press.

1937   ALBERT EIDE PARR reported on the time variations of temperature, salinity, and velocity of flow at a line of stations across the Straits of Florida.

A. E. Parr, 1937, "Report on Hydrographic Observations at a Series of Anchor Stations across the Straits of Florida," *Bull. Bingham oceanogr. Coll.*, 6(3): 62 pp.

1937   CARL-GUSTAF A. ROSSBY presented a view of the mutual adjustment
(and   of pressure and velocity distributions in idealized current systems.
1938)

C.-G. Rossby, 1937/1938, "On the mutual adjustment of pressure and velocity distributions in certain simple current systems," *J. Mar. Res.*, 1: 15–28, 239–263.

1937   ATHELSTAN F. SPILHAUS made an experimental study of the behavior of a jet stream in a rotating tank as suggested by Rossby's wake-stream theory of 1936.

A. F. Spilhaus, 1937, "Note on the flow of streams in a rotating system," *J. Mar. Res.*, 1: 29–33.

1938   MILUTIN MILANKOVITCH proposed a Pleistocene chronology based on the changes of insolation associated with cyclic changes in the orbit and relative position of the earth's axis of rotation.

M. Milankovitch, 1938, "Die Chronologie des Pleistocäns," *Bull. Acad. Sci. Math. et Nat., Belgrade*, 4: 49.

1938   RAYMOND B. MONTGOMERY calculated the annual variation in the total transport of the Gulf Stream system from the evidence of sea-surface slopes contained in tide gauge records.

R. B. Montgomery, 1938, "Fluctuations in monthly sea level on eastern U.S. coast as related to dynamics of western North Atlantic Ocean," *J. Mar. Res.*, 1: 175–176.

1939   CARL-GUSTAF A. ROSSBY and collaborators introduced the concept of the conservation of potential vorticity and applied it to long waves in the atmospheric westerlies.

C.-G. Rossby and collaborators, 1939, "Relations between variations in the intensity of the zonal circulation of the atmosphere and the displacements of the semi-permanent centers of action," *J. Mar. Res.*, 2: 38–55.

1940   COLUMBUS O'DONNELL ISELIN discussed the variation in the transport of the Gulf Stream on the basis of 15 dynamic sections made between Montauk Point and Bermuda.

C. O'D. Iselin, 1940, "Preliminary report on long-period variations in the transport of the Gulf Stream system," *Pap. phys. Oceanogr.*, 8(1): 40 pp.

1940   GEORGE CLARKE SIMPSON showed that an increase in the solar constant can account for terrestrial glaciation if the intensified storminess of the earth's atmosphere results in an increase in the earth's albedo and a consequent cooling of its surface.

Sir George C. Simpson, 1940, "Possible causes of change in Climate and their limitations," *Proc. Linn. Soc. Lond.*, 152: 190–219.

1942   HARALD ULRIK SVERDRUP suggested that the salt content and density of Antarctic bottom water might be increased in winter as a consequence of freezing of ice adjacent to the Antarctic continent, in addition to the effects noted by Brennecke (1921).

H. U. Sverdrup, 1942, p. 155, *Oceanography for Meteorologists*, New York: Prentice-Hall.

1942   HARALD ULRIK SVERDRUP, with MARTIN W. JOHNSON and RICHARD H. FLEMING, published a general treatise on oceanography, *The Oceans, Their Physics, Chemistry and General Biology*. New York: Prentice-Hall.

1946   HARRY H. HESS described the flat-topped sea mounts of the southern North Pacific, giving them the name *guyot*.

H. H. Hess, 1946, "Drowned Ancient Islands of the Pacific Basin," *Amer. J. Sci.*, 244: 772–791.

1946   BÖRJE KULLENBERG made tests of his core sampler, which is forced into sea bottom sediments by external hydrostatic pressure maintained in excess of the inside pressure by the lifting force applied to an internal piston.

B. Kullenberg, 1947, "The Piston Core Sampler," *Svenska hydrogr.-biol. Komm. Skr.*, 3, Hydro 1(2): 46 pp.

1946   WILLARD FRANK LIBBY began investigations leading to a method of age determination based on measurements of the carbon-14 isotope, which has a half-life of about 5600 years and is therefore useful for Pleistocene chronology. The method was confirmed in 1949.

W. F. Libby, E. C. Anderson, and J. R. Arnold, 1949, "Age Determination by Radiocarbon Content: World-Wide Assay of Natural Radiocarbon," *Science*, 109: 227–228.

1947   HARALD ULRIK SVERDRUP proposed a theory for wind-driven currents in the central portions of a baroclinic ocean.

H. U. Sverdrup, 1947, "Wind-Driven Currents in a Baroclinic Ocean; with Application to the Equatorial Currents of the Eastern Pacific," *Proc. Nat. Acad. Sci., Wash.*, 33: 318–326.

1947   HARALD ULRIK SVERDRUP and WALTER H. MUNK developed empirical formulae to be used in the forecasting of the sea state associated with geostrophic winds computed from synoptic charts of the atmospheric pressure distribution at sea level.

H. U. Sverdrup and W. H. Munk, 1947, *Sea and Swell: Theory of Relations for Forecasting*, U.S. Navy Hydrographic Office Publication 601.

1948   WALTER HANSEN made a computation of ocean tide by relaxation methods, based on coastal observations.

W. Hansen, 1948, "Die Ermittlung der Gezeiten beliebig gestalteter Meeresgebiete mit Hilfe des Randwertverfahrens," *Dtsch. hydrogr. Z.*, 1: 157–163.

1948   HENRY STOMMEL proposed an explanation of the westward intensification of the wind-driven surface circulation of the ocean.

H. Stommel, 1948, "The Westward Intensification of Wind-Driven Ocean
Currents," *Trans. Amer. geophys. Un.*, 29: 202–206.

1949   MICHITAKA UDA traced the development and hydrography of a
persistent loop in the Kuroshio southeast of Honshu.

M. Uda, 1949, "On the Correlated Fluctuation of the Kuroshio Current and
the Cold Water Mass," *Oceanogr. Mag.*, 1: 1–12.

1950   DIRK BROUWER and A. J. J. VAN WOERKOM recalculated Milankovitch's solar radiation curves on the basis of more recent solutions for
the planetary masses and orbital coefficients, and showed, among
other things, how sensitive such computations can be to small errors
when extended over periods of geologic time.

D. Brouwer and A. J. J. van Woerkom, 1950, *Astr. Papers Amer. Ephemeris*,
13(2).

1950   WALTER H. MUNK obtained a solution, based on the vertically integrated vorticity equation, for the pattern of streamlines in a wind-
driven ocean, and from the curl of the estimated wind stress computed the mass transports.

W. H. Munk, 1950, "On the Wind-driven Ocean Circulation," *J. Met.*, 7:
79–93.

1950   WALTER H. MUNK and GEORGE F. CARRIER developed the theory of
the wind-driven ocean circulation to apply to ocean basins of simple
nonrectangular shape, noting the changes of the circulation pattern
which accompany a change in lateral eddy viscosity.

W. H. Munk and G. F. Carrier, 1950, "The Wind-driven Circulation in
Ocean Basins of Various Shapes," *Tellus*, 2: 158–167.

1950   WALTER H. MUNK and R. L. MILLER estimated the changes in the
earth's angular velocity of rotation produced by the fluctuations in
the circulations of the oceans and atmosphere.

W. H. Munk and R. L. Miller, 1950 "Variation in the Earth's Angular
Velocity Resulting from Fluctuations in Atmospheric and Oceanic Circulation," *Tellus*, 2: 93–101.

1950   ALFRED C. REDFIELD applied Fjeldstad's theory (1929) to estuaries
of the Atlantic coast of North America, treating the problem of

damped, progressive waves in long, narrow estuaries with reflection from the bay head.

A. C. Redfield, 1950, "The Analysis of Tidal Phenomena in Narrow Embayments," *Pap. phys. Oceanogr.*, 11(4): 36 pp.

1950   ALFRED H. WOODCOCK published the first of his measurements of the concentration of sea-salt condensation nuclei in the atmosphere.

A. H. Woodcock, 1950, "Condensation Nuclei and Precipitation," *J. Met.*, 7: 161–162.

1951   FREDERICK C. FUGLISTER published his interpretation of the multiple current structure of the Gulf Stream in the region of the Canadian maritime provinces.

F. C. Fuglister, 1951, "Multiple Currents in the Gulf Stream," *Tellus*, 3: 230–233.

1951   FREDERICK C. FUGLISTER and L. VALENTINE WORTHINGTON gave an account of a synoptic study of the meander pattern of the Gulf Stream off the coasts of New England and the Maritime Provinces.

F. C. Fuglister and L. V. Worthington, 1951, "Some Results of a Multiple Ship Survey of the Gulf Stream," *Tellus*, 3: 1–14.

1951   WALTER H. MUNK and ERIK PALMÉN suggested that lateral friction does not balance the wind stress on the Antarctic circumpolar current, arguing that bottom friction is developed as the current is "deepened" by meridional circulation.

W. H. Munk and E. Palmén, 1951, "Note on the Dynamics of the Antarctic Circumpolar Current," *Tellus*, 3: 53–55.

1951   GORDON A. RILEY studied the distribution of nonconservative properties in the Atlantic Oceans in relation to the conservative properties to determine the biological rates of change.

G. A. Riley, 1951, "Oxygen, Phosphate, and Nitrate in the Atlantic Ocean," *Bull. Bingham oceanogr. Coll.*, 13(1): 126 pp.

1951   CARL-GUSTAF A. ROSSBY attempted to characterize in terms of simple analytical models the narrow and rapid flows that develop in the oceans and atmosphere.

C.-G. Rossby, 1951, "On the Vertical and Horizontal Concentration of Momentum in Air and Ocean Currents," *Tellus*, 3: 15–27.

1952   BRUCE C. HEEZEN and MAURICE EWING suggested that the orderly sequence of breaks in telegraph cables on the ocean floor south of

Newfoundland was caused by a suspension flow associated with the earthquake of November 18, 1929.

B. C. Heezen and M. Ewing, 1952, "Turbidity Currents and Submarine Slumps, and the 1929 Grand Banks Earthquake," *Amer. J. Sci.*, 250: 849–873.

1952   ROGER REVELLE and ARTHUR E. MAXWELL found that the heat flow through the ocean floor is nearly the same as that through the continental crust (1.2×10^{-6} cal/cm^2/sec).

R. Revelle and A. E. Maxwell, 1952, "Heat Flow Through the Floor of the Eastern North Pacific Ocean," *Nature, Lond.*, 170; 199–200.

1954   TOWNSEND CROMWELL, R. B. MONTGOMERY, and E. D. STROUP described the equatorial undercurrent in the Pacific, now called the Cromwell Current, an eastward flow centered on the equator with the core of maximum velocity below the surface in the presence of easterly winds.

T. Cromwell, R. B. Montgomery, and E. D. Stroup, 1954, "Equatorial Undercurrent in Pacific Ocean Revealed by New Methods," *Science*, 119: 648–649.

1954   BERNARD LUSKIN, BRUCE C. HEEZEN, MAURICE EWING, and MARK LANDISMAN described an echo sounder having sufficiently high time resolution to measure a depth change of less than one fathom in depths of 3000 fathoms.

B. Luskin, B. C. Heezen, M. Ewing, and M. Landisman, 1954, "Precision Measurement of Ocean Depth," *Deep-Sea Res.*, 1: 131–140.

1955   FREDERICK C. FUGLISTER showed the variety of interpretations possible in contouring synoptic charts based on inadequate data.

F. C. Fuglister, 1955, "Alternative Analyses of Current Surveys," *Deep-Sea Res.*, 2: 213–229.

1955   P. C. LINEIKIN discussed the theory of the effects of wind stress on a baroclinic ocean and defined the Lineikin spiral, which differs from the Ekman spiral (1905).

P. C. Lineikin, 1955, "On the Determination of the Thickness of the Baroclinic Layer in the Sea," *C. R. Acad. Sci. U.R.S.S.*, 101: 461–464.

1955   JOHN C. SWALLOW described a neutrally buoyant float and an acoustic signaling system for tracking the motions of deep currents in the open sea.

J. C. Swallow, 1955, "A Neutral Buoyancy Float for Measuring Deep Currents," *Deep-Sea Res.*, 3: 74–81.

This chronological summary has been cut off quite arbitrarily at the end of 1955 because it becomes increasingly difficult to distinguish, at close range, between those contributions of lasting importance and others which have only passing influence on contemporary thought. Readers who wish to explore recent developments in physical oceanography (in English) will find most of it in the following journals:

Tellus

Deep-Sea Research

Journal of Meteorology

Journal of Marine Research

Limnology and Oceanography

Journal of Geophysical Research (formerly *Transactions of the American Geophysical Union*)

Philosophical Transactions of the Royal Society (London)

Geophysical Journal (formerly *Geophysical Supplements to the Monthly Notices of the Royal Astronomical Society*).

Appendix B

MISCELLANEOUS ASTRONOMICAL CONSTANTS*

Angular velocity of earth rotation	-0.729211×10^{-4} sec^{-1}
Mean solar day	86,400 sec = 1.0027379 sidereal day
Sidereal day	86,164.09054 mean solar sec (23 hr 56 min 4.09054 sec)
Solar parallax	8.80 sec of arc
Mean distance of sun	1.4945×10^8 km
Mean density of sun	1.41 gm/cm^3
Solar diameter	1.393×10^6 km
Obliquity of the ecliptic	23°26′59″
†Tidal constituents, mean increase in hour angle	
of sun	15°/hr
celestial longitude of sun	0.0411°/hr
celestial longitude of moon	0.5490°/hr
celestial longitude of lunar perigee	0.0046°/hr
Mean distance of moon from earth	384,393 km
Moon's sidereal period	27.322 days
Moon's synodic period	29.5 ± 0.5 days
Earth's mass	5.975×10^{27} gm
Sun's mass	1.987×10^{33} gm (330,000 earth masses)
Moon's mass	7.343×10^{25} gm (1/81.56 earth mass)
Gravitation constant G	$(6.670 \pm 0.005) \times 10^{-8}$ dyne · cm^2/gm^2
§Acceleration due to gravity g	
sea level, 0° latitude	978.039 cm/sec^2
sea level, 90° latitude	983.217 cm/sec^2

---

* Adapted from *Smithsonian Physical Tables*, 1954 (Smithsonian Miscellaneous Collections, Vol. 120), 9th revised ed., prepared by W. E. Forsythe, Washington, D.C.: The Smithsonian Institution.

† P. Schureman, 1940, *Manual of Harmonic Analysis and Prediction of Tides*, 2nd ed., U.S. Coast and Geodetic Survey Special Publication No. 98, Washington, D.C.: U.S. Government Printing Office; A. T. Doodson and H. D. Warburg, 1941, *Admiralty Manual of Tides*, London: H.M. Stationery Office.

§ Based on a U.S. Coast and Geodetic Survey formula.

Appendix C

MISCELLANEOUS TERRESTRIAL CONSTANTS*

Area of the earth........................510.100 × 10⁶ km²

Area of the land........................148.847 × 10⁶ km²

Area of the oceans....................361.254 × 10⁶ km²

Mean depth of oceans..................3.790 km

Volume of the oceans..................1.369 × 10⁹ km³

Percent ocean volume colder than 10°C....93%

Percent ocean volume colder than 4°C.....76%

Mean depth of thermocline..............0.2 km

Volume of lower atmosphere.............4 × 10⁹ km³

Standard sea level pressure of
atmosphere...........................1.01325 × 10⁶ dynes/cm²

Approximate pressure gradient of
oceans..............................0.1 atm/m or 0.5 lb/in²/ft

Maximum insolation at top of
atmosphere...........................2 ly/min

24-hr mean insolation at top of
atmosphere...........................0.5 ly/min

24-hr mean insolation at surface...........0.25 ly/min

Mean temperature of earth surface........287°K

Mean albedo of earth...................0.34

†Heat flux through sea floor..............1.2 × 10⁻⁶ ly/sec

§Evaporation rate from oceans............93 — 106 cm/year

Ellipticity of earth......................1/297

Equatorial radius......................6378 km

---

* *Smithsonian Meteorological Tables*, 1951 (Smithsonian Miscellaneous Collections, Vol. 114), 6th revised ed., prepared by Robert J. List, Washington, D.C.: The Smithsonian Institution.

† R. Revelle and A. E. Maxwell, 1952, *Nature, Lond.*, 170: 199–200.

§ G. Wüst, 1936, *Länderkundliche Forschung*, Festschrift Norbert Krebs: 347–359; H. Mosby, 1936, *Ann. Hydrogr., Berl.*, 64: 281–286.

Appendix D

GROSS PROPERTIES OF SEA WATER

(For physical units and chemical definitions see B. Helland-Hansen, J. P. Jacobsen, and T. G. Thompson, 1948, *Un. géod. géophys. int., Ass. Océanogr. phys. Publ. Sci.*, 9: 28 pp.)

Characteristic density, $f(s, t, p)$ . . . . .1.025 gm/cm^3
Velocity of sound, $f(s, t, p)$ . . . . .at surface $= 1448.6$ m/sec
Specific heat, $Cp, f(s)$ . . . . .0.932 cal/gm/°C at 35 $^0/_{00}$
Adiabatic lapse rate, $f(s, p)$ . . . . .approx. 0.1°C/km
Maximum surface temperature . . . . .32°C
Minimum surface temperature . . . .-2°C
Median surface temperature . . . . .20°C
Average temperature . . . . .3.8°C
Latent heat of fusion and vaporization . . . . .same as pure water

SPECIFIC HEAT AT CONSTANT PRESSURE FOR VARIOUS SALINITIES*

S ($^0/_{00}$)	0	5	10	15	20	25	30	35	40
C_p (cal/gm/°C)	1.000	0.982	0.968	0.958	0.951	0.945	0.939	0.932	0.926

EFFECTS OF PRESSURE ON THE PROPERTIES OF SEA WATER AT 0°C AND SALINITY OF 35 $^0/_{00}$*

Pressure, decibars	Density, gm/cm^3	Speed of sound, m/sec	Adiabatic temperature change, °C/1000 decibars
0	1.02813	1448.6	0.035
2,000	1.03748	1484.4	0.072
4,000	1.04640	1519.7	0.104
6,000	1.05495	1554.2	0.133
8,000	1.06315	1587.7	0.159
10,000	1.07104	1620.0	0.181

* Adapted from R. B. Montgomery, 1957, "Oceanographic Data," pp. 2–115 to 2–124, *American Institute of Physics Handbook*, New York: McGraw-Hill.

Appendix E

GLOSSARY OF SYMBOLS

Latin

A	area, arbitrary point, station, coefficient of eddy diffusion
a	acceleration
B	secondary point, station, magnetic flux density
$\mathbf{C}$	velocity vector
C_D	drag coefficient
Cl	chlorinity
C_p	specific heat at constant pressure
c, c_h	horizontal velocity
c_g	geostrophic velocity
c_m	meander velocity
D	dynamic height, layer thickness, depth of frictional influence
d	depth, electrical thickness, distance, radius of equivalent circle
d_r	depth ratio
e	constant $= 2.71838$, eccentricity, vapor pressure, coefficient of thermal expansion
Ek	Ekman number
Eu	Euler number
F	force, friction
f	Coriolis parameter, planetary vorticity
Fr	Froude number
G	constant of universal gravitation
g	local gravity
g_0	standard gravity
$\mathbf{H}$	magnetic field vector
h, h'	depth or fractional depth
i	electric current density
$\mathbf{i, I}$	unit vector in x direction
$\mathbf{j, J}$	unit vector in y direction
J	dimensionless number, mechanical equivalent of heat
K	Kelvin temperature

k, K   unit vector in z direction

   k dimensionless ratio $|\mathbf{C}_h \times \mathbf{K}H_z/(-\rho\mathbf{i})_h|$

   L heat of vaporization, wavelength, horizontal distance $\overline{AB}$

  L_r length ratio

   l wavelength

  ly   langley $=$ cal/cm^2

M, m mass

N, n integer

   n normal horizontal component

   p pressure, arbitrary point

  Q heat content

R, r radius, electrical resistance

  R Bowen ratio

 Re   Reynolds number

 Ro   Rossby number

  S interelectrode distance, salinity

  s increment of length

  T temperature, duration in time, period of oscillation, transport

  T_r time ratio

  t time

  U wind speed

  u east component, velocity

  V potential difference, velocity

  V_r velocity ratio

  v north component, velocity

  w down component, velocity

  x east coordinate

  y north coordinate

  z down coordinate

Z, z depth, zenith angle

Greek

  α specific volume

  β df/dy (variation of the Coriolis parameter f with latitude on an xy-plane)

  γ specific weight

  Δ difference

  δ finite difference, dynamic anomaly

  ζ relative vorticity

  η refractive index, surface elevation

  θ angle, potential temperature

  μ molecular viscosity, magnetic permeability

μ_e eddy viscosity
ν kinematic viscosity μ/ρ
λ wavelength, longitude
π constant $= 3.14159$
ρ density, electrical resistivity
σ conventionalized density
τ stress
ϕ latitude, electric potential
Ω earth's angular velocity
ω angular velocity relative to the earth

Arbitrary symbols

∂ partial variation
∇ del (a vector operator), gradient
grad_h horizontal component of gradient
div_h horizontal divergence
$^0/_0$ parts per hundred
$^0/_{00}$ parts per thousand

Appendix F

SMALLER HANDBOOKS AND TABLES OF CONSTANTS

Handbook of Physical Constants, 1942, F. Birch, J. F. Schairer, and H. C. Spicer, editors, Geological Society of America Special Paper No. 36.

Smithsonian Meteorological Tables, 1951 (Smithsonian Miscellaneous Collections, Vol. 114), 6th revised ed., prepared by Robert J. List, Washington, D.C.: The Smithsonian Institution.

Smithsonian Physical Tables, 1954 (Smithsonian Miscellaneous Collections, Vol. 120), 9th revised ed., prepared by W. E. Forsythe, Washington, D.C.: The Smithsonian Institution.

American Institute of Physics Handbook, 1957, D. E. Gray, coordinating editor, *et al.*, New York: McGraw-Hill.

Tables of Physical and Chemical Constants and Some Mathematical Functions, 1957, G. W. C. Kaye and T. H. Laby, 11th ed., London: Longmans, Green.

Handbook of Chemistry and Physics, 1958, C. D. Hodgman, R. C. Weast, and S. M. Selby, editors, 40th ed. (frequently revised), Cleveland: Chemical Rubber Publishing Co.

Name Index

Page numbers in parentheses refer to listings in Appendix A.

Subject Index

411